BENEFICIAL MICROBES FOR SUSTAINABLE AGRICULTURE AND ENVIRONMENTAL MANAGEMENT

Current Advances in Biodiversity, Conservation and Environmental Sciences

BENEFICIAL MICROBES FOR SUSTAINABLE AGRICULTURE AND ENVIRONMENTAL MANAGEMENT

Edited by
Jeyabalan Sangeetha
Devarajan Thangadurai
Saher Islam

Apple Academic Press Inc.
4164 Lakeshore Road
Burlington ON L7L 1A4
Canada

Apple Academic Press, Inc.
1265 Goldenrod Circle NE
Palm Bay, Florida 32905
USA

First issued in paperback 2021

Exclusive worldwide distribution by CRC Press, a member of Taylor & Francis Group

No claim to original U.S. Government works

ISBN 13: 978-1-77463-509-4 (pbk)
ISBN 13: 978-1-77188-818-9 (hbk)

Library and Archives Canada Cataloguing in Publication

Title: Beneficial microbes for sustainable agriculture and environmental management / edited by
    Jeyabalan Sangeetha, Devarajan Thangadurai, Saher Islam.
Names: Sangeetha, Jeyabalan, editor. | Thangadurai, D., editor. | Islam, Saher, editor.
Description: Series statement: Current advances in biodiversity, conservation and environmental
    sciences book series | Includes bibliographical references and index.
Identifiers: Canadiana (print) 2019022360X | Canadiana (ebook) 20190223634 |
    ISBN 9781771888189 (hardcover) | ISBN 9780429284137 (ebook)
Subjects: LCSH: Agricultural microbiology. | LCSH: Sustainable agriculture. |
    LCSH: Microorganisms. | LCSH: Microbial ecology. | LCSH: Environmental management.
Classification: LCC QR51 .B46 2020 | DDC 630.2/79—dc23

CIP data on file with US Library of Congress

Apple Academic Press also publishes its books in a variety of electronic formats. Some content that appears in print may not be available in electronic format. For information about Apple Academic Press products, visit our website at **www.appleacademicpress.com** and the CRC Press website at **www.crcpress.com**

ABOUT THE CURRENT ADVANCES IN BIODIVERSITY, CONSERVATION AND ENVIRONMENTAL SCIENCES BOOK SERIES

Series Editors:
Jeyabalan Sangeetha, PhD
Assistant Professor, Central University of Kerala, Kasaragod, Kerala, India

Devarajan Thangadurai, PhD
Assistant Professor, Karnatak University, Dharwad, Karnataka, India

Biodiversity and Conservation: Characterization and Utilization of Plants, Microbes, and Natural Resources for Sustainable Development and Ecosystem Management

Editors: Jeyabalan Sangeetha, PhD, Devarajan Thangadurai, PhD, Hong Ching Goh, PhD, and Saher Islam, MPhil

Beneficial Microbes for Sustainable Agriculture and Environmental Management

Editors: Jeyabalan Sangeetha, PhD, Devarajan Thangadurai, PhD, and Saher Islam, MPhil

ABOUT THE EDITORS

Jeyabalan Sangeetha, PhD
Assistant Professor, Central University of Kerala at Kasaragod,
South India

Jeyabalan Sangeetha, PhD, is an Assistant Professor at the Central University of Kerala at Kasaragod, South India. She earned her BSc in Microbiology and PhD in Environmental Science from Bharathidasan University, Tiruchirappalli, Tamil Nadu, India. She holds an MSc in Environmental Science from Bharathiar University, Coimbatore, Tamil Nadu, India. She is the recipient of the Tamil Nadu Government Scholarship and Rajiv Gandhi National Fellowship of University Grants Commission, Government of India for her doctoral studies. She served as Dr. D.S. Kothari Postdoctoral Fellow and UGC Postdoctoral Fellow at Karnatak University, Dharwad, South India during 2012–2016 with funding from the University Grants Commission, Government of India, New Delhi. Her research interests are in the fields of environmental toxicology, environmental microbiology, environmental biotechnology, and environmental nanotechnology.

Devarajan Thangadurai, PhD
Assistant Professor, Karnatak University at Dharwad, South India

Devarajan Thangadurai, PhD, is Senior Assistant Professor at Karnatak University in South India and Editor-in-Chief of international journals *Biotechnology, Bioinformatics and Bioengineering,* and *Acta Biologica Indica.* He received his PhD in Botany from Sri Krishnadevaraya University in South India as CSIR Senior Research Fellow with funding from the Ministry of Science and Technology, Government of India. Dr. Thangadurai served as a postdoctoral fellow at the University of Madeira, Portugal, University of Delhi, India, and ICAR National Research Centre for Banana, India. He is the recipient of a Best Young Scientist Award with a Gold Medal from Acharya Nagarjuna University, India, and a VGST-SMYSR Young Scientist Award of the Government of Karnataka, Republic of India. He has authored/edited over twenty books with publishers of national/international reputation. He has also visited twenty countries in East, South, Southeast and West Asia, Eastern, Southern, and Western

Europe, North Africa, and the Middle East for academic visits, scientific meetings, and international collaborations.

Saher Islam, MPhil
IRSIP Scholar, Cornell University, Ithaca, New York, USA

Saher Islam, MPhil, is an IRSIP Scholar at Cornell University, Ithaca, New York, USA, a PhD Candidate in Molecular Biology and Biotechnology and HEC Scholar of the Islamic Republic of Pakistan at the University of Veterinary and Animal Sciences, Lahore. She received her MPhil in Molecular Biology and Biotechnology in 2014 and BS (Hons.) in Biotechnology and Bioinformatics from the University of Veterinary and Animal Sciences at Lahore, Pakistan. She worked as Research Associate in projects funded by Grand Challenges, Canada, and served as an Internee in Pakistan Council of Scientific and Industrial Research, Lahore. She has visited the USA, the UK, Singapore, Germany, the UAE, Egypt, Italy, Russia, Maldives, Malaysia, and Sri Lanka for trainings, courses, and meetings. She is the recipient of EMBL Boehringer Ingelheim Fonds Travel Grant and a Travel Grant from the Wildlife Conservation Society. She has research interests in biodiversity, conservation, genetics, molecular biology, biotechnology, and bioinformatics.

CONTENTS

CONTRIBUTORS

Özlem Akkaya
Department of Molecular Biology and Genetics, Gebze Technical University, Kocaeli–41400, Turkey

Prasad Andhare
PD Patel Institute of Applied Sciences, Charotar University of Science and Technology,
Changa, Gujarat–388421, India

Abdullah Adil Ansari
Department of Biology, University of Guyana, Georgetown, Guyana

Priya Ashrit
Department of Biotechnology, MS Ramaiah Institute of Technology, Bengaluru–560054,
Karnataka, India

Navneet Batra
Department of Biotechnology, GGDSD College, Sector-32-C, Chandigarh–160030, India

Yelda Özden Çiftçi
Department of Molecular Biology and Genetics, Gebze Technical University, Kocaeli–41400, Turkey

Selvakumar Gopal
Department of Microbiology, Allagappa University, Karaikudi–630003, Tamil Nadu, India

Dweipayan Goswami
Department of Biochemistry and Biotechnology, St. Xavier's College (Autonomous),
Ahmadabad, Gujarat–380009, India

Manivannan Govindasamy
Department of Microbiology and Biotechnology, NMSS Vellaichamy Nadar College,
Madurai–625019, Tamil Nadu, India

Aparna Gunjal
Department of Microbiology, Savitribai Phule Pune University, Pune–411007, Maharashtra, India

Ravichandra Hospet
Department of Botany, Karnatak University, Dharwad–580003, Karnataka, India

Bakulranjan Jana
ICAR-RCER, Research Center on Makhana, Darbhanga, Bihar–846005, India

Sudeshna Menon
Department of Biochemistry and Biotechnology, St. Xavier's College (Autonomous),
Ahmadabad, Gujarat–380009, India

Mohammed Abdul Mujeeb
Department of Microbiology and Biotechnology, Karnatak University, Dharwad–580003,
Karnataka, India

Abhishek Mundaragi
Department of Microbiology, Davangere University, Davangere–577002, Karnataka, India

Sivakumar Natesan
Department of Molecular Microbiology, School of Biotechnology, Madurai Kamaraj University, Madurai–625021, Tamil Nadu, India

Neelu Nawani
Dr. D.Y. Patil Vidyapeeth's, Dr. D. Y. Patil Biotechnology and Bioinformatics Institute, Tathawade, Pune–411033, Maharashtra, India

Dhaval Patel
Department of Biochemistry and Biotechnology, St. Xavier's College (Autonomous), Ahmadabad, Gujarat–380009, India

Neha Patil
Department of Microbiology, Waghire College, Saswad, Pune–412301, Maharashtra, India

Nittaya Pitiwittayakul
Department of Agricultural Technology and Environment, Faculty of Sciences and Liberal Arts, Rajamangala University of Technology Isan, Nakhon Ratchasima Campus, Nakhon Ratchasima 30000, Thailand

Prathima Purushotham
Department of Botany, Karnatak University, Dharwad–580003, Karnataka, India

Md. Maksudur Rahman
Biomass Energy Engineering Research Center, School of Agriculture and Biology, Shanghai Jiao Tong University, 800 Dongchuan Road, Shanghai–200240, PR China

Preeti Ranawat
Department of Botany and Microbiology, Hemvati Nandan Bahuguna Garhwal University, Srinagar (Garhwal)–246174, Uttarakhand, India

Seema Rawat
School of Life Sciences, Central University of Gujarat, Gandhinagar, Gujarat–382030, India

Bindu Sadanandan
Department of Biotechnology, MS Ramaiah Institute of Technology, Bengaluru–560054, Karnataka, India

Jeyabalan Sangeetha
Department of Environmental Science, Central University of Kerala, Periye, Kasaragod–561716, Kerala, India

Shanmugapriya Saravanabhavan
Department of Microbiology and Biotechnology, NMSS Vellaichamy Nadar College, Madurai–625019, Tamil Nadu, India

Manobendro Sarker
Department of Food Engineering and Technology, State University of Bangladesh, Dhanmondi, Dhaka–1205, Bangladesh; Biomass Energy Engineering Research Center, School of Agriculture and Biology, Shanghai Jiao Tong University, 800 Dongchuan Road, Shanghai–200240, PR China

Mine Gül Şeker
Department of Molecular Biology and Genetics, Gebze Technical University, Kocaeli–41400, Turkey

Neetu Sharma
Department of Biotechnology, GGDSD College, Sector-32-C, Chandigarh–160030, India

Megha Ramachandra Shinge
Department of Microbiology and Biotechnology, Karnatak University, Dharwad–580003,
Karnataka, India

Abhinashi Singh
Department of Biotechnology, GGDSD College, Sector-32-C, Chandigarh–160030, India

Simmi Maxim Steffi
Department of Environmental Science, Central University of Kerala, Periye,
Kasaragod–561716, Kerala, India

Somboon Tanasupawat
Department of Biochemistry and Microbiology, Faculty of Pharmaceutical Sciences,
Chulalongkorn University, Bangkok 10330, Thailand

Devarajan Thangadurai
Department of Botany, Karnatak University, Dharwad–580003, Karnataka, India

Shivasharana Chandrabanda Thimmappa
Department of Microbiology and Biotechnology, Karnatak University, Dharwad–580003,
Karnataka, India

Sebastian Vadakan
Department of Biochemistry and Biotechnology, St. Xavier's College (Autonomous),
Ahmadabad, Gujarat–380009, India

V. Vijayalakshmi
Department of Biotechnology, MS Ramaiah Institute of Technology, Bengaluru–560054,
Karnataka, India

Meghmala Waghmode
Department of Microbiology, Annasaheb Magar Mahavidyalaya, Hadapsar, Pune–411028,
Maharashtra, India

ABBREVIATIONS

$(NH_4)_2SO_4$	ammonium sulfate
1,3,6,8-THN	1,3,6,8-tetrahydroxynaphthalene
2,4-D	2,4-dichlorophenoxyacetic acid
ABA	abscisic acid
ABR	anaerobic baffled reactor
ABTS	2,2-azino-bis-3-ethylbenzothiazoline-6-sulfonic acid
ACC	1-aminocyclopropane-1-carboxylase
ACDS	1-aminocyclopropane-1-carboxylate deaminase
AHLs	acyl-homoserine lactones
$AlPO_4$	aluminum phosphate
AM	arbuscular mycrrhizal
AMF	arbuscular mycorrhizal fungi
ARDRA	amplified rDNA restriction analysis
Arg	arginine
As	arsenic
ASTM	American Society of Testing and Materials
ATP	adenosine triphosphate
atpD gene	gene encoding the β-subunit of the ATP synthase
AzoA	azoreductases A
AzoB	azoreductases B
BAF	biological aerated filter
B-ARISA	bacterial automatic ribosomal intergenic spacer analysis
BCA	biological control agents
bchL gene	gene encoding bacteriochlorophyll
BenA	β-tubulin
BGA	blue green algae
BOD	biological oxygen demand
C	carbon
$CaCl_2$	calcium chloride
$CaCO_3$	calcium carbonate
CaM	calmodulin
Cd	cadmium
CD	compact disc

c-di-GMP	cyclic-di-guanosine monophosphate
CHH	crude horn hydrolysate
CMN	common mycelia network
CMV	cucumber mosaic virus
CO	carbon monoxide
CO_2	carbon dioxide
COD	chemical oxygen demand
COGs	cluster of orthologs genes
CPM	capsaicin in the medium
Cr	chromium
CSA	continuously stirred aerobic
Cu	copper
DAPG	2,4-diacetylphloroglucinol
DCIP	dichloroindophenol oxidoreductase
Dgeo	ortholog genes in *Deinococcus geothermalis*
*Dge*RecA	recA protein of *Deinococcus geothermalis*
DGGE	denaturing gradient gel electrophoresis
DHI	5, 6-dihydroxyindole
DHICA	5, 6-dihydroxyindole-2-carboxylic corrosive
DHN	1,8-dihydroxynaphthalene
*Dmu*RecA	recA protein of *Deinococcus murrayi*
DNA	deoxyribonucleic acid
DO	dissolved oxygen
DOE	department of energy
DOPA	*O*-dihydroxyphenylalanine
DP	direct pathways
*Dra*RecA	recA protein of *Deinococcus radiodurans*
DSBs	double-strand breaks
DW	distilled water
EC	electric conductance
EGSB	expanded granular sludge bed
EMOs	effective microorganisms
EPA	Environmental Protection Agency
EPS	exopolysaccharides
ERM	extra-radical mycelium
EtOH	ethanol
E-waste	electronic waste
$FADH_2$	flavin adenine dinucleotide

FAO	Food and Agriculture Organization
FBR	fluidized bed reactor
Fe	iron
FeCl$_3$	iron (III) chloride
FePO$_4$	ferric phosphate
FISH	fluorescence *in situ* hybridization
GAs	giberellins
GDHB	glutaminyl-3,4-dihydroxybenzene
GEM	genetically engineered microorganisms
GFP	green fluorescent protein
Glc	glycogen
glnII	gene encoding glutamine synthetase II
Gy	gray
H$_2$	hydrogen
H$_2$O$_2$	hydrogen peroxide
HEP-2	human epidermoid larynx carcinoma cell line
HO	hydroxyl radicals
HRDC	helicase and RNAse D C-terminal
HRT	hydraulic retention time
IAA	indole-3-acetic acid
ICPA	International Commission of *Penicillium* and *Aspergillus*
IL	interleukins
IMO	indigenous microorganisms
IR	ionizing radiations
IRM	intra-radical mycelium
ISR	induced systematic resistance
IST	induced systemic tolerance
ITS	internal transcribed spacer
JA	jasmonic acid
K	potassium
K$_2$HPO$_4$	dipotassium phosphate
KB	ketobutyrate
KEGG	Kyoto encyclopedia of genes and genomes
kGy	kilogray
KH$_2$PO$_4$	monopotassium phosphate
LA	Luria agar
L-DOPA	L-dihydroxyphenylalanine
LiP	lignin peroxidase

LPS	lipopolysaccharides
MAAS	microalgae activated sludge
MAMPs	microbe-associated molecular patterns
MDA	malondialdehyde
ME	malt remove
MEL1	melanin-overproducing mutant
$MgSO_4$	magnesium sulfate
MIC	minimum inhibitory concentration
mM	millimolar
MMR	mismatch repair
MN	melanin-shrouded nanoparticle
MnP	manganese peroxidase
MP	mycorrhizal pathway
MPTP	1-methyl-4-phenyl-1,2,5,6-tetrahydropyridine
MR	methylenetetrahydrofolate reductases
MtEnod11	MtEnod11 (pour early nodulin 11 gene of *Medicago truncatula*)
MTI	MAMP-triggered immunity
Myc	mycorrhiza
N	nitrogen
NA	nutrient agar
$Na_2HPO_4·H_2O$	disodium hydrogen phosphate
NaCl	sodium chloride
NAD/NADH	nicotinamide adenine dinucleotide
NADH-DCIP	nicotinamide adenine dinucleotide-2,6-dichloroindophenol
NADPH	nicotinamide adenine dinucleotide phosphate
$NaH_2PO_4·2H_2O$	sodium dihydrogen phosphate dihydrate
$NaMoO_4$	sodium molybdate
NaOCl	sodium hypochlorite (household bleach)
NAs	naphthenic acids
NFA	nitrogen free agar
NH_3-N	ammonia nitrogen
NH_4	ammonium
NHEJ	non-homologous end joining
nif genes	gene encoding enzyme involved nitrogen fixation
NO	nitric oxide
NO_3	nitrate

nod gene	gene encoding nod protein involved in the induction of root hair curling
Nod	nodulation
NWS	non-water stressed
O_2^-	superoxide
OH^-	hydroxyl radical
OTUs	operational taxonomic units
P	phosphorus
PC	personal computer
PCA	plate count agar
PCBs	printed circuit boards
PCP	pentachlorophenol
PCR	polymerase chain reaction
PCR-DGGE	polymerase chain reaction denaturing gradient gel electrophoresis
PD	potato dextrose
PDA	potato dextrose agar
PGPB	plant growth promoting bacteria
PGPF	plant growth promoting fungi
PGPR	plant growth promoting rhizobacteria
PHS	paralana hot springs
PKS	polyketide synthase
PMSRs	peptide methionine sulfoxide reductases
PPM	parts per million
PPOs	polyphenol oxidases
PR	pathogenesis-related
PSM	phosphate-solubilizing microorganisms
pufM gene	gene encoding the M subunit of the reaction center of anoxygenic photosynthesis
QS	quorum sensing
R2A	Reasoner's 2A agar
RBC	rotating biological contactors
RCBD	randomized complete block design
RDBR	rotating drum bioreactor
rDNA	ribosomal DNA
RDR	radiation-desiccation response
RDRM	radiation-desiccation response motif
recA gene	gene encoding recA protein essential for repair and maintenance of DNA

RFLP	restriction fragment length polymorphism
RMs	redox mediators
ROS	reactive oxygen species
RPB2	RNA polymerase II second largest subunit
SA	salicylic acid
SAR	systemic acquired resistance
SBR	sequencing batch reactor
SCP	single cell protein
SLP	salts low-phosphate
SODs	superoxide dismutases
SOR	superoxide reductase
spp.	species
SR	systemic resistance
SSA	single strand annealing
SSBs	single strand breaks
SSF	solid state fermentation
STE	simulated textile effluent
Su	sucrose
Suc	succinate
TCE	trichloroethylene
TDS	total dissolved solids
THN	1,3,8-trihydroxynaphthalene
TKN	total Kjeldahl nitrogen
TM	thiophanate methyl
TNF-α	tumor necrosis factor-α
TOC	total organic carbon
TSA	tryptic soy agar
TSS	total soluble solids
TTSS-SPII	type III secretion system
TV	television
UASB	up-flow anaerobic sludge blanket reactor
UFC	up-flow fixed film column
UFCR	up-flow column reactor
ups	UV-inducible pili operon of *Sulfolobus*
UV	ultraviolet
UV-R	ultraviolet radiation
VAM	vesicular arbuscular mycorrhizae
VOCs	volatile organic compounds

WEEE	waste electrical and electronic equipment
WS	water stress
WWTPs	wastewater treatment plants
Zn	zinc

PREFACE

Microbes can be found in any kind of habitats from normal to extreme conditions. Microbes are playing a major role in ecosystem functioning, and some of the microbial functions in the ecosystem are decomposition, biogeochemical cycling, biocontrol agents, bioremediation, and bioaugmentation. Microbial biosynthetic abilities have been highly influenced in the search for solutions to problems faced by mankind in maintaining a quality environment. Microbes have enabled constructive and cost-effective responses, which would have been impossible through physical or chemical methods. Nowadays, microbial technologies have been applied to a wide area of environmental problems, with considerable solutions. Interaction among microbial populations in natural environments and their potential roles in the food web, biogeochemical cycles, and evolution of life makes microbial ecology as an essential study area.

Microbes are the most abundant organisms in the biosphere and regulate many critical elemental and biogeochemical phenomena. They are the key players in the carbon cycle and related biological reactions, and microbial ecology is a vital research area for understanding the contribution of the biosphere in global warming and the response of the natural environment to climate variations. This book provides basic to advanced information in environmental microbiology by exploring fascinating insights into microbial diversity.

In this context, this book briefly discusses the diversity of microbes and their potential application in nutrient cycling, environmental stress, plant growth, biocontrol, melanin production, bioremediation of xenobiotics, wastewater treatment, radiation resistance, and vermicomposting of organic wastes with updated information and useful illustrations. The potential of applying mycorrhiza and endophytic bacteria as bioenhancers, biofertilizers, and bioprotectors in commercial agricultural and horticultural systems and its impact on nutrient cycling, hormone signaling, environmental stress, plant growth, siderophore production, and biocontrol are discussed in Chapters 1 to 3. Chapter 4 discusses the interaction of fungi with plants and its effect on agriculture with special reference to *Penicillium* as phytopathogen and phytoaugmentor. A comprehensive discussion has been attempted on the ill effects of wastes, effluents, and dyes on the environment and human health and their degradation using microbes and biofilms in Chapters 5 through 8. Bioremediation of metal contamination using soil microbes, treatment of wastewater by indigenous and effective microbes, and resistance against radiation through thermophiles are discussed in detail in Chapters 9 to 11. Finally, biodegradation of organic wastes using microbe and earthworm mediated vermicomposting is briefly discussed in the last chapter of this book.

This book contains chapter contributions from well-established experts in their respective fields and is currently reviewed by many academic professionals. The editors wish to specially thank the authors and coauthors for their valuable contributions and extraordinary effort in revising and finalizing respective chapters towards the successful publication of this book. We would like to express our gratitude to Sandy Jones Sickels, Vice President, and Ashish Kumar, Publisher and President, Apple Academic Press, Inc., USA, for bringing out this book with excellent quality and for timely production.

—**Jeyabalan Sangeetha, PhD**
Devarajan Thangadurai, PhD
Saher Islam, MPhil

CHAPTER 1

MYCORRHIZA: A POTENTIAL BIO-ENHANCER IN THE AGRICULTURE PRODUCTION SYSTEM

BAKULRANJAN JANA

ICAR-RCER, Research Center on Makhana, Darbhanga, Bihar–846005, India

1.1 INTRODUCTION

The agriculture industry is confronted today with the pressure of bourgeoning population, depleting, and degrading natural resources of soil and water, climate change, and shortage manpower and the non-availability of optimal technology. Due to rapid urbanization, we are losing the fertile soil for agriculture, and on the other hand, infertile and degraded soils have been taken for cultivation with a view to area expansion. In such a scenario, research on abiotic stresses is gaining momentums in India. This has been termed as a gray revolution, which may play a vital role for future agriculture in tropical and subtropical areas. In abiotic stress conditions, mycorrhizae are economically important symbionts to plant in arable conditions (Table 1.1). Mycorrhizae are simply a symbiotic association between numerous fungi and roots of higher plants/vascular host plants (Kirk et al., 2001). The term mycorrhizae mean fungus root, and it was derived from the Greek words '*mykós*' means fungus, and 'rhiza' means root' (Frank, 1885). This type of plant and fungus association was first observed in certain forest plant species, now is widespread and to affect most of the agricultural plant species, including many agronomic or field crops (Brady, 1995). From an economic point of view, the mycorrhizal associations have great importance because it significantly increases the availability of water and nutrients to the plants. It has been found that several essential nutrients especially from arable degraded soil to plant, resulted in better crop stand and thereby minimizing the irrigation facility and increasing productivity of several horticultural and agronomic crops in the agriculture production system. In a broad sense, fungal plant symbiotic association makes available the fungi with current photosynthates, i.e., sugar and other organic metabolites in the form of plant root exudates to use as food for VAM fungi. On the other hand, the fungi help in enhanced availability of water and other essential micronutrients, including phosphorous, which are essential for plant growth and development. The most common type of mycorrhizal association in the plant is VAM fungi, which are recently known as arbuscular mycorrhizal (AM) fungi, and its distribution in the plant kingdom is substantial covering 2/3 of land plants (Hodge, 2000).

Although several kinds of mycorrhizae present in the entire plant kingdom, with regard to practical importance in agriculture, there are mainly two types of *mycorrhiza*, like *Ectomycorrhiza* and *Endomycorrhiza*, which are actively engaged with host plants in a natural ecosystem. The former group includes hundreds of different fungal associations, primarily with tree species of horticulture and forestry systems, such as pine-oak, fir, and hemlock. These fungi are actively involved in colony formation with fungal mantle forming common mycelia network (CMN) on the root surface and stimulated by the root exudates of the plant. The fungal hyphae do not penetrate the cell wall. They only penetrate the feeder root and develop around the cells of the cortex. On the other hand, the economically most important group of fungi are *endomycorrhizae*, which penetrate the root cells and forms two types of hyphal masses within the root cells, which are known as vesicles and arbuscules (Figure 1.1). The cell walls of the root cortex of the host plant are accessed by fungal hyphae of the vesicular-arbuscular mycorrhizae (VAM), which are presently known as AM fungi (AMF). Inside the root cells, highly branched small structures are formed sporadically, which are known as arbuscules formed by fungi and act as a medium that transfers the mineral nutrients from fungi to the host plant species. Vesicles are another structure, which is served as a storage organ for plant nutrients and other organic metabolites. A vast range of agronomical crops like wheat, maize, bajra, cotton, potatoes, sugarcane, cassava, and dry land rice have the AMF association with their root systems. Many horticultural crops, including apple, grapes, citrus, cocoa, coffee, and rubber, also have AMF associations. Many

TABLE 1.1 Different Groups of Economically Important Mycorrhizae

Types of Fungi	Scientific Names/ Groups	Host Plant Species	Media pH and Rhizosphere	References
Endomycorrhiza	*Glomus* spp.	Cultivable crops, cereals, pulses, fiber crops, and horticultural crops	Alternative dry and moist friable, acid and alkaline, pH 5.5–7.5	Redecker et al., 2000; Fitter, 2005; Santos-González et al., 2006
Ericoid mycorrhizal fungi	*Rhizoscyphus ericae*	Mediterranean climate zones in chaparral vegetation systems	Muck and sandy pH >5.8, rocky, nutrient poor	Selosse et al., 2007
Orchid fungi	Myco-heterotrophic	Orchids	Deep sand sandy loam clay loam, Dry and friable	Bidartondo et al., 2002; Leake, 2004, 2005
Achlorophyllous	Monotropoid mycorrhiza	Achlorophyllous monotropoid plants	Forest soil, leaf mould, pH 4.3–8.0	Leake, 2004; Bidartondo, 2005
Arbutoid mycorrhizae	*Arctostaphylos virgata* and *Arbutus* spp.	This type of mycorrhizae involves plants of the Ericaceae subfamily Arbutoideae. It is, however, different from ericoid mycorrhizae and resembles ectomycorrhizae. Hyphae of the arbutoid do actually penetrate the outer cortical cells and fill them in coils.	Peat and loam/sand-peat Moist and dry forest soil, pH > 7.2	Leake, 2004
Ectomycorrhiza	Basidiomycota and Ascomycota	Trees or shrubs from cool, temperate boreal or montane abundantly found in forests, arctic-alpine, and dwarf shrub communities	Dry and friable, acid, and alkaline pH 4.5–8.5	Wallander et al., 2001; Taylor and Alexander, 2005

economically important trees like maple, yellow poplar, and redwood have been found to exist with AMF associations for mutual benefits.

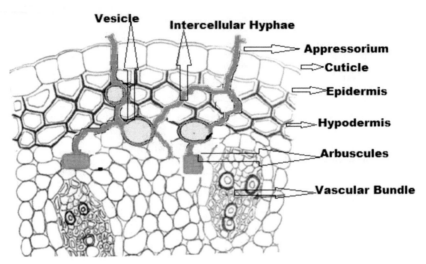

FIGURE 1.1 Mycorrhizae association with host plant at the cellular level.

1.2 ROLE OF MYCORRHIZAE

Due to rapid urbanization and increasing population, people are searching for underexploited areas to grow more food. Even to mitigate the climate change scenario and to maintain soil fertility, symbiotic microorganisms played a significant role in the agricultural production system. An interest in soil microbes and their active role and use in agriculture flared a few decades ago. Despite of that, most of the commercial products quickly flickered out, and laboratory successes had not translated to the field application to revive the microbial contribution to agricultural development. One of the few agricultural microbes that did catch hold was AMF for crop growth in almost all agricultural crops and the bacterium *Rhizobium*, which helps in nitrogen fixation through legumes in soils.

The major challenges of the new millennia are to grow food and other economic crops at adverse weather conditions to feed the billions of rural poor and to sustain food security. Under the climate change scenario, agricultural crops are exposed to the extent of biotic or abiotic stresses such as drought, flood attenuation, freezing stress, and salt loading that influence plant growth and development and thereby productivity. Water stress (WS) is one of the major abiotic stresses that directly influence plant productivity. Agriculture production also affected by drought and salinity at a major part of the world (Gueta-Dahan et al., 1997) as well as in India. Wang et al. (2003) stated that WS alone is responsible for crop loss worldwide, decimating average yields by 50%. Water deficit in plant tissue resulted in inhibition of photosynthesis, thus leading to a negative effect on yield. Some plants are drought tolerance, i.e., the ability to maintain the photosynthetic activity under WS. Cornic (1994) explained that plants response to water deficit through affecting transpiration and rapid closure of stomata. Apart from photosynthesis, respiration, metabolism, and translocation of growth promoters and nutrient ions, the colony of AMF prevent to develop other fungi related to plant diseases in the rhizosphere (Alam, 1999; Jaleel et al., 2008). Smith and Read (2008) observed that the hyphae of the AMF not only uptake the water but also absorb nutrients like phosphorus (P), nitrogen (N), potassium (K), zinc (Zn) or copper (Cu). In another study, Brachmann and Parniske (2006) also reported that the symbiotic association between plant roots and AMF was known to be one of the beautiful and widespread plant strategies to increase nutrient and water uptake to cope with adverse weather condition for mutual benefit. The soil fungi which had intra-radical mycelium (IRM), which elongated in the root cortex, absorb the nutrient for the host plant. Extra-radical mycelium (ERM) of AMF spread in the soil around the root and provide the surface area by which the AM fungus absorbs nutritional elements such as for transport and transfer to the host. Zaidi et al. (2003) reported that the presence of mycorrhizal fungi in the roots of chickpea or Bengal gram, which improves the growth and yield, especially in phosphorus-deficient soils. Even AMF imparts a positive role in various leguminous crops, particularly the enhancement of phosphate uptake and growth through their symbiotic association (Atimanav and Adholeya, 2002). In abiotic stress condition, drought reduces both nutrient and water uptake by the roots and decrease the transport of metabolites and ions from the roots to the shoots due to restriction of transpiration rates and impaired membrane permeability and thereby active transport. Mycorrhizal fungi generally increase fertilizer use efficiency. The research studies about mycorrhizal fungi in a wide range of conditions, plants, and soils were studied since 1981 to find out a viable symbiotic association for crop growth and soil health. The studies were also performed in diverse climatic conditions from field to greenhouse and even in nurseries of horticultural importance. The vesicular-arbuscular mycorrhizal fungi (AMF), i.e., *Glomus deserticola*, *G. intraradices*, *G. fasciculatum*, *G. mosseae*, and *G. etunicatum* have been identified as promising in arable agriculture world till now. Aguilera-Gomez et al. (1999) tried to demonstrate the importance of mycorrhizae for phosphate solubilization and more efficient utilization of P in a sustainable agricultural production system, particularly in the paper industry. The useful effect of mycorrhizal colonization increased leaf number, large leaf area, and greater shoot: root ratio and fruit mass as compared to non-VAM plants through higher water absorption and nutrient uptake. At high P levels, the increment of reproductive growth was 4.5 folds in mycorrhizal plants. Due to greater uptake of P from the soil and greater extra radical hyphae of fungi, enable the plants to become more vigorous and even stay strong and stout at adverse weather conditions or soil moisture stress. This was supported by Al-Karaki (1998), who found that AMF inoculation on growth and water-use efficiency (grams dry matter produced per kilogram water evapo-transpired) while studied with two wheat genotypes grown under (drought-sensitive and drought-tolerant) under stress and normal condition. The result of this study revealed that biomass production was higher in well water plants than stress plants. The water-use efficiency of mycorrhizal plants was higher than that of non-mycorrhizal plants. So far, shoot dry matter is concerned; differences between non-mycorrhizal and mycorrhizal plants were due to the positive effect of arbuscular mycorrhizal fungal-root associations. Total P mobilization of AMF colonization increased by both genotypes, in spite of their water-stress level. Mycorrhizal plants due to symbiosis had increased root growth, which paved the way for a greater absorption area for nutrient uptake. In addition to P nutrition, the absorption of water and other nutrients by mycorrhizal plants possessed better growth and development as compared to the non-mycorrhizal plant. The benefit/cost ratio was higher in even AMF, which increased host plant dry matter under WS condition in wheat than that of well-watered plants.

Absorption rates of AMF treated plant on growth and uptake of phosphorus (P) and other micronutrients like zinc (Zn), copper (Cu), and manganese (Mn) for barley plant grown with and without drought stress. The different rates of AMF inoculums were highly depended on soil moisture. Rate of root AMF colonization depended on inoculum rates resulted in even higher growth rates grown with WS than that of non-water stressed (NWS), unlike wheat. Plant biomass and nutrient content of P, Zn, Cu, and Mn increased due to an increase in inoculums rate up to 240 spores of *Glomus mosseae* per 100 g dry soil where soil moisture is an apparent character. Al-Karaki et al. (2004) believed that mycorrhizal colonization was the maximum in well-watered inoculated plants with AMF as compared to water-stressed plants. Biomass and grain yields were higher in AM mycorrhizal plots irrespective of soil moisture condition. Plants inoculated with *G. eutunicatum* had higher biomass and grain yields than that of colonized by *G. mosseae*. Shoot P and Fe concentration were higher in mycorrhizal plants as compared to non-mycorrhizal (plants) where soil moisture effect was nil. VAM fungal inoculation act as a bio-enhancer in terms of plant growth and yield by regulating uptake of P and Cu, especially under stress of moisture in the soil. Soil pore space is perfectly used by AMF and improves nutrient and water uptake in adverse conditions. In wheat crop, mycorrhizal application paves the way for a reduction in the effects of drought stress in semi-arid areas of the world.

Bethlenfalvay et al. (1988) found that the dry weights of AMF inoculated plants were greater at severe stress than that of the non-treated plant. *G. mosseae* was a very useful AMF, and its colonization at roots is insensitive to stress. The biomass of fungi along with the length of the ERM, was also greater in stressed condition rather than in non-stressed soil conditions to plants. Growth enhancement by AMF treated plants with relation to P-fertilization was attributed to increased uptake of water as well as several minerals along with efficient P absorption. The survival of AMF treated the plant to soil WS than that of non-VAM plants, even in lower wilting points are the beauty of the association of mycorrhizae. Brejeda et al. (1998) reported that seedlings inoculated with rhizosphere microflora showed positive improvement of root and shoot growth by 15-fold greater shoot and root yields, recovered up to 6-fold more N and 36-fold more P than seedlings inoculated with rhizosphere microflora. These responses were consistent for all four cultivars of switchgrass and were mainly due to AMF. Seedlings inoculated with rhizosphere populations from seeded switchgrass stands averaged 1.5-fold greater shoot and root yields than seedlings inoculated with rhizosphere populations from native prairies. AMF induced plant growth-promoting fungi (PGPF) and other rhizosphere fungi, due to symbiotic association, may be responsible for the substantial increase in shoot and root weights, P, and N uptake. This result was further confirmed by Fidelibus et al. (2001). They found that water-use characteristics of AMF treated plants differed from those of non-AM plants regarding well supplied with P. They used AM fungal isolates of different geographic origins. Inoculation of citrus seedlings with *Glomus* isolates from arid and semi-arid areas resulted in different patterns of water uptake and thereby plant growth and development. AMF induced plants and non-AMF plants had similar shoot size but different in root growth (dry weight and length). Leaf nutrient analysis of the AMF plant showed that P concentration was 12–56% higher as compared to non-AM plants. There was a positive correlation between enhanced root growth and leaf P concentration. They also found that a faster recovery from moisture stress by AM plants and AMF induced plants had lower leaf conductance than non-AM plants in case of gradual soil drying. *Glomus* mycorrhizae had a beneficial effect on Citrus growth and also in rootstock survival after grafting by providing greater P concentration to rootstock and minimizing stress effect to the grafted rootstock mitigating desiccation.

Similarly, Ruiz-Lozano et al. (1995) observed that seven *Glomus* species had effects on plant growth characters like mineral uptake of P and stomatal conductance. They also recorded additional characters like transpiration, the CO_2 exchange rate, water use efficiency, and proline accumulation under drought and irrigated condition. Seven AM fungal species showed a wide array of host plant drought tolerance. They thought that the alleviating stress appeared to be based on physiological processes rather than nutrient uptake by AM responsive host plant. AMF association protected mycorrhizal plants against WS by the endophytes. Plant physiological processes like leaf conductance and transpiration and P and K uptake were enhanced by *Glomus* spp. *Glomus deserticola*, a potential AMF mainly responsible for the aggressive colonizer and effective species, which mitigated drought enormously as compared to *G. occultum*, which was least in colonization and ultimately less water and nutrient absorption.

The study of Subramanian and Charest (1999) reflected the effect of mycorrhizal fungi on maize plants when the external hyphae of an arbuscular mycorrhizal helped in the uptake of nitrogen in fungus (*Glomus intraradices*). Rapid mycorrhizal colonization by *G. intraradices* augmented nutritional status and N uptake and its assimilation in maize plants under drought stress. N acquisition and increased capacity of its assimilation resulted to sustain the host plant in drought stress (moderate). The protein enzymatic activity like glutamine synthetase activity was increased by 30% under drought conditions, which might be due to the hyphal transport of N (NO_3 or NH_4). P status under AM colonization rendered positive root growth at stress conditions. This finding was further confirmed with the study of Tobar et al. (1994b), which entailed us the importance of the external mycelium with regard to transport of N from

15N-labeled nitrate responsible for plant nutrition in case of well-irrigated or water-stressed inoculated plants. They confirmed that AMF provided transport of N from nitrate source though the hyphal network important for dry agricultural soils (predominant with NO_3). Similarly, Tobar et al. (1994a) stated that mycorrhizal activity directly related to N uptake under water-stressed conditions and had positive effects on plant growth. They grew *Glomus mosseae* and *G. fasciculatum* in a neutral agricultural soil and found both the fungal species increased the 15N enrichment of plant tissues under water-stressed. This indicated in relatively dry soil, where a direct effect of AMF on N acquisition is essential. Both the fungi improved biomass production of the host plant. It had been found that *G. mosseae* had more effect on N uptake than that of *G. fasciculatum* under water-stressed conditions in agricultural field. This result has also corroborated the findings of Tarafdar (1996), where AMF tested on the crop *Vigna aconitifolia*, *Prosopis juliflora*, and *Cenchrus ciliaris* in marginal land. In poor fertility sandy soil and indigenous mycorrhizal fungi, they found more N assimilation shoot biomass, and other nutrients uptake like, P, K, Fe, Zn, and Cu. In arid conditions, the more pronounced effect was observed in *Prosopis juliflora* with *Glomus fasciculatum*. The effect of mycorrhizae under semi-arid condition is very much conducive to the growth and development of host plants and thereby enhanced agriculture production in arable crops.

The horticultural crops like fruits and plantation crops, which are perennial in nature thrive best under arable and marginal land. The mycorrhizas play a vital role for yield and quality production of many fruits plantation crops and medicinal plants. The fruit crops like litchi, VAM fungi association is necessary for plant growth and development and, ultimately the yield.

Subramanian et al. (2006) established a positive relationship between tomato plant with VAM fungi and recorded higher growth in terms of fruit yield, and quality attributes nutrient content, reproductive behavior, and water status under stress conditions. They found that drought tolerance was higher when inoculation with *Glomus intraradices* was done. It also showed enhanced nutritional status of the host plant, with regard to N and P. They further added that Vitamin C and reduced acidity from AM associated tomato fruit. Mycorrhizal colonization also lessened the deteriorating fruit quality caused by drought. In general, mycorrhizal fungi improved the nutritional status and water content of the host plant to withstand in field conditions facing different intensities of drought. Runjin

(1989) determined the influence of AMF on nutrient and water uptake of apple plants. They gave special emphasis to establish a relationship of water uptake at the seedling stage with VAM colonization. In sterilized soil medium, the apple plant inoculated with *Glomus versiforme* and *G. macrocarpum* enhanced drought tolerance, elements uptake, improved available water status, overall growth, and development of the plant. AM colonization increased the transpiration rate of the apple leaves and reduced the stomatal resistance. The plant is also experienced with the rate of recovery from the WS. External hyphae growth of AMF paved the way for enhancing the absorption and translocation of water. Zn and Cu absorption was strengthening by the roots and weakened the P-Cu and P-Zn interactions. Plant growth, mineral nutrition, especially nutrients uptake and water relations, are interlinked with the effects AMF. Ishii et al. (2005) reported that (AM) fungi had better fruit quality with larger trees when applied within the rhizosphere of Satsuma mandarin trees, than non-VAM control trees. Under low concentrations of applied phosphorus (P), AMF showed higher growth and development of citrus tree (Shrestha et al., 1995). Reduction of P enhances the number of AM spores in the soil, thereby, AMF infection in the roots increased. The juice content was markedly increased with AMF, even sugar-acid ratio, and TSS (total soluble solids) were increased, and chlorophyll b and the b value of peel color get affected by -P and -P+CH increased as compared to control. Carotenoids in the peel were affected by the reduction of P in soil. In particular, the reduction of P fertilizer positively affected the growth of AMF. It also increased the percentage of AM infection in citrus roots in the preliminary stage, which gave a boost to the yield of citrus trees. Thus, insoluble P fertilizer in the form of bone dust might be useful for the propagation and maintenance of AM inoculums. Waterer and Coltman (1989) stated that inoculation timing on the development and yields of AMF colonized bell peppers increased nutrient and water availability. They gave a treatment of *Glomus aggregatum* under filed conditions and greenhouse. Tissue P concentrations when low (P soil inoculation), which ultimately gave better plant weights and fruit yields than that of non-inoculated plants. Seedling inoculation was better than transplanted later inoculation. AMF association with bell pepper plant gave better P utilization, which may be a substituted for a greater portion of phosphate fertilizer. Regarding plant biomass and dry matter, Yano-Melo et al. (1999) stated the effects of three native AM fungal species collected from the Brazilian semi-arid region on *in vitro* growth of banana plantains. A significant difference in plant growth characters like height, leaf area, the weight of shoot and root, and fresh and dry matter of shoot between inoculated and non-inoculated plants were observed after three months of acclimatization. Leaf area and height of inoculated plants were approximately 57% and 32% higher. Dry matter of shoots was increased by 45–64% in mycorrhizal plants. Plants inoculated with *Glomus clarum* showed an increment of around 45% in the fresh weights of shoots and roots. Inoculation with AMF increased the growth of micro-propagated banana plantlets during the acclimatization period. This might be due to higher rates of photosynthesis and nutrient transport from AM to plantain.

Further study by Jin et al. (2005) determined the nitrogen path in AMF developed by *Glomus intraradices* when carrot roots were grown *in vitro*. After providing 15N in NH_4 to the system, the fungal synthesized predominantly arginine (Arg). It indicated that NH_4 was the most likely form of N transferred to host cells after Arg breakdown. Extrametrical mycelium, formed by AMF, played an important role in N nutrition to the carrot. While studied with cadmium (Cd) toxicity, Yu et al. (2005) found that the importance of earthworms and AMF on the phytoremediation of soils. Earthworms and mycorrhizal fungi survived to decompose Cd in ryegrass shoot biomass. Inoculation of both earthworms and mycorrhizal fungi increased ryegrass shoot Cd uptake. The earthworms, mycorrhizal fungi, and their interaction paved the way for alleviating the process of eliminating the metal contamination in soil.

1.3 FUTURE OF AMF AND SYMBIOTIC ASSOCIATION

Researchers across the world, particularly from North America, Europe, and Africa, strongly believed that AMF is very useful by saving water, and improving nutrition to plant could help the stakeholder belongs to rural poor from arable farmland. A lot of field trials have been conducted using AMF inoculums, principally on many economically sound crops like apple, plum, and cherry in temperate regions. Due to different types of soil availed in these regions, the application of phosphate fertilizer was very effective (acid soils of mountain and terrace). In developing countries, phosphate fertilizer was relatively cheaper. Therefore, farmers in developing countries belong to tropical and subtropical climate have no great motivation to switch on to commercial mycorrhizal inoculums for yield maximization. The farmers and stakeholders are not interested in AMF due to the lack of awareness. This situation might be changing with increasing phosphate prices, which might influence the decision of the farmers to switch on mycorrhizae application in their agricultural production system. The most of the tropical soils are phosphate deficient or where exceptionally low phosphate levels and phosphate retaining capacities. Use of AMF in future cases may be encouraged if degraded land brought under cultivation and prices of the phosphate fertilizer undergone high. Tropical region is where the judicious use of AMF mycorrhizal inoculums could potentially bring about a substantial amount of saving for the farmers by decreasing the amount of phosphate fertilizer and other fertilizers with phosphate combinations. Therefore, technology developed by highly facilitated laboratories of a developed country is looking for applications of AMF through biotechnological intervention to launch its products in tropical agriculture system of the developing county like Africa, Asia, and other sub-continents. It is a fact that enormous potential for mycorrhizal application in the tropics could pave the way for further research and development on AMF and could establish its scientific importance in agriculture as well as in plant kingdom.

1.4   MYCORRHIZAE AND CARBON ASSIMILATION

In soil, plant, mycorrhiza interface there was significant carbon flow from host plant to mycorrhizal fungi (Ho and Trappe, 1973; Bevege et al., 1975) and arguments from analogy with saprophytic and other symbiotic fungi (Lewis and Harley, 1965a, b; Smith et al., 1969) made sugars strong candidates (mycorrhizae) for the forms of carbon transferred. Woolhouse (1975) postulated that the host root cortex may release sugars and other metabolites to the symbiotic association (plant to fungal interfaces) by passive efflux that could be stimulated by the presence of native as well as in symbionts fungus. It has been observed that no plant transporters are involved in such type of carbon efflux (Sauer et al., 1994). Further study is essential to have better identification of any carriers (Harrison, 1999). Shachar-Hill et al. (1995) studied with nuclear (atomic) magnetic resonance spectroscopy using isotopic labeling in AMF roots and intra-radical hyphae (Solaiman and Saito, 1997). Results revealed that fungal symbionts efficiently used hexose within the root cortex. On the contrary, there was no evidence to the uptake of Glycogen (Glc), fructose (Fru), mannitol, or succinate (Suc) by the ERM (Pfeffer et al., 1999). Intra-radical hyphae utilized the modest amount of Suc (Solaiman and Saito, 1997) and raised the possibility based on fractional enrichments to make less probable *in vivo* for carbon utilization (Shachar-Hill et al., 1995). In diverse biotrophic symbiosis, including AM, host extracellular (acid) invertase imparts (Dehne, 1986; Farrar and Lewis, 1987; Snellgrove et al., 1987) a considerable amount of hexose that was dominantly taken up. For disease in powdery mildew/wheat association, recent results by Sutton et al. (1999) reveal that Suc is indeed hydrolyzed before uptake. In the case of ectomycorrhizal fungi (*Amanita muscaria* and *Hebeloma crustuliniforme*) Suc utilization depended on invertase activity in the root cortex of the host spruce (Salzer and Hager, 1991). Fungi might have active or passive carbon flow systems (Blumenthal, 1976; Lagunas, 1993), and the probable happening in AMF cannot be ruled out. The concentration gradient at the interface might be passive, which propels fungal carbon uptake and conversion (hexose to trehalose and glycogen) easy and rapid (Shachar-Hill et al., 1995). In another way, there may be active transport together with the active carrier, which penetrates the cortical cell wall easily. In fungi, H^+-hexose co-transport is a common phenomenon (Sanders, 1988).

1.5   METABOLISM IN THE MYCORRHIZAL ROOT

In the case of AMF, root carbohydrate pools are significantly observed as compared to uncolonized plants (Douds et al., 2000). Expression of hexose transporters were enhanced at the level of gene expression, probably due to the involvement of uptake of sugar in cortical cells in intraradical hyphae (Harrison, 1996). Changes in the expression of invertase had also been reported by Blee and Anderson (2000) in arbuscules of AMF fungi. As the respiration rate in mycorrhizal roots was substantially higher than that of non-mycorrhizal fungi (Shachar-Hill et al., 1995; Douds et al., 2000; Graham, 2000), the transfer of carbohydrate to the fungus and with mycorrhizal roots showed stronger sink for photosynthates than another fungi (Douds et al., 2000). Martin et al. (1998), the acquired Glc had direct incorporation in cells of AMF, and again these converted into trehalose and glycogen via mannitol. Glycogen, trehalose, and tetrahalose were detected successfully in the intra and extra-radical (Shachar-Hill et al., 1995) mycelium and germinated fungal spore (Bago et al., 1999; Pfeffer et al., 1999). Their presence might be due to cytoplasmic hexose, which may act as a buffer.

1.6   NUTRIENT UPTAKE AND TRANSPORT BY MYCORRHIZAE

The major function of mycorrhizae is the ability to uptake the nutrients such as inorganic phosphorus, mineral or organic nitrogen, and amino acids from the soil and transported to host plant and sometimes specialized transporters located on their membrane adsorb nutrient efficiently (Table 1.2). P and N transporters located in the root hairs and epidermis, uptake the nutrients directly from the soil-root interface due to high affinity is termed as direct pathways (DP) (Smith et al., 1969). On the other hand, the mycorrhizal pathway (MP) involves the nutrients absorption from the soil by the ERM and its translocation to the IRM thereby; finally, the nutrients are utilized by the host from the fungal-plant interface (Harrison et al., 2002). It is interesting to speculate that the AM fungus could use the downregulation of the DP to increase its C availability. A higher dependence on the MP for nutrient uptake has been shown to stimulate the C allocation to the root system (Nielson et al., 1998). The plant provides sucrose, which is broken down by invertase or sucrose synthase (plant-derived) into hexoses, which the fungus takes up through high-affinity monosaccharide transporter (Helber et al., 2011). Induction of plant acid invertase in the mycorrhizal interface is essential for the AMF because it is not able to utilize sucrose as a C source. It is a fact that an increase in the C availability, the AM fungus stimulates the P transport in the AM symbiosis (Bucking et al., 2005). It has been found that C also acts as a trigger for fungal N uptake and transport. Stimulation in N transport is driven by changes in fungal gene expression (Fellbaum et al., 2012).

1.7   AMF FOR NODULATION AND DISEASE RESISTANCE

In natural ecosystems, active symbiosis paves the way for the nitrogen and phosphate and potash economy, particularly on infertile soils or in marginal land (Wheeler and Miller, 1990). All bacteria and fungi present in soil did not help to the potential symbiotic association through mycorrhizae. Some bacteria help mycorrhizae for nutrient transfer and disease protection. Garbaye (1994) acknowledged about bacterial help for mycorrhizal establishment. Kosuta et al. (2003) reported that the AMF have been found to release *myc* (Mycorrhiza), responsible for accelerating the activities of the nodulation factor's inducible gene *MtEnod11* (Pour early nodulin 11 of *Medicago truncatula*). AMF also helped to the rhizobial bacteria through help in nodulation. Symbiotic association of AMF with the

TABLE 1.2 Different Functions of Mycorrhizae in Soils and Plants

Functions	Detailed Mechanism	Plant Types	References
Mobilization of nutrient from organic substrates	In mycorrhizal mycelia, enzymatic activities lead to mobilize nutrients from complex organic sources to available nutrient sources to plants in ecosystems with low nutrient availability. Mycorrhizal fungi dominated media/soil having more decomposed litter and humus from the forest or cultivable species, where they apparently mobilized N, and P to host plant.	Agricultural crops (80%)	Read and Perez-Moreno, 2003; Lindahl et al., 2005
Mycorrhizal effects with bacteria	Mycorrhizae help in nitrogen fixation by giving enhancing effect in rhizobium nodule formation in soil.	Leguminous crops	Newsham et al., 1995; Finlay, 2004; Johansson et al., 2004; Frey-Klett et al., 2007
Weathering of minerals	Mycorrhizal mycelia, either alone or in association with bacteria or other fungi in the soil, release nutrients from mineral particles and rock surfaces. The effect of climate, along with mycorrhizal association, may bring the spectacular change in the rhizosphere.	Forest plants, Arid plants	Landeweert et al., 2001; Finlay and Rosling, 2006; Wallander, 2006
Carbon cycling	Transfer of energy-rich carbon compounds from the host plant to soil microbial populations by decomposition, and here is carbon flow to the soil have been established. Production of glomalin, a wide range of glycoproteins are involved in the stability of soil aggregates, and maintaining soil health is only associated with the arbuscular mycorrhizal mycelium.	Agricultural crops	Finlay and Söderström, 1992; Johnson et al., 2002, 2004
Bioremediation	Degradation of two chlorinated aromatic herbicides 2,4-D and Atrazine by ericoid and ectomycorrhizal fungi (*Paxillus involutus* and *Suillus variegatus*) and their functional complementarities with AMF community have been established. AMF have a range of effects that contribute to the amelioration of different types of biotic and abiotic stresses experienced by their host plant, including metal	Rye, *B. campestris*, *B. napus* and *N. tabacum*	van Der Heijden et al., 1998; Ruis-Lozano et al., 2006; Colpaert, 2008; Finlay et al., 2008

TABLE 1.2 *(Continued)*

Functions	Detailed Mechanism	Plant Types	References
	toxicity (Cd), oxidative stress, water stress, and effects of soil acidification. Research studies confirm that enhanced tolerance of AM symbiotic plants to water deficit may involve modulation of drought-induced plant genes.		
Detoxification of soil	Present studies provide the evidence of a functional AMF to Cu, Zn superoxide dismutase, which may provide protection against localized host defense responses involving ROS. Solubilization of toxic metal minerals and metal tolerance by ericoid and ectomycorrhizal fungi has been proved.	*Helianthus annuus, Glycine max*	Fomina et al., 2005; Lanfranco et al., 2005
Biotic interaction	Endosymbiotic bacteria have been reported in both AM fungi and the ectomycorrhizal fungus *Laccaria bicolor*. AM fungi may influence bacterial assemblages in the rhizosphere, and its mycelia clearly play a significant role in the microbial processes influencing ecosystem functioning.	Agricultural crops (Particularly Leguminous crops)	Bertaux et al., 2003; Jargeat et al., 2004; Singh et al., 2008

economically important plant's changes in root exudation patterns induced by AM colonization (Soderberg et al., 2002; Marschner and Baumann, 2003), which also augment the influence for the composition of bacterial communities in the plant rhizosphere or mycorrhizosphere. With the help of PGPR, the manipulation of crop rhizosphere is unique and easy, and thereby bio-control of plant pathogens had remarkable results (Nelson, 2004; Ren et al., 2007). In soil, changes in the bacterial community were influenced by complex interactions between plant and fungal genotypes (Marschner and Baumann, 2003; Marschner and Timonen, 2005). In another study, Toro et al. (1997) demonstrated that there was a synergic effect between both *Enterobacter* sp. and *Bacillus subtilis* and *Glomus intraradices*, which promoted the establishment of the mycorrhizal association and increased the host plant growth.

With regard to plant disease control, Grandmaison et al. (1993) stated that there was increasing deposition of phenolic compounds in AMF root and then bound to cell walls showed resistance of AMF root to pathogenic fungi. Several mechanisms of plant disease control by mycorrhizal fungi have been developed from time to time. Mycorrhizae develop microbial networks at the rhizosphere of the host plants, which imparts the development against disease infection. Mycorrhizae act as a mechanical barrier for the pathogen infection and subsequent spread at the plant root system. During the growth process via lignifications, thickening of the cell wall happened, which in turn prevent the entry of soil born plant pathogens (Dehne and Schoenbeck, 1979). Host root also gathers or retains different metabolites (terpenes and phenols), by the signal of mycorrhizal root and propel defense system of the host plant (Krupa et al., 1973; Sampangi, 1989). AMF also helps in increased accumulation of ortho-dihydroxy phenols in roots and deter the activity of plant-pathogen (Krishna et al., 1985). An AM fungus produces antifungal and antibacterial antibiotics which prevent pathogenic infection (Marx, 1972). Mycorrhizae prevent uptake of essential nutrients in the rhizosphere for other pathogens (Reid, 1990). AMF may also harbor more actinomycetes, which are antagonistic to root pathogens (Secilia and Bagyaraj, 1987).

1.8   COMMERCIAL USE OF AM FUNGI (AMF)

Although the separation or isolation of AMF is easy, but difficult to culture in the laboratory where it did not grow with the symbiotic association. So far, AMF thrives best in a natural habitat with the association in higher plants. In most of the cases, mycorrhizal fungal propagules are used to produce commercial symbionts in a non-sterile medium, either soil or some other non-sterile substrate or organic waste. The growth of the mycorrhizal inoculums was very slow in the laboratory as compared to many other microorganisms and propagule numbers (spores, hyphae, and mycelium); in many products, which reflects their presence was in low concentration. To establish AMF in the field, therefore, requires large volumes/weights of inoculums, which are very costly. Hence, the production of AMF in the laboratory and its dispersal to the farmers at the commercial level did not suffice the cost of inorganic phosphate fertilizers. In some laboratories, scientists have developed AMF inoculums efficiently in a gel-based carrier, which is highly concentrated, but these products are yet to come into the market as commercial products. Such types of technological breakthroughs are required for AMF proliferation/propagation to be realistically used in the agriculture production system. In almost all soils, mycorrhizae are present in high or low amount in association with important field and horticultural crops. Hence, the use of AMF symbionts/inoculums may give a boost for production by maintaining soil and plant health.

In the past four years, worldwide fertilizer cost (total) has gone up to more than 400%, and farmer's dependency on the use of chemical fertilizer is dwindling. The cost of DAP, nitrogen, and other fertilizers and combined costs per acre now exceed, and farmers are trying to find a viable alternative to reduce their fertilizer and water costs (up to 30%). For them, AMF could be a good option. The market intervention of true to type mycorrhizal products could change the science and farmers' interface with profit and soil health for sustainable agriculture, which is long been expected by the country like India and other counties of arid and semi-arid world.

1.9   CONCLUSION

Of late, the use of inorganic/synthetic fertilizer in agriculture has been substantially increased across the world and across the different farming communities to mitigate the growing demand for food for people. Due to the over demand for fertilizers and plant nutrients, there are escalating cost of inputs in agriculture production system. It has been found that excessive use of fertilizers resulted in a worse impact on soil, water, ecosystem, and total environment. In such a situation, the concept of using AMF as a bio-fertilizer is cost-effective, energy-saving, eco-, and environment-friendly, and promising and perspective for rural and poor farmers of the developing countries. The practical utility of AMF in agriculture food production system is well established. To keep the soil healthy and fertility, status intact, and augment of production by supporting sound plant health through the decomposition of organic matter, AMF could be successfully used in sustainable agriculture or low externally input agriculture. In India, mycorrhizae are successfully used in fruit crops like apple, cherry, plum, citrus, banana, litchi, medicinal, and aromatic plants and grain crops like wheat, maize, pulses (mung) and oilseeds (sunflower and mustard). In addition to these, it is widely used for soil amelioration from cadmium toxicity through the cultivation of rapeseed. In Africa, Cassava is also one of the top crops, and AMF is well known for enhancing the yield and growth of the above-said crop. Cassava, the crop of Columbia, is a major staple for nearly a billion rural poor living in more than 100 countries, from South America-Africa-Asia (Brazil to Nigeria to Thailand). Producing more cassava means poor communities can eat more food or give livelihood security to the rural poor. In semiarid areas, farmers are suffered from low yield and paying more price to fertilizers. Scientists are trying to get more yields or grow more crops for the sustainable manner in the underexploited areas through the use of soil microorganisms, especially using different kinds of mycorrhizae in the agriculture production system.

KEYWORDS

- AM fungi
- arid and semi-arid climate
- carbon assimilation
- defense action
- *Glomus* spp.
- metabolism
- mycorrhizae
- nodulation
- nutrient uptake
- phosphate acquisition
- water uptake

REFERENCES

Aguilera-Gomez, L., Davies, F. T., Olalde-Portugal, V., Duray, S. A., & Phavaphutanon, L. Influence of phosphorus and endomycorrhiza (*Glomus intraradices*) on gas exchange and plant growth of Chile Ancho pepper (*Capsicum annum* L. cv. San Luis). *Photosynthetica*, **1999**, *36*(3), 149–174.

Alam, S. M. Nutrient uptake by plants under stress conditions. In: Pessarakli, M., (ed.), *Handbook of Plant and Crop Stress* (pp. 285–314). Marcel Dekker, New York, **1999**.

Al-Karaki, G. N. Benefit, cost and water-use efficiency of arbuscular mycorrhizal durum wheat grown under drought stress. *Mycorrhiza*, **1998**, *8*(1), 41–45.

Al-Karaki, G. N., McMichael, B., & John, Z. Field response of wheat to arbuscular mycorrhizal fungi and drought stress. *Mycorrhiza*, **2004**, *14*(4), 263–269.

Atimanav, G., & Adholeya, A. AM inoculations of five tropical fodder crops and inoculums production in marginal soil amended with organic matter. *Biol. Fertil. Soil*, **2002**, *35*, 214–218.

Bago, B., Pfeffer, P. E., Douds, D. D., Brouillette, J., Bécard, G., & Shachar-Hill, Y. Carbon metabolism in spores of the arbuscular mycorrhizal fungus *Glomus intraradices* as revealed by nuclear magnetic resonance spectroscopy. *Plant Physiol.*, **1999**, *121(1),* 263–271.

Bertaux, J., Schmid, M., Hutzler, P., Hartmann, A., Garbaye, J., & Frey-Klett, P. Occurrence and distribution of endobacteria in the plant-associated mycelium of the ectomycorrhizal fungus *Laccaria bicolor* S238N. *Environmental Microbiology*, **2005**, *7*(11), 1786–1795.

Bethlenfalvay, G. J., Brown, M. S., Ames, R. N., & Thomas, R. S. Effects of drought on host and endophyte development in mycorrhizal soybeans in relation to water use and phosphate uptake. *Physiologia Plantarum*, **1988**, *72*, 565–571.

Bevege, D. I., Bowen, G. D., & Skinner, M. E. Comparative carbohydrate physiology of ecto-endomycorrhizas. In: Sanders, F. E., Mosse, B., & Tinker, P. B., (eds.), Endomycorrhizas (pp. 149–174). Academic Press, London, UK, **1975**.

Bidartondo, M. I. The evolutionary ecology of myco-heterotrophy. *New Phytologist.*, **2005**, *167*(2), 335–352.

Bidartondo, M. I., Redecker, D., Hijri, I., Wiemken, A., Bruns, T. D., Domínguez, L., Sérsic, A., Leake, J. R., & Read, D. J. Epiparasitic plants specialized on arbuscular mycorrhizal fungi. *Nature*, **2002**, *419*, 389–392.

Blee, K. A., & Anderson, A. J. Defense responses in plants to arbuscular mycorrhizal fungi. In: Podila, G. K., & Douds, D. D., (eds.), Current Advances in Mycorrhizae Research (pp. 27–44). APS Press, St. Paul, **2000**.

Blumethal, H. J. Reserve carbohydrates in fungi. In: Smith, J. E., & Berry, D. R., (eds.), *The Filamentous Fungi, Biosythesis and Metabolism* (Vol. 2, pp. 292–307). Edward Arnold, London, **1976**.

Brachmann, A., & Parniske, M. The most important symbiosis on earth. *PLoS Biol.*, **2006**, *4*(7), e239, https://doi.org/10.1371/journal.pbio.0040239 (Accessed on 5 October 2019).

Brady, N. C. Organisms of the soil. In: *Nature and Properties of Soils* (pp. 253–277). Prentice-Hall of India, Ltd., New Delhi, **1995**.

Brejeda, J. J. Moser, L. E., & Vogel, K. P. Evaluation of switchgrass rhizosphere microflora for enhancing seedling yield and nutrient uptake. *Agron. J.*, **1998**, *90*, 753–758.

Bücking, H., & Shachar-Hill, Y. Phosphate uptake, transport, and transfer by the arbuscular mycorrhizal fungus *Glomus intraradices* is stimulated by increased carbohydrate availability. *New Phytologist*, **2005**, *165*(3), 899–912.

Cornic, G. Drought stress and high effects on leaf photosynthesis. In: Baker, N. R., & Boyer, J. R., (eds.), *Photo Inhibition of Photosynthesis: From Molecular Mechanisms to the Field* (pp. 297–313). Bios Scientific Publishers, Oxford, **1994**.

Dehne, H. W. Improvement of the VA mycorrhiza status in agriculture and horticulture. *Trans XIII Int. Cong. Soil Sci. Hamburg*, **1986**, *6*, 817–825.

Dehne, H. W., & Schoenbeck, F. The influence of endotrophic mycorrhizae on plant disease. I. Colonization of tomato plants by *Fusarium oxysporum* F. sp. *lycopersici*. *Phytopathology*, **1979**, *95*, 105–105.

Douds. D. D., Pfeffer, P. E., & Shachar-Hill, Y. Carbon partitioning, cost and metabolism of arbuscular mycorrhizas. In: Kapulnick, Y., & Douds, D. D., (eds.), *Arbuscular Mycorrhizas: Physiology and Function* (pp. 107–129). Kluwer Academic Publishers, Dordrecht, The Netherlands, **2000**.

Farrar, J. F., & Lewis, D. H. Nutrient relations in bio-trophic infections. In: Pegg, G. F., & Ayres, P. G., (eds.), *Fungal Infection of Plants* (pp. 104–113). Cambridge University Press, Cambridge, **1987**.

Fellbaum, C. R., Gachomo, E. W., Beesetty, Y., Choudhari, S., Strahan, G. D., Pfeffer, P. E., Kiers, E. T., & Bucking, H. Carbon availability triggers fungal nitrogen uptake and transport in the arbuscular mycorrhizal symbiosis. *Proceedings of the National Academy of Sciences USA*, **2012**, *109*, 2666–2671.

Fidelibus, M. W., Martin, C. A., & Stutz, J. C. Geographic isolates of *Glomus* increase root growth and whole-plant transpiration of *Citrus* seedlings grown with high phosphorus. *Mycorrhiza*, **2001**, *10*, 231–236.

Finlay, R. D. Mycorrhizal fungi and their multifunctional roles. *Mycologist*, **2004**, *18*(2), 91–96.

Finlay, R. D. Söderström, B., & Allen, M. F. Mycorrhiza and carbon flow to soil. In: *Mycorrhizal Functioning* (pp. 134–160). Chapman and Hall, London, **1992**.

Finlay, R. D., Lindahl, B. D., & Taylor, A. F. S. Responses of mycorrhizal fungi to stress. In: Avery, S., Stratford, M., & Van West, P., (eds.), *Stress in Yeasts and Filamentous Fungi* (pp. 201–220). Elsevier, Amsterdam, **2008**.

Finlay, R. D., Rosling, A., & Gadd, G. M. Integrated nutrient cycles in forest ecosystems, the role of ectomycorrhizal fungi. In: *Fungi in Biogeochemical Cycle* (pp. 28–50). Cambridge University Press, Cambridge, UK, **2006**.

Fitter, A. H. Darkness visible, reflections on underground ecology. *J. Ecol.*, **2005**, *93*, 231–243.

Fomina, M. A., Alexander, I. J., Colpaert, J. V., & Gadd, G. M. Solubilization of toxic metal minerals and metal tolerance of mycorrhizal fungi. *Soil Biology and Biochemistry*, **2005**, *37*, 851–866.

Frank, A. B. Über die auf Würzelsymbiose beruhende Ehrnährung gewisser Bäum durch unterirdische Pilze. *Berichte der Deutschen Botanischen Gesellschaft (in German)*. **1885**, *3*, 128–145.

Frey-Klett, J., Garbaye, J., & Tarkka, M. The mycorrhiza helper bacteria revisited. *New Phytologist*, **2007**, *176*(1), 22–36.

Garbaye, J. Helper bacteria: A new dimension to the mycorrhizal symbiosis. *New Phytol.*, **1994**, *128*, 197–210.

Graham, J. H. Assessing costs of arbuscular mycorrhizal symbiosis agroecosystems fungi. In: Podila, G. K., & Douds, D. D., (eds.), *Current Advances in Mycorrhizae Research* (pp. 127–140). APS Press, St. Paul, **2000**.

Grandmaison, J., Olah, G. M., Van Calsteren, M. R., & Furlan, V. Characterization and localization of plant phenolics likely involved in the pathogen resistance expressed by endomycorrhizal roots. *Mycorrhiza*, **1993**, *3*, 155–164.

Gueta-Dahan, Y., Yaniv, Z., Zilinskas, A., & Ben, H. G. Salt and oxidative stress; similar and specific responses and their relation to salt tolerance in citrus. *Planta*, **1997**, *203*(4), 460–469.

Harrison, M. J. A sugar transporter from *Medicago truncatula*: Altered expression pattern in roots during vesicular-arbuscular (VA) mycorrhizal associations. *Plant J.,* **1996**, *9*(4), 491–503.

Harrison, M. J. Molecular and cellular aspects of the arbuscular mycorrhizal symbiosis. *Annu. Rev. Plant Physiol.,* **1999**, *50,* 361–389.

Helber, N., Wippel, K., Sauer, N., Schaarschmidt, S., Hause, B., & Requena, N. A versatile monosaccharide transporter that operates in the arbuscular mycorrhizal fungus *Glomus* sp. is crucial for the symbiotic relationship with plants. *Plant Cell*, **2011**, *23*, 3812–3823.

Hodge, A. Microbial ecology of the arbuscular mycorrhiza. *Microbial Ecology*, **2000**, *32*(2), 91–96.

Ishii, T., Kirino, S., Zeng, M., Aikawa, J., Matsumoto, I., & Kadoya, K. *Effect of the Reduction of Phosphorus Fertilizer for Citrus Iyo Orchards on the Development of Vesicular-Arbuscular Mycorrhizae and the Quality of Fruit.* Ehime University, Japan, **2005**.

Jaleel, C. A., Gopi, R., Sankar, B., Gomathinayagam, M., & Panneerselvam, R. Differential responses in water use efficiency in two varieties of *Catharanthus roseus* under drought stress. *Comp. Rend. Biol.*, **2008**, *331*, 42–47.

Jargeat, P., Cosseau, C., Ola'h, B., Jauneau, A., Bonfante, P., Batut, J., & Bécard, G. Isolation, free-living capacities, and genome structure of "*Candidatus* Glomeribacter gigasporarum," the endocellular bacterium of the mycorrhizal fungus *Gigaspora margarita*. *J. Bact.*, **2004**, *186*, 6876–6884.

Jin, H., Pfeffer, P. E., Douds, D. D., Piotrowski, E., Lammers, P. J., & Sachar-Hill, Y. The uptake, metabolism, transport and transfer of nitrogen in an arbuscular mycorrhizal symbiosis. *New Phytologist*, **2005**, *168*, 687–696.

Johansson, J., Paul, L., & Finlay, R. D. Microbial interactions in the mycorrhizosphere and their significance for sustainable agriculture. *FEMS Microbiol. Ecol.*, **2004**, *48*(1), 1–13.

Johnson, D., Leake, J. R., Ostle, N., Ineson, P., & Read, D. J. *In situ* $^{13}CO_2$ pulse-labeling of upland grassland demonstrates that a rapid pathway of carbon flux from arbuscular mycorrhizal mycelia to the soil. *New Phytologist*, **2002**, *153*(2), 327–334.

Kirk, P. M., Cannon, P. F., David, J. C., & Stalpers, J. *Ainsworth and Bisby's Dictionary of the Fungi.* CAB International, Wallingford, UK, **2001**.

Kosuta, S., Chabaud, M., Lougnon, G., Gough, C., Denarie, J., Barker, D. G., & Becard, G. A diffusible factor from arbuscular mycorrhizal fungi induces symbiosis-specific MtENOD11 expression in roots of *Medicago truncatula*. *Plant Physiol.*, **2003**, *131*(3), 952–962.

Krishna, K. R., Shetty, K. G., Dart, P. J., & Andrews, D. J. Genotype dependent variation in mycorrhizal colonization and response to inoculation of pearl millet. *Plant Soil*, **1985**, *86*(1), 113–125.

Krupa, S., Anderson, J., & Marx, D. H. Studies on ecto-mycorrhizae of pine. IV. Volatile organic compounds in mycorrhizal and nonmycorrhizal root systems of *Pinus echinata* Mill. *Eur. J. Plant Pathol.*, **1973**, *3*, 194–200.

Lagunas, R. Sugar transport in *Saccharomyces cerevisiae*. *FEMS Microbiol. Rev.*, **1993**, *10*(3/4), 229–242.

Landeweert, R., Hofflund, E., Finlay, R. D., & Van Breemen, N. Linking plants to rocks: Ectomycorrhizal fungi mobilize nutrients from minerals. *Trends Ecol. Evol.*, **2001**, *16*(5), 248–254.

Lanfranco, L., Novero, M., & Bonfante, P. The mycorrhizal fungus *Gigaspora margarita* possesses a Cu Zn superoxide dismutase that is up-regulated during symbiosis with legume hosts. *Plant Physiol.*, **2005**, *137*, 1319–1330.

Leake, J. R. Myco-heterotroph/epiparasitic plant interactions with ectomycorrhizal and arbuscular mycorrhizal fungi. *Curr. Opi. Plant Biol.*, **2004**, *7*(4), 422–428.

Leake, J. R. Plants parasitic on fungi, unearthing the fungi in myco-heterotrophs and debunking the 'saprophytic' plant myth. *Mycologist*, **2005**, *19*(3), 113–122.

Lindahl, B. D., Finlay, R. D., Cairney, J. W. G., Dighton, J., Oudemans, P., & White, J. Enzymatic activities of mycelia in mycorrhizal fungal communities. In: Dighton, J., Oudemans, P., & White, J., (eds.), *The Fungal Community, its Organization and Role in the Ecosystem* (pp. 331–348). Marcel Dekker, New York, **2005**.

Marschner, P., & Baumann, K. Changes in bacterial community structure induced by mycorrhizal colonization in split-root maize. *Plant Soil*, **2003**, *251*(2), 279–289.

Marschner, P., & Timonen, S. Interactions between plant species and mycorrhizal colonization on the bacterial community composition in the rhizosphere. *Appl. Soil. Ecol.*, **2005**, *28*(1), 23–36.

Martin, F., Boiffin, V., & Pfeffer, P. E. Carbohydrate and amino acid metabolism in the *Eucalyptus globulus-Pisolithus tinctorius* ectomycorrhiza during glucose utilization. *Plant Physiol.*, **1998**, *118*(2), 627–635.

Marx, D. H. Ectomycorrhizae as biological deterrents to pathogenic root infections. *Ann. Rev. Phytopathol.*, **1972**, *10*, 429–441.

Nelson, L. M. Plant growth promoting rhizobacteria (PGPR): Prospects for new inoculants. *Crop Manag.*, **2004**, 3, doi: 10.1094/CM-2004-0301-05-RV.

Newsham, K. K., Fitter, A. H., & Watkinson, A. R. Multi-functionality and biodiversity in arbuscular mycorrhizas. *Trends Ecol. Evol.*, **1995**, *10*(10), 407–411.

Nielsen, K., Sørensen, P. G., Hynne, F., & Busse, H. G. Sustained oscillations in glycolysis: An experimental and theoretical study of chaotic and complex periodic behavior and of quenching of simple oscillations. *Biophys. Chem.*, **1998**, *72*(1/2), 49–62.

Pfeffer, P. E., Douds, D. D., Bécard, G., & Shachar-Hill, Y. Carbon uptake and the metabolism and transport of lipids in and arbuscular mycorrhiza. *Plant Physiol.*, **1999**, *120*(2), 587–598.

Read, D. J., & Perez-Moreno, J. Mycorrhizae and nutrient cycling in ecosystems: A journey towards relevance? *New Phytologist*, **2003**, *157*, 475–492.

Redecker, D., Kodner, R., & Graham, L. E. Glomalean fungi from the Ordovician. *Science*, **2000**, *289*(5486), 1920–1921.

Reid, C. P. P. Mycorrhizas. In: Lynch, J. M., (ed.), *The Rhizosphere* (pp. 281–315). John Wiley and Sons Ltd., Chichester, UK, **1990**.

Ren, T. R., Xie, Y. H., Zhu, W. W., Li, Y. H., & Zhang, Y. Activities and toxicity of a novel plant growth regulator 2-furan-2-yl-[1,3]dioxolane. *J. Plant Growth Regul.*, **2007**, *26*, 362–368.

Ruiz-Lozano, J. M., Azcon, R., & Gomez, M. Effects of arbuscular-mycorrhizal *Glomus* species on drought tolerance: Physiological and nutritional plant responses. *App. Environ. Microbiol.*, **1995**, *61*, 456–460.

Ruiz-Lozano, J. M., Porcel, R., & Aroca, R. Does the enhanced tolerance of arbuscular mycorrhizal plants to water deficit involve modulation of drought induced plant genes? *New Phytologist*, **2006**, *171*(4), 693–698.

Runjin, L. Effects of vesicular-arbuscular mycorrhizas and phosphorus on water status and growth of apple. *J. Plant Nutr.*, **1989**, *12*, 997–1017.

Salzer, P., & Hager, A. Sucrose utilization of the ectomycorrhizal fungi *Amanita muscaria* and *Hebeloma crustuliniforme* depends on the cell wall invertase activity of their host *Picea abies*. *Bot. Acta.*, **1991**, *104*, 439–445.

Sampangi, R. K. Some recent advances in the study of fungal root diseases. *Ind. Phytopth.*, **1989**, *22*, 1–17.

Sanders, D. Fungi. In: Baker, D. A., & Hall, J. L., (eds.), *Solute Transport in Plant Cells and Tissues* (pp. 106–165). Longman Scientific and Technical, Harlow, **1988**.

Santos-González, J. C., Finlay, R. D., & Tehler, A. Seasonal dynamics of arbuscular mycorrhizal root colonization in a semi-natural grassland. *App. Environ. Microbiol.*, **2007**, *73*, 5613–5623.

Sauer, N., Baier, K., Gahrtz, M., Stadler, R., Stolz, J., & Truerait, E. Sugar transport across the plasma membranes of higher plants. *Plant Mol. Biol.*, **1994**, *26*, 1671–1679.

Secilia, J., & Bagyaraj, D. J. Bacteria and actinomycetes associated with pot cultures of vesicular-arbuscular mycorrhizas. *Can. J. Microbiol.*, **1987**, *33*(12), 1069–1073.

Selosse, M. A., Setaro, S., Glatard, F., Richard, F., Urcelay, C., & Weiss, M. Sebacinales are common mycorrhizal associates of Ericaceae. *New Phytologist*, **2007**, *174*(4), 864–878.

Shachar-Hill, Y., Pfeffer, P. E., Douds, D., Osman, S. F., Doner, L. W., & Ratcliffe, R. G. Partitioning of intermediate carbon metabolism in VAM colonized leek. *Plant Physiol.*, **1995**, *108*, 7–15.

Shrestha, Y. H., Ishii, T., & Kadoya, K. Effect of vesicular-arbuscular mycorrhizal fungi on the growth, photosynthesis, transpiration and the distribution of photosynthates of bearing satsuma mandarin (*Citrus reticulata*) trees. *J. Jpn. Soc. Hortic. Sci.*, **1995**, *64*, 517–525.

Singh, B. K., Naoise, N., Ridgway, K. P., McNicol, J., Young, J. P. W., Daniell, T. J., Prosser, J. I., & Millard, P. Relationship between assemblages of mycorrhizal fungi and bacteria on grass roots. *Environ. Microbiol.*, **2008**, *10*(2), 534–541.

Smith, D. C., Muscatine, L., & Lewis, D. H. Carbohydrate movement from autotroph to heterotroph in parasitic and mutualistic symbioses. *Biol. Rev.*, **1969**, *44*, 17–90.

Smith, S. E., & Read, D. J. *Mycorrhizal Symbioses*. Academic Press, UK, **2008**.

Snellgrove, R. C., Stribley, D. P., & Hepper, C. M. Host-endophyte relationships: Invertase in roots. *Rothamsted Exp. Stn. Rep.*, **1986**, *1*, 142.

Soderberg, K. H., Olsson, P. A., & Baath, E. Structure and activity of the bacterial community in the rhizosphere of different plant species and the effect of arbuscular mycorrhizal colonization. *FEMS Microbiol. Ecol.*, **2002**, *40*(3), 223–231.

Solaiman, M. D., & Saito, M. Use of sugars by intraradical hyphae of arbuscular mycorrhizal fungi revealed by radiorespirometry. *New Phytol.*, **1997**, *136*, 533–538.

Subramanian, K. S., & Charest, C. Acquisition of N by external hyphae of an arbuscular mycorrhizal fungus and its impact on physiological response in maize under drought-stressed and well-watered conditions. *Mycorrhiza*, **1999**, *9*(2), 69–75.

Subramanian, K. S., Santhanakrishnan, P., & Balasubramanian, P. Responses of field grown tomato plants to arbuscular mycorrhizal fungal colonization under varying intensities of drought stress. *Scientia Horticulturae*, **2006**, *107*(3), 245–253.

Sutton, P. N., Henry, M. J., & Hall, J. L. Glucose, and not sucrose, is transported from wheat to wheat powdery mildew. *Planta,* **1999**, *208*(3), 426–430.

Tarafdar, J. C. The role of vesicular arbuscular mycorrhizal fungi on crop, tree and grasses grown in an arid environment. *J. Arid. Environ.*, **1996**, *34*(2), 197–203.

Taylor, A. F. S., & Alexander, I. The ectomycorrhizal symbiosis: Life in the real world. *Mycologist,* **2005**, *19*, 102–112.

Tobar, R. M., Azcon, R., & Barea, J. M. The improvement of plant N acquisition from an ammonium-treated, drought-stressed soil by the fungal symbionts in arbuscular mycorrhizae. *Mycorrhiza,* **1994a**, *4*(3), 105–108.

Tobar, R. R., Azcon, R., & Barea, J. M. Improved nitrogen uptake and transport from 15N labeled nitrate by external hyphae of arbuscular mycorrhizal under water-stressed conditions. *New Phytologist,* **1994b**, *126*(1), 119–122.

Toro, M., Azcon, R., & Barea, J. Improvement of arbuscular mycorrhiza development by inoculation of soil with phosphate-solubilizing rhizobacteria to improve rock phosphate bioavailability [^{32}P] and nutrient cycling. *Appl. Environ. Microbiol.*, **1997**, *63*, 4408–4412.

Van Der Heijden, M. G. A., Klironomos, J. N., Ursic, M., Moutoglis, P., Streitwolf-Engel, R., Boller, T., Wiemken, A., & Sanders, I. R. Mycorrhizal fungal diversity determines plant biodiversity, ecosystem variability and productivity. *Nature,* **1998**, *396*, 69–72.

Wallander, H. Mineral dissolution by ectomycorrhizal fungi. In: Gadd, G. M., (ed.), *Fungi in Biogeochemical Cycles* (pp. 28–50). Cambridge University Press, Cambridge, UK, **2006**.

Wallander, H., Nilsson, L. O., Hagerber, D., & Bååth, E. Estimation of the biomass and seasonal growth of external mycelium of ectomycorrhizal fungi in the field. *New Phytologist,* **2001**, *151*(3), 752–760.

Wang, W., Vinour, B., & Altman, A. Plant responses to drought, salinity and extreme temperatures: Towards genetic engineering for stress tolerance. *Planta,* **2003**, *218*(1), 1–14.

Waterer, D. R., & Coltman, R. R. Response of mycorrhizal bell peppers to inoculation timing, phosphorus, and water stress. *HortScience,* **1989**, *24*(4), 688–690.

Wheeler, C. T., & Miller, I. M. Uses of actinorhizal plants in Europe. In: Schwintzer, C. R., & Tjepkema, J. D., (eds.), *The Biology of Frankia and Actinorhizal Plants* (pp. 365–390). Academic Press, New York, USA, **1990**.

Woolhouse, H. W. Membrane structure and transport problems considered in relation to phosphorus and carbohydrate movement and the regulation of the endotrophic mycorrhizal associations. In: Sanders, F. E., Mosse, B., & Tinker, P. B., (eds.), *Endomycorrhizas* (pp. 209–223). Academic Press, London, UK, **1975**.

Yano-Melo, A. M., Saggin. O. J., Lima-Filho, J. M., Melo, N. F., & Maia, L. C. Effect of arbuscular mycorrhizal fungi on the acclimatization of micropropagated banana plantlets. *Mycorrhiza,* **1999**, *9*, 119–123.

Yu, X., Cheng, J., & Wong, M. H. Earthworm-mycorrhiza interaction on Cd uptake and growth of ryegrass. *Soil Biology and Biochemistry,* **2005**, *37*(2), 195–201.

Zaidi, A., Khan, M. S., & Amil, M. Interactive effect of rhizotrophic micro organisms on yield and nutrient uptake of chickpea (*Cicer arietinum* L.) *Eur. J. Agron.,* **2003**, *19*(1), 15–21.

CHAPTER 2

PLANT GROWTH-PROMOTING MICROBIOME NETWORK

ÖZLEM AKKAYA, MINE GÜL ŞEKER, and YELDA ÖZDEN ÇIFTÇI

Department of Molecular Biology and Genetics,
Gebze Technical University, Kocaeli–41400, Turkey

2.1 INTRODUCTION

Bacteria and plants have a huge relevance stemming from symbiotic relationships that continue for a long time and have an influence on their evolutionary process. To date, endophytes have been pervasively found to exist almost in every plant tissue (Guerin et al., 1898; Zhang, 2006; Staniek et al., 2008). Bacterial endophytes not only appear in plant microbiome just as rhizospheric and epiphytic microbiota but also internal tissues (endophytic) (Hardoim et al., 2015). Regardless of the number of bacteria in a special soil sample, bacteria may affect plants in three ways; beneficial, harmful, or neutral (when viewed from the plant standpoint).

The bacteria that can promote plant growth, i.e., plant growth-promoting rhizobacteria (PGPR), include those that free-living and form certain symbiotic relationships with plants (e.g., *Rhizobia* spp. and *Frankia* spp.) while bacterial endophytes can colonize some or a part of the internal tissues of the plant. Despite the differences between these bacteria, they all use the same mechanisms. Plant growth-promoting bacteria (PGPB) can generally promote plant growth by either facilitating direct source or modulating plant growth regulator levels, or indirectly by acting as biocontrol bacteria via reducing the inhibitory effects of various pathogenic agents on plant growth and development (Glick et al., 1995).

Since the beginning of 21st century, the probability to investigate possible advantages and applications of PGPB's or endophytic microbes in the fields of medicine, pharmacy, and agriculture has become practicable due to intensive research and exploration of plant-microbe interaction systems as they produce a diverse range of biologically active secondary metabolites (Strobel et al., 2004; Gunatilaka, 2006; Zhang et al., 2006; Staniek et al., 2008) that supply host plant tolerance against various biotic and abiotic (heat, salt, disease, and drought) stresses (Stone et al., 2000; Redman et al., 2002; Rodriguez and Redman, 2008). However, up to now, commercially 'sustained production' of these pharmaceutically valuable metabolites could not be achieved (Kusari et al., 2014). Thus, the interaction between endophyte and host organism needs more attention. With these aims, endophytes starting from its definition, impact, and localization in its host plant together with its potential, will be reviewed in this chapter.

2.2   THE TERM OF ENDOPHYTIC BACTERIA

Endophytes can be defined as "a group of microorganism that infects the internal tissues of the plant regardless of any important symptom of infection and/or any visible sign of disease and lives in mutualistic relationships with plants at least in a part of their life cycle" (Bacon and White, 2000). The term of "endophyte" was introduced to the literature initially by de Bary (1866), and since then this term is used to describe a large group of organisms including bacteria, fungi, and insects (Feller, 1995; Marler et al., 1999; Kobayashi and Palumbo, 2000; Stone et al., 2000).

Plant-associated bacteria include endophytic, rhizospheric, and phyllospheric bacteria. Since these endophytic bacteria can proliferate inside the plant tissue, they interact closely with their hosts; they compete less for nutrients and are protected more than the negative changes in the environment when compared to the bacteria in the rhizosphere or in the phyllosphere (Reinhold-Hurek and Hurek, 1998; Beattie, 2007). Phyllosphere can be called the external regions of plant parts such as stalks, flowers, fruits, and leaves. Rhizospheric bacteria live directly vicinity of the roots and, therefore, root exudates are thought to be a major influence on the diversity of microorganisms in the rhizosphere (Kloepper et al., 1991).

Endophytic colonization of bacteria could be observed in whole plant tissues or organs (Turner et al., 2013), starting from the root to meristematic cells (Pirttilä et al., 2000), including pollen (Madmony et al., 2005). Plants containing a large number of microbial species are complex microecosytems (McInroy and Kloepper, 1995), some of which may predominantly be found (Van Peer et al., 1990). Differently, in some cases, only one endo-phyte could be found in its host plant (Seker et al., 2017), which may be evaluated as species-specific. Sometimes endophytic species cannot be isolated because they are present in plant tissues at very low concentrations or unculturable.

Different methods have been used by many investigators to isolate bacterial endophytes (Hallmann et al., 1997). Endophytes are isolated from the initial surface sterilization after culturing from a ground tissue extract or by direct culturing of plant tissues or by using appropriate media for bacteria (Rai et al., 2007). Traditionally, to define endophytic bacteria, the morphological characteristics of them have been examined, and biochemical tests have been carried out. Reinhold-Hurek co-workers (1998) reported the criteria for identifying "true" endophytic bacteria, and accordingly, it should be isolated from surface-sterilized tissues together with its microscopic evidence that shows "tagged" bacteria in the inner part of the plant. When the second criterion cannot be fulfilled, it is proposed to use the term 'putative endophyte.' It is also possible to identify true endophytes by their capability to re-infect seedlings that have been surface sterilized.

Ribosomal DNA (rDNA) internal transcribed spacer (ITS) sequence analysis is generally utilized for the determination of microorganisms. rDNA ITS offers considerable evidence for revealing phylogenetic asso-ciations as genera or species (Youngbae et al., 1997). Along with recent advances in biotechnology, more molecular studies have been carried out with endophytes, by using DNA markers, DNA cloning, and expression studies. The denaturing gradient gel electrophoresis (DGGE) profiles of the 16S rRNA gene sequencing were utilized to settle on various endophytic bacteria that could not be cultured compared to the band profiles obtained from culturable endophytes from the citrus plant (Araújo et al., 2002). Bacterial automatic ribosomal intergenic spacer analysis (B-ARISA) and pyrosequencing were also used to study bacterial endophytic populations of potato (Manter et al., 2010).

Metagenomic studies could also be utilized to find microorganisms that could not be easily detected from diverse environments. For instance, a 1-aminocyclopropane-1-carboxylate deaminase gene (*acdS*) operon was determined from an uncultured endophytic microorganism colonizing potato by this approach. In addition, metagenomic analysis can be comple-mentary to PCR-based methods and supply information on whole operon profiles (Nikolic et al., 2011). The complete genome of endophytic plant growth-promoting gamma-proteobacterium *Enterobacter* sp. 638 isolated from the stem of poplar, was sequenced (Taghavi et al., 2010). According to sequencing result, it has 4,518,712 bp chromosome and a 157749 bp plasmid (pENT638-1) that contain genes expressed for adaptation (i.e., colonization/establishment inside the plant, root adhesion, biotic stress protection, and enhanced plant growth). The techniques such as genome sequencing, comparative genomics, microarray, next-generation gene sequencing, metagenomics, and metatranscriptomics are used to find out the host-endophyte relationship (Kaul et al., 2016).

Thanks to molecular biology techniques, more than 200 bacterial genera from 16 phyla including *Actinobacteria, Aquificae, Acidobacteria, Bacteroidetes, Chloroflexi, Cyanobacteria, Deinococcus-Thermus, Cholorobi, Fusobacteria, Verrucomicrobiae, Firmicutes, Nitrospira, Proteobacteria, Gemmatimonadetes, Planctomycetes,* and *Spirochaetes* (Berg and Hallmann, 2006; Sessitsch et al., 2012) have been indicated as endophytes since their reliable initial isolation (Samish et al., 1959; Mundt and Hinkle, 1976). Among these phyla, Actinobacteria, Firmicutes, and Proteobacteria contain the members of *Enterobacter, Herbaspirillum, Azoarcus, Bacillus, Burkholderia, Pseudomonas, Streptomyces, Gluconobacter, Stenotrophomonas,* and *Serratia* (Suzuki et al., 2005; Krause et al., 2006; Bertalan et al., 2009; Ryan et al., 2009; Taghavi et al., 2009, 2010; Deng et al., 2011; Pedrosa et al., 2011; Weilharter et al., 2011), were notified as predominant. As species of these genera omnipresent in the soil/rhizosphere, they could be nominated as the main source of endophytic colonizers (Hallmann and Berg, 2006). Phyllosphere, the anthosphere, and seeds could also represent the other possible endophyte sources (Compant et al., 2010). Moreover, endophytic bacteria can also be used as a vector for the transfer of interested genes into plants. In genetic studies, plasmids of endophytic bacteria can be used as transmission vectors instead of the whole organism (Berde et al., 2010).

Endophytes can accelerate the plant growth by (i) supplying the necessary nutrients, e.g., nitrogen fixation, phosphate solubility or iron chelation, (ii) preventing pathogenic infections through antifungal or antibacterial agents, (iii) outcompeting pathogens for nutrients by siderophore production, or (iv) establishing the plant's systemic resistance (SR) through four interrelated mechanisms: phytostimulation, biofertilization, bioremediation, and biocontrol (Bloemberg and Lugtenberg, 2001) (Figure 2.1).

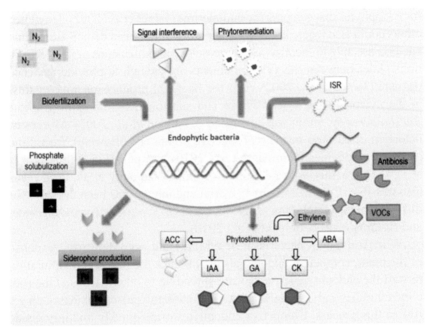

FIGURE 2.1   Schematic diagram of the common mechanisms in plant-bacterial endophyte interactions.

2.3   LOCALIZATION AND DISTRIBUTION OF ENDOPHYTIC BACTERIA INSIDE THE PLANTS

Endophytes are mostly intercellular in terms of location in host cells and tissues. It has been found that endophytic bacteria can colonize different plant parts, including roots, tubers, stems, leaves, ovules, and seeds, and also immature flower buds (Mishagi and Donndelinger, 1990; Benhizia et al., 2004). However, most plants have a higher number of endophytes in their roots than other tissue parts (Rosenblueth and Martínez-Romero, 2004). They can become intracellular and enter the cytoplasm of the host cell or placed on the periplasmic surface (Thomas and Sekhar, 2014; White et al., 2014a). Passive and active several colonization routes have been identified according to the strain (Hallmann, 2001). The progression of these pathways allows the bacteria to migrate from the rhizoplane to the cortical cell layer (Mercado-Blanco et al., 2014). Xylem vascular system is the principal transport way for systemic colonization of inner plant parts for endophytes that can pass the endodermis (James et al., 2002). However, others could also locate in intercellular areas. It has been demonstrated that bacteria are able to localize xylem vessels, and dimensions of the perforations plates between the xylem elements are enough to provide bacterial passage (James et al., 2002) even the bacterial progression was so slow in the xylem. In the xylem tissues and substomal chambers of grape, a *Burkholderia* sp. strain was identified (Compant et al., 2005). Moreover, detection of bacteria inside reproductive organs (i.e., flowers), fruits, and seeds of grapevines (Compant et al., 2010) and pumpkin (Fürnkranz et al., 2012), together with the pollen of pine (Madmony et al., 2005) was also possible. Colonization by bacterial endophytes has been declared for different plants, including *Angiospermae, Bryophytes, Gymnospermae,* and *Pteridophytes* (Compant et al., 2010).

When endophytic bacterial are inside the plant, according to the response of the plant, competent endophytes may induce the necessary stimulation to start the endophytic life cycle and spreading to other tissues of the root cortex and beyond. Endoglucanases and endopolygalacturonidases play a role in this process. Competent endophytes often quickly multiply inside the plant as soon as they reach high cell numbers (Barraquio et al., 1997). Endophytic population sizes correlate positively with the developmental stage of the plant and progressively increase and reach maximal stages of the seedling stage (Van Overbeek et al., 2008). However, it is remained unknown whether endophytes need to reach a specific organ or tissue for optimal performance of their functions.

As previously mentioned, the isolation of bacteria from surface-sterilized plant tissues does not prove that it is a "true endophyte," so it is necessary to demonstrate that the bacteria are in plants by means of various labeling techniques such as immunological detection of bacteria, fluorescence tags, confocal laser scanning microscopy and immuno-gold labeling in combination with transmission electron microscopy, specific oligonucleotide probes (Chelius and Triplett, 2000; Hartmann et al., 2000) or bacterial marking with green fluorescent protein (GFP) (Verma et al., 2004). It has been observed that bacterial cells initially colonize the rhizosphere via microscopic visualization that permits the analysis of strains labeled by (i) gfp- or gusA (Charkowski et al., 2002), (ii) immu-nomarkers, or (iii) fluorescence *in situ* hybridization (FISH) (Gamalero et al., 2003; Loy et al., 2007). Transformation of microorganisms with plasmids bearing GFP or insertion of GFP into their genomes ensures a helpful experimental tool for the determination of the behavior of specific microbes when interacting with host tissues and cells (Chalfie et al., 1994; Valdivia et al., 1996). GFP is nontoxic; therefore, it does not interfere with cell function, and it provides a unique and visual phenotype for studying the behaviors of microorganisms inside the plant tissues. For example, after the introduction of GFP-tagged *Sinorhizobium meliloti* into rice, it was reported that its growth under sterile conditions and the endophyte population densities within plants were very high (Chi et al., 2005). After inoculation with poplar trees, Germaine et al. (2004) found that various GFP-labeled endophytes were determined in the inner tissues of the poplar tree. *Pantoea agglomerans* 33.1, tagged with the GFP gene, was observed to have colonized *Eucalyptus* roots, primarily in intercellular spaces, stems, and xylem vessel (Ferreira et al., 2008).

The presence of bacterial endosymbionts was demonstrated using electron microscopy in the cells of *in vitro* peach palm shoots (De Almeida et al., 2009). In another study, the presence of non-culturable endosymbiont bacteria located in plant tissues were revealed by molecular biologic analysis such as DGGE and 16S rDNA sequencing (Abreu-Terazi et al., 2010). That showed a high similarity of *Moraxella* sp., *Brevibacillus* sp., and cyanobacterium. Following surface sterilization, Abreu-Terazi et al. (2010) isolated DNA from axenic cultures of pineapple (five-year-old) and identified bacteria belonging to families of *Actinobacteria*, *Alphaproteobacteria*, and *Betaproteobacteria* by similar techniques. In the same study, it was discovered that different bacterial communities could be found in various plant organs. Moreover, microbiological analysis, visualization techniques, and sequencing methods revealed out the evidence of vast microbiome including known, unknown bacteria and fungi within regenerated *Atriplex* sp. leaves and roots (Lucero et al., 2011). In meristematic explants of *Aglaonema* cultivars, bacteria that belong to 13 different genera were identified by Fang and Hsu (2012), and 30% of these isolates were *Pseudomonas aeruginosa*. In the contaminant plant cultures detected from many laboratories in Poland, 108 isolates were observed from many genera, most commonly *Bacillus*, *Methylobacterium*, and *Pseudomonas* (Kaluzna et al., 2013). All of the isolates in this study, except *Staphylococcus*, were located inside plant tissues without showing any detrimental effect.

2.4 INTERACTIONS OF ENDOPHYTIC BACTERIA AND *IN VITRO* PLANT CULTURES

In the last 20 to 25 years, many researches are published on the coloniza-tion of endophytic bacteria inside *in vitro* plant cultures (Reed et al., 1995; Reed and Tanprasert, 1995; Dunaeva and Osledkin, 2015). These are associated with free-living bacteria, plant-inhabiting, animal-related, and processed foods, sewage, or human pathogens (Cassells and Tahmatsidou, 1996; Fletcher et al., 2013). There were 14 isolates of bacteria belonging to 9 genera, including in the animal- or human-related ones, found at the 1 cm long shoot apices of papaya after the surface sterilization (Tomas et al., 2007). Despite the surface sterilization of the initial explants, the plant cultures were not necessarily free of bacteria. While the presence of some bacteria may be evident at the beginning, sometimes the contaminants may not be detected until the proliferation stage, and often even after the suboptimal conditions or acclimation (Pirttilä et al., 2008; Thomas, 2011). For instance, pathogenic *Xanthomonas axonapadis* was symptomless in *Anthurium* cultures for one year (Norman and Alvarez, 1994), and patho-genic *Agrobacterium vitis* was present in the latent stage of *in vitro Vitis vinifera* shoot cultures for 14 weeks (Poppenberger et al., 2002).

Many isolates from plant tissue culture have promoted plant growth rates or influenced morphogenesis *in vitro*. Regulation of optimum condi-tions, growth, and organogenesis of plantlets are easier in *in-vitro* compared to *ex vitro* plant production. These are used in both research and industry. Some plant varieties, such as recalcitrant genotypes do not effectively multiply or regenerate rooting *in vitro*. However, if they are inoculated with endophytic beneficial bacteria and supported by exogenous regulators and metabolites, these plant varieties can overcome culturing challenges. For example, rodestrine, a phytohormone produced by *Rhodobacter sphaer-oides*, promote the rooting of mulberry microshoots (Sunayana et al., 2005) and *Bacillus* spp. produce indole-3-acetic acid (IAA) that enhances rooting of strawberry (Dias et al., 2009). Endogenous bacteria related with micropropagated *Prunus avium* genotypes were found to influence plant growth rate (Quambusch et al., 2014).

The beneficial effects of several bacterial strains on the explants *in vitro* have led to the biotization of plant cultures with useful bacteria (Nowak, 1998). One of the best-known examples for biotization is *Burk-holderia phytopharmans* PsJNTM strain, previously known as *Pseu-domonas* spp. PsJN, and it does not grow on the plant medium without plant explants (Sessitsch et al., 2005). It stimulates the growth of micro-shoots and microroots as colonized on both surface and internal tissues (Nowak, 1998), allows for more effective use of water and enhances the resistance of the plant to pathogens (Theocharis et al., 2012) and cold stress (Fernandez et al., 2012). Another bacteria, *Pseudomonas* spp., has been reported to produce polysaccharides that can prevent the overhydra-tion of oregano (Shetty et al., 1995), raspberry (Ueno et al., 1998) and anise (Bela et al., 1998) cultures. *Methylobacterium* sp. D10 and *Methy-lophilus glucoseoxidans* elicit the morphogenic callus production from wheat embryos (Kalyaeva et al., 2003). The stimulation of callus somatic embryogenesis, which is derived from geranium hypocotyls, is mediated by the *Bacillus circulans* strain (Murthy et al., 1999); *Curtobacterium citreum* promotes the development of axillary shoots in geranium cultures (Panicker et al., 2007). *Bacillus* spp. promotes root growth, *Azotobacter chroococcum* enhances the number of shoots in wheat (Andressen et al., 2009), *Sphingomonas* spp. facilitates the adaptation to climate change of micropropagated strawberries in greenhouse conditions (Dias et al., 2009), *Azospirillum brasilense* 243 increases the compatibility of micropropa-gated fruit rootstocks (Vettori et al., 2010). Scherling et al. (2009) reported that the interaction between *Paenibacillus* P22 and poplar shoot explants, which can assimilate atmospheric nitrogen, causes significant changes in plant metabolism, consistent with experiments on bacterial interaction with plant growth during *in vitro* conditions (Russo et al., 2012).

Moreover, methylobacteria generally found in soil and plant surfaces; several beneficial interactions of these group bacteria with plants as endo-phytes were identified (Madhaiyan et al., 2011). For example, *M. extorquens* that produces a pink pigment (Christoserdova et al., 2003) was determined in meristematic cells of *Pinus silvestris*. When this strain was inoculated on the plant callus, it affected the growth and regeneration of that plant through diverse mining (Pirtilä et al., 2008). It is known that many strains isolated from different places such as soil, rhizosphere water, plant parts (Lorentz et al., 2006), and *in vitro* plant cultures (Ulrich et al., 2008) secrete different metabolites and enzymes with plant growth regulators (Timmusk et al., 1999). For instance, *P. glucanolyticus* isolated from the black pepper roots is able to solubilize potassium (Sangeeth et al., 2012). In addition, Lorentz et al. (2006) reported that many *Paenibacillus* strains, which have antimicrobial activity, belonging to the various species.

Beneficial effects of bacteria can occur under stress conditions. For instance, in harmony with the greenhouse conditions, those bacteria can protect plants from losing water, microbial threats; also they can activate the growth of conductive tissues in plants to supply the absorption of water and nutrients (Chandra et al., 2010). One of the most potential applications for using beneficial bacteria in *in-vitro* plant propagation is, of course, the improvement of acclimatization (Digat et al., 1987). In potato and strawberry, inoculated microshoots before rooting with *Pseudomonas aureofaciens* strain have been shown to grow better (Zakharchenko et al., 2011). Moreover, *Azospirillum brasilense* is a strain on *Prunus cerasifera* promoted rooting and climate adaptation (Russo et al., 2008).

Furthermore, some contaminants can also aid the maintenance of the plant tissues without the addition of any plant growth regulators. Habituated cultures can continue to grow at a specific point of their growth without adding the exogenous plant growth regulators. Compared to the transcripts of habituated and non-habituated *Arabidopsis thaliana* (*A. thaliana*) cells, it showed various expressions of 800 genes, including up-regulation of genes encoding cytokine receptors (Pischke et al., 2006). At this point, up to current knowledge, it can be stated that the auxin and cytokinin balance in plant tissues could be changed by endophytic bacteria with their metabolic activity. The balance is influenced by the activity of endophytes producing auxins, gibberellins, cytokinins, and other plant hormones (Arshad and Frankenberger, 1991).

Also, recently in our study, the presence of plant growth-promoting putative endophytic bacterium associated with the microshoots of Fraser photinia has been identified (Seker et al., 2017). This plant-bacteria relationship has ensured healthy and vigor microshoots together with the promotion of rooting without supplying auxin to the plant medium in *in-vitro* conditions.

2.5 MECHANISMS UNDERLYING BACTERIAL ENDOPHYTES-PLANT INTERACTIONS

2.5.1 PHYTOSTIMULATION

Phytostimulation is the promotion of plant growth via the production of phytohormones (Bloemberg and Lugtenberg, 2001). They have a complex, diverse, and significant role that combines both assumed developmental pathways and environmental dynamic responses in plant growth (Durbak et al., 2012). Certain bacteria produce phytohormones, i.e., auxins (i.e.,

IAA), gibberellins (GAs), and cytokinins (Bottini et al., 2004; Tsavkelova et al., 2006; Kudoyarova et al., 2014). It could be assumed that phytohormones (IAA and ethylene biosynthesis) have been exploited as signal molecules between bacteria and plants (Yuan et al., 2008). Moreover, phytohormone production can be also regulated by bacterial action on the plant. The production of jasmonic acid (JA) (Forchetti et al., 2007) and salicylic acid (SA) (De Meyer et al., 1999) were reported in endophytic bacteria isolated from sunflower. The auxin synthesis that plays a key role in plant root development has been identified in many bacterial strains belong to the genera of *Azospirillum*, *Pseudomonas*, and *Bacillus* (Dodd et al., 2010). Cytokinins can be produced by *Arthrobacter*, *Azospirillum*, *Azotobacter*, *Bacillus*, *Rhizobium*, and *Pseudomonas* strains. When the plants inoculated with *B. subtilis* that produce cytokinin showed enhanced chlorophyll content and cytokine accumulation. Finally, this inoculation increased the weight of shoots and roots (Arkhipova et al., 2007). PGPBs such as *Azospirillum*, *Brevibacterium, Bacillus*, *Lysinibacillus*, and *Pseudomonas* have been shown to synthesize abscisic acid (ABA) and affect its level in plants, especially under stressful conditions like salinity (Belimov et al., 2014). For instance, under drought conditions, when maize inoculated with ABA producer strain *Azospirillum lipoferum* USA 59b, this interaction support to the plant growth by inducing high ABA production in the plants (Cohen et al., 2009).

The optimal functioning of PGPBs includes the synergistic interaction between ACC deaminase and both plant and bacterial auxin, IAA. These bacteria favor to plant growth and help to plants against biotic and abiotic stress conditions. ACC deaminase, responsible for the cleavage of the plant ethylene precursor (Honma and Shimomura, 1978), is closely related to the stimulation of plant growth by microorganisms (Glick et al., 1998). Therefore, the PGPB have a role as a sink for ACC, thus lowering plant ACC levels, decreasing the amount of ACC within the plant that can be converted into ethylene (Belimov et al., 2009). *Medicago truncatula* (*M. truncatula*) and *A. thaliana* mutants were used to analyze the effect of plant defense/response pathways in controlling the number of endophytic bacteria (Iniguez et al., 2005). They reported that the amount of *Klebsiella* sp. strain Kp342 and *Salmonella typhimurium* (*S. typhimurium*) strain decreased by ethylene, a signal molecule which induces plant SR. While the ethylene precursor, ACC, reduced colonization of endophytic bacteria, an ethylene-insensitive *M. truncatula* mutant was hyper-colonized by *Klebsiella* sp. strain Kp342. Various endophytes, including *Arthrobacter* spp. and *Bacillus* spp. in pepper plants (Sziderics et al., 2007), as well as *Pseudomonas putida* and *Rhodococcus* spp. in peas (Belimov et al., 2001) release ACC deaminase to increase plant growth. Even though the mechanism of plant growth promotion is unknown, ACC deaminase production may reduce abiotic stress by balancing the level of ethylene in the plant since the elevated amount of ethylene inhibit cell division, DNA synthesis, and root/shoot growth (Burg, 1973). Moreover, the production of other plant hormones, including IAA, JA, and ABA by bacterial strains may also stimulate plant growth (Forchetti et al., 2007). When *Miscanthus* seedlings were inoculated with *Herbaspirillum frisingense* (*H. frisingense*) GSF30T, elicited the root and the shoot growth. Transcriptomic analysis has provided detailed information on ethylene and jasmonate signaling regulation and showed that the induction of plant growth is promoted by the activation of phytohormones (Straub et al., 2013).

Rothballer et al. (2008) showed that *H. frisingense* GSF30T produce IAA and similar positive effect of IAA was observed in wheat inoculated with *B. subtilis* (Egorshina et al., 2012). In the case of *Azospirillum* spp., while root stimulation is increased by auxin, production of the other phytohormones had a slightly effect on nitrogen fixation and production (Steenhoudt and Vanderleyden, 2000). This strain can also be used in the field, for example, re-inoculation of seedlings with *Azospirillum* sp. strain B510 that was previously isolated from surface-sterilized stems of rice, enhanced root proliferation and mass production in field conditions (Isawa et al., 2010). In addition, three *Pseudomonas* strains stimulated wheat growth and its spike length both in laboratory and field conditions (Iqbal and Hasnain, 2013). In both cases, these effects were come up with phytohormone production rather than nitrogen fixation. Thus, PGPBs alter the phytohormone level, affecting the hormonal balance of the plant and its resistance to stress.

2.5.2   PHYTOREMEDIATION

Some bacterial endophytes promote plant growth by increasing plant tolerance or resistance capacity to high concentrations of pollutants. The remediation of organic compounds and toxic metals is probably through effective plant-bacterial interactions, and the cleaning of contaminated soils with plants is called phytoremediation. The abilities of the PGPBs mentioned under these conditions are to absorb heavy metal ions from the extracellular space, to accumulate them in the cell wall, or to convert them into less toxic forms. The inclusion of bacteria interacting with plants in the phytoremediation process is increasingly accepted as an alternative to combat the inherent weaknesses of the plants (Abhilash et al., 2012). The microbial community in the rhizosphere has been shown to play a part in the degradation of trichloroethylene (TCE) contaminant in groundwater (Weyens et al., 2009a). In this sense, it has been proposed to use various endophytic bacteria for the degradation of hydrophobic compounds (Stenuit et al., 2010) or for the delivery of the plant N-feed or P, Fe, or hormones (Marchand et al., 2010).

There are some studies published on the degradation ability of endophytic bacteria on endophytes (Mitter et al., 2013b). The first report on bacteria isolated from plant inner tissue that had the ability to degrade hydrocarbons was published by Siciliano et al. (2001). Those bacteria were intensely located on the root interior rather than bulk soil. Density of endophytic bacteria was correlated with amount of hydrocarbons in soil. Besides, some hydrocarbon degradable bacteria isolated from plant growing soil contaminated with crude oil were also identified. These bacteria have enormous potential to degrade alkane and aromatic hydrocarbons (Phillips et al., 2008; Yousaf et al., 2010). In addition, different bacterial strains that have the ability to degrade 2,4-dichlorophenoxyacetic acid (2,4-D) were also isolated from poplar internal tissues (Porteous Moore et al., 2006; Taghavi et al., 2011).

Endophytes like *Burkholderia cepacia* have been reported to improve both remediation and biomass production in the host (Weyens et al., 2009a, b). At the same time, some of the organic pollutants degrading endophytic bacteria have resistance to heavy metals, and they enhanced the capacity of phytoremediation of soil and water (Weyens et al., 2010b). For instance, five endophytic bacteria had ACC deaminase activity, solubilized phosphorus, and produced IAA and siderophore were isolated from *Sedum plumbizincicola*, the Cd/Zn hyperaccumulator plant (Ma et al., 2015). Moreover, it was found that the strains have high Cd, Pb, and Zn resistance. In addition, when plants were inoculated with *Bacillus pumilus* (*B. pumilis*) strain E2S2, increased Cd uptake and physiological properties (i.e., root and shoot length, fresh, and dry biomass) were observed in comparison to non-inoculated plants. Recently, *Sinorhizobium meliloti* (*S. meliloti*) strain CCNWSX0020, isolated from *Medicago lupulina* under copper stress was reported (Kong et al., 2015). According to the report, it elevated nitrogen content, Cu accumulation, and plant growth. Moreover, when plants inoculated with *S. meliloti* in high Cu levels, genes responsible for antioxidant response were also stimulated.

In pot experiments, *Bacillus* sp. SLS18 increased the biomass of sorghum grown in the presence of either manganese or cadmium. Also, the existent of a similar effect in dicotyledon species indicates broad host range applicability (Luo et al., 2012). For example, the inoculation of barley with bacterial strains *Arthrobacter myosorens* and *Flavobacterium* sp. reduced the mobility of Cd ions in soil and prevented contamination of the grain with heavy metal ions. On the other hand, the interaction of metallophilic bacteria, which function as accumulators or hyperaccumulators of heavy metals, with plants in cleaning up the contaminated soils will provide useful information for the phytoextraction of metals (Hu et al., 2007), such as the isolates from the rhizosphere of *Alyssum murale* plants that contained Ni-resistant bacteria *Sphingomonas* and *Microbacterium* which were capable of decreasing the content of mobile Ni in soil and inoculated plants. Moreover, the association of mustard (*Brassica juncea*) plants with *Xanthomonas* sp., *Pseudomonas* sp., *Azomonas* sp., and *Bacillus* sp. strains elevated the mobility of Cd ions (Belimov et al., 2011; Luo et al., 2012). Therefore, it is important to note that the capability of microorganisms for the accumulation of heavy metal ions may give optional or different ways to chemical methods of soil amendment.

2.5.3   BIOFERTILIZATION

The promotion of plant growth through the provision of basic nutrients or by increasing availability is called biofertilization (Bashan, 1998). Nitrogen fixation is the conversion of atmospheric nitrogen to ammonia, a well-studied form of biofertilization. Numerous PGPBs, including *Azospirillum*, *Pantoea agglomerans*, *Acetobacter*, *Azoarcus*, *Herbaspirillum* spp., *Azoarcus* spp. and *Aeromonas* have been extensively studied for their nitrogen fixation ability (Bloemberg and Lugtenberg, 2001; Verma et al., 2001; Hurek et al., 2002; Dobbelaerae et al., 2003).

Utilization of the potential of biological nitrogen fixation is assumed as a priority target for the development of novel cropping systems in 21st-century agriculture (Roesch et al., 2008). The difficulty underlying the more widespread use of biological nitrogen fixation in agricultural production is due to the complexity of nitrogenase, an ancient enzyme used by prokaryotes to convert atmospheric N_2 to NH_3, responsible for atmospheric nitrogen fixation. Free-living endophytes with biological nitrogen fixation ability utilize the low oxygen environment created by the plant as it optimizes nitrogen activity (Elliott et al., 2009). Some endophytes also produce compounds such as triterpenes, while another plant-associated bacterium may require input from the host plant for the reaction to occur (Doty et al., 2009).

Several endophytic *B. pumilus* strains were isolated, which could reduce the acetylene from sunflower tissues and were able to make nitrogen fixation (Forchetti et al., 2007). Similar features have also been reported for *Bacillus* spp. strain isolated from the cactus *Pachycereus pringlei* (Puente et al., 2009). Also, various bacilli with the *nifH* encoding the small subunit of nitrogenase have been reported (Ding et al., 2005; Rashedul et al., 2009). The rate of nitrogen fixation from *Bacillus* species such as *B. azotofixans*, *B. coagulans*, *B. holimixa*, and *B. macerans* can be as high as 18.8% of the number of spore-forming bacteria in the soil (Melentev et al., 2007). There is evidence that free-living and diatrophic bacteria from the genera *Azotobacter*, *Azospirillum*, *Rhizobium*, and *Bradirhizobium* as well as *Rhizobacteria*, *Pseudomonas*, and *Bacillus* can also be elicite nitrogen fixation (Dobbelaere et al., 2003).

Diazothropy is a symbiotic relationship between bacteria and their host plant. Conception of diazotrophy is important to understand the mechanisms of plant growth promotion. It is known that rhizobia fix the atmospheric nitrogen and serve to plants. Some PGPBs are diazotrophic and can fix nitrogen (Lodewyckx et al., 2002), but not all PGPBs are diazotrophic as they cannot fix atmospheric nitrogen or cannot fix the amount enough for their host (Hong et al., 1991). There are a large number of diazotrophic bacteria such as *Acetobacter diazotrophicus*, *Herbaspirillum* sp. and *Azospirillum* sp. in some agronomically important plants such as sugarcane (*Saccharum* sp.), rice (*Oryza sativa*), wheat (*Triticum aestivum*), and maize (*Zea mays*). At this point, endophytic diazotrophs have the advantage as compared to root-associated diazotrophs. For example, *Azospirillum* sp. and *Azobacter* sp. are located in the depths of the plant roots and can grow in the atmosphere with little oxygen, which is suitable for the study of the nitrogenase (McInroy and Kloepper, 1995; Triplett, 1996; Dobbelaere et al., 2003).

Furthermore, endophytic bacteria that fix nitrogen in the internal tissues of the plants without damaging them (Govindarajan et al., 2006, 2008) can provide up to 65% of plant nitrogen in poplar clones (Knoth et al., 2014). Endophytes identified in tomato grown in the field, in sorghum

(Wong-Villarreal and Caballero-Mellado, 2010) and in mimosa (Elliott et al., 2009) were generally found as *Burkholderia* spp. while free-living and nitrogen-fixing *Sphingomonas* spp. were identified in cottonwood and willow growing in nutrient-poor conditions (Doty et al., 2009). The effectiveness of nitrogen fixation varies between different bacteria, such as seven free-living rhizospheric and endophytic *Burkholderia* spp. (*B. unamae, B. tropica, B. silvatlantica, B. xenovorans, B. vietnamiensis, B. kururiensis,* and *B. sacchari*) (Martinez-Aguilar et al., 2008).

Phosphorus ranks second important key element among the mineral nutrients the plant needs after nitrogen. Although both organic and inorganic forms of phosphorus are abundant in the soil, its availability is limited because mostly it is found in an insoluble form. Despite the P content in average soil is about 0.05% (w/w), only 0.1% of the total P is available to plant because of poor solubility and its fixation in soil (Illmer and Schinner, 1995). Certain PGPBs can increase the solubility and availability of phosphate. With the release of low molecular weight acids, chelating of the metal cation that adheres to phosphorus can be provided hence making it more accessible for plants (Kpomblekou and Tabatabai, 2003). For elimination of the phosphorus starvation of plants, endophytes are becoming important since they can dissolve phosphorus-containing substances with organic acids, phosphatases, or other metabolites (Dobbelaere et al., 2003). The phosphate dissolution capacity of the *Pseudomonas fluorescens* CHA0 strain depends on its ability to produce gluconic acid (De Werra et al., 2009). *Achromobacter xiloxidans* and *Bacillus pumilus* in sunflower (*Helianthus annuus*) were identified as having the highest chelating capabilities (Forchetti et al., 2007). Yazdani and Bahmanyar (2009) showed that the use of PGPB in fertilizer treatments for corn (*Zea mays*) decreased the need for phosphorus treatment by 50% without significant loss in grain yield.

2.5.4   *BACTERIAL SIGNALING MOLECULES AND BIOCONTROL*

Bacteria can regulate gene expression through "quorum sensing" (QS), a term that refers to gene expression controlled by the response to signal molecules in a cell-density-dependent manner (Fugua et al., 1994). Some bacteria use these molecules to create their own pathogenicity; others living in the same environment could break them down. So far, it has been indicated that the production of antibiotics, virulence factors, and plant cell wall-degrading exo-enzymes is activated in respond to QS (Von

Bodman et al., 2003). Noticeably, plants induce QS-regulated signals in bacteria and could also control its level (Bauer and Mathesius, 2004). For this reason, it would be interesting to investigate whether endophytes produce QS compounds that affect the growth of plants and their adaptation properties.

Some mechanisms employed by phytogenic bacteria (Vant Slot and Knogge, 2002) are also possible to be used by other endophytes. Many candidate genes with unknown functions were expressed differently during plant bacterial interaction (Rocha et al., 2007), especially inoculation of sugarcane with endophytic nitrogen-fixing bacteria altered the expression profile of *shr5* (a plant receptor kinase) gene (Vinagre et al., 2006). In the presence of beneficial endophytes, the genes of the ethylene-signaling pathway were also differently expressed (Cavalcante et al., 2007). Genes such as *carAB*, which is required for degradation of a fatty acid signaling molecule, can be produced by recombinant biocontrol strains and they reduce virulence caused by *Xanthomonas* sp. and *Xylella fastidiosa* (Newman et al., 2008).

Plants produce secondary metabolites that mimic or inhibit QS molecules (Gao et al., 2003). For example, tobacco plants were engineered to synthesize acyl-homoserine lactones (AHLs) in the chloroplast (Scott et al., 2006), and it was shown that AHL was transported and secreted on the phyllosphere and in the rhizosphere. The virulence through *Erwinia carotovora* was reduced by pretreatment of potato slices with *Bacillus thuringiensis* (Dong et al., 2004). Due to deterioration of QS molecules of *E. carotovora* with AHL lactonase, potato slices pretreated with *B. thuringiensis* lacked the ability to produce AHL-lactonase, but it did not reduce maceration. On the other hand, the plants that are manipulated to express AHL lactonase have exhibited a positive effect in the protection of plants against pathogens (Zhang, 2003). Recently, a gene encoding a new type of AHL-lactonase, which is nominated *qsdA* from *Rhodococcus erythropolis*, has been characterized. This gene was able to gain quorum quenching capacity to *P. fluorescens* that led to enhanced protection of potato tuber against the soft rot pathogen *Pectobaterium carotovorum* (Uroz et al., 2008). Rosmarinic acid, a highly antioxidant phenolic compound, has a potential antimicrobial activity against a range of soil-borne microorganisms, and it was induced in sweet basil hair root cultures after contact with *Pythium ultimum* (*P. ultimum*) (Bais et al., 2002). In addition, recently endophytic actinobacteria has great potential, as it is the source of novel bioactive compounds, including antibiotics, antifungals, and antitumor compounds

(Qin et al., 2011). Endophytic bacteria inhibit bacterial communication and biofilm formation, and virulence in this way, without suppressing bacterial growth, both by producing specific bioactive products and by blocking pathogenic QS. Accordingly, it has been reported that cell-free lysates of endophytic bacteria suppress QS molecules and prevent biofilm formation in *P. aeruginosa* PAO1 (Rajesh and Ravishankar, 2013). Therefore, this kind of "quorum quenching" gains importance as an alternative innovative approach in dealing with drug-resistant bacteria (Kusari et al., 2014).

Biocontrol is the promotion of plant growth through protection from plant pathogens. Endophytic bacteria are emerging as promising biological methods to control plant pathogens (Koumoutsi et al., 2004). Since the ultimate goal of the action mechanism of PGPBs is not pathogen destruction, they are different from the classical plant protection chemicals. It is used to stimulate SR in order to control the pathogens, which are present in large quantities in the environment. The majority of PGPBs are active to trigger the cascade of defense responses due to the production of various metabolites (Maksimov et al., 2011). Bacterial endophytes may interfere with the development of phytophagous insects and nematodes through the synthesis of biologically active compounds, termed antipathogenic action (Azevedo et al., 2000). Commonly used mechanisms of biocontrol mediated by PGPBs are antibiosis (Haas and Défago, 2005; Lugtenberg and Kamilova, 2009), induced systematic resistance (ISR) (Van Loon, 1998, 2007; Kloepper et al., 2004), competition for niches and nutrients (Kamilova et al., 2005; Validov, 2007) and predation and parasitism (Harman et al., 2004). The ISR associated with PGPBs is effective against a broad spectrum of plant pathogens including oomycetes, fungi, bacteria, and viruses and even insects and herbivores (Van Oosten et al., 2007, 2008). This resistance was observed in a variety of plant species including Arabidopsis, beans, carnation, eucalyptus, radish, tobacco, tomato, rice, maize, and wheat (Kloepper et al., 2004; Van Der Ent et al., 2009; Beneduzi et al., 2012; Yi et al., 2013). Development of PGPB-induced ISR in plants may involve the generation of reactive oxygen species (ROS) as a critical event in the formation of priming effect and after inoculation of a pathogen, plants treated with PGPB show early ROS formation (Conrath et al., 2006).

Pathogen-induced systemic acquired resistance (SAR) co-exists with coordinated activation of genes associated with pathogenesis (PR), and many of them encode proteins with antimicrobial activity (Van Loon et al., 2007). This resistance is controlled by the redox-regulated protein NPR1, which is activated by SA, and functions as a transcriptional co-activator for a large part of the PR genes (Pieterse et al., 2012). SAR induced by endophytic bacteria is phenotypically similar to SAR associated pathogenesis, except activation of JA and ethylene synthesis (Pieterse et al., 2014). *B. pumilus* SE 34 induces SR against Fusarium wilt on tomatoes (Benhamou et al., 1998). After pre-inoculation of Arabidopsis seedlings with two strains of *Streptomyces* sp., seedlings were inoculated with *Erwinia caratovora* (*E. caratovora*). The plants found to be protected from disease symptoms as endophyte-free plants decayed within 5 days. ISR is conducted by one strain and SAR is achieved by the other strain due to the fact that gene stimulation in Arabidopsis was specific to strain (Conn et al., 2008).

Several microbial components, which are referred to as microbe-associated molecular patterns (MAMPs), are recognized by plants and act as elicitors, triggering a generalized MAMP-triggered immunity (MTI). Despite generally described in the context of pathogenicity, MAMPs are highly conserved in whole classes of microbes, including endophytes. MTI includes the production of reactive oxygen and nitrogen species (Newman et al., 2013). The antibiotic iturin A is produced by *Bacillus* sp. CY22 and it suppresses the root rot of balloon flower caused by *Rhizoctonia solani* (*R. solani*) (Cho et al., 2003). Also, *Pseudomonas fluorescens* carrying the chitinase-encoding gene *chiA*, can control the phytopathogenic fungus *R. solani* on bean seedlings (Downing and Thomson, 2000).

Endophytes offer tremendous promise to discover natural products with therapeutic value (Xiong et al., 2013). One of the interesting examples of bioactive plant-associated metabolites by endophytes is hypericin, which originally isolated from the plant *Hypericum perforatum* (Nahrstedt and Butterweck, 1997). This compound has many pharmacologically interesting properties, such as its distinctive antidepressant properties, anti-inflammatory, antimicrobial, and antioxidant activities (Tammaro and Xepapadakis, 1986). Recently a *Streptomyces* sp. was isolated from an annual plant *Lolium perenne* (Guerney and Mantle, 1993). This endophyte contains a diketopiperazine called methylalbonoursin and consists of leucine and phenylalanine. Moreover, a number of new antibiotics identified as munumbicins A, B, C, and D, were isolated from a streptomycete within a snake vine plant (Castillo et al., 2002). These are broad-spectrum antibiotics that can be active against several humans as well as plant pathogenic fungi and bacteria.

Endophytic bacteria isolated from rice seeds showed vast fungal activity against *Gaeumannomyces graminis, P. myriotylum, Heterobasidium annosum,* and *R. solani* (Mukhopadhyay et al., 1996). *Enterobacter cloaca,* an endophyte isolated from corn also has antibiotic ability against *Fusarium moniliforme* (Hinton and Bacon, 1995). This ability was confirmed by Chen et al. (1995) on cotton plants in which it reduced wilt disease symptoms caused by a *Fusarium* sp. Van Buren et al. (1993) reported that pathogen-endophyte antagonistic activity on *Clavibacter michiganensis* subsp. *sepedonicum* (Van Buren et al., 1993) and *P. fluorescens* 89B-27 and *Serratia marcescens* 90–166 activity against to *Pseudomonas syringae* pv. *lachrymans* (Liu et al., 1995). Additionally, control of plant-parasitic nematodes (Hallmann et al., 1995) and insects (Dimock et al., 1988) via rhizospheric and endophytic bacteria were also observed (Kloepper et al., 1991).

Iron (Fe) is an essential element that plays a role as a catalyst in enzymatic processes, oxygen metabolism, electron transfer, and DNA/RNA synthesis (Barry et al., 2009). Since it regulates surface motility and stabilizes the polysaccharide matrix, Fe is also essential for biofilm formation. Under iron-deficient growth conditions, the microbial surface hydrophobicity is reduced, which causes the surface protein composition to be altered to limit the biofilm formation. Due to the low Fe availability in the environment, microorganisms have developed specific uptake mechanisms such as the production of siderophores. Siderophores such as pyrethroids and SA produce iron, compete with phytopathogens for trace metals, and indirectly contribute to disease control (Duffy and Défago, 1999). One of the most studied siderophores with a strong antifungal effect is pseudobacin produced by *P. putida* B10, and it suppressed the development of oxysporum in iron-depleted soils (Kloepper et al., 1980). In addition, pseudobacin produced by *P. putida* WCS 358 inhibited the growth of *Ralstonia solanacearum* in eucalyptus, the growth of *Erwinia carotovora* in tobacco and of *B. cinerea* in tomatoes (Bakker et al., 2007). Antibiotics can affect microorganisms through not only the inhibition of cell wall synthesis and respiratory enzymes but also the alteration of membrane functions and protein synthesis. As an example, *P. fluorescens* CHA0 produces 2,4-diacetylphloroglucinol (DAPG) and this compound destructs the membranes and suppress the germination of zoospores of the oomycete *Pythium* spp. (Melentev et al., 2007). Therefore, PGPBs produce antimicrobial metabolites such as DAPG and enhance the disease suppression in plants. For example, when seeds were inoculated with

DAPG-producing endophytic isolates, eggplant wilt caused by *R. sola-nacearum* was reduced by 70% (Ramesh et al., 2009). Similarly, endophytic bacteria belonging to genus *Bacillus* are known to produce various antifungal and antibacterial lipopeptides, including iturins, bacillomycins, fengycins, and surfactins (Gond et al., 2014). It is known that antibiotics from *Bacillus* spp. stimulate the swellings of hyphal apices in *Sclerotinia sclerotiorum, Alternaria alternata, Drechslera oryzae, Fusarium roseum,* and pathogenic *Puccinia graminis* in hyphal apicals (Duffy et al., 2003).

Why *Bacillus* endophytes frequently occur in the natural populations of plants? One of the possible reasons can be the disease protection of *Bacillus* endophytes by using its lipopeptides, compliments in several wild populations (White et al., 2014b). It was known that endophyte infection might induce the expression of disease resistance genes in host plants. With regard to this, *Bacillus* endophytes in corn resulted in enhanced expression of defense-related genes (Gond et al., 2015). Therefore, it is known that the efficiency of antibiotic production in many endophytic bacterial strains has been shown to increase the resistance of plants to pathogens. However, the exact details of the interaction between endophytes and host plants leading to the induced expression of resistance genes are presently unknown.

Volatile organic compounds (VOCs) evaporate under high vapor pressures, and under normal conditions, they can enter the atmosphere. Low molecular weight (<300 g/mol^{-1}) compounds such as alcohols, aldehydes, ketones, and hydrocarbons take part in this class. VOCs help the plant growth and response ISR and IST in plants (Ryu, 2015). Volatile substances such as acetoin and 2,3-butanediol released from some PGPBs, can change plant-bacteria interactions and promote plant growth. In several numbers of mechanisms that PGPBs use to interact with plants, VOC emissions are an important participant. VOCs have mainly play role on antibiosis and biocontrol of plant pathogens. Recent studies using gas chromatography and mass spectrometry have revealed the capacity of bacteria to produce outstanding VOCs. Bacteria can produce different volatiles such as ammonia, alcohols, butyrolactones, phenazine-1-carboxylic acid, HCN, and some of them have antifungal activity on various fungal species (Trivedi et al., 2008). Different kinds of VOCs can be produced based upon growth conditions and environment (Schulz and Dickschat, 2007). Moreover, *B. subtilis* GB03 emitted VOCs, which promoted various hormone signaling pathways in *A. thaliana*, including auxin, cytokinins, brassinosteroids, gibberellins, and SA. Therefore, this study led to new opportunities on the role of volatiles during plant-microbe interaction accompanied by plant growth (Zhang et al., 2007).

As defensive mechanism of plants is triggered ISR occurs and leads to resistance against to pathogen infection. Regarding this, PGPB VOCs could behave as bioprotectants by ISR (Ryu et al., 2014). In addition, ISR is used for enhancing tolerance to abiotic stress by chemical and physical changes elicited by PGPBs. Zhang et al. (2008) showed the role of VOCs emitted from *B. subtilis* GB03 on ISR to salt stress in *Arabidopsis*. In another hand, endophytic bacteria have been detected in English oak (*Quercus robur*), common ash (*Fraxinus excelsior*), and can degrade VOCs such as TCE (Weyens et al., 2010a; Kang et al., 2012) that could have a role in phytoremediation. Recent studies on these mechanisms are shown in Table 2.1.

2.6   A MODEL HOST TO STUDY PLANT-ENDOPHYTE INTERACTION

A. thaliana is an ideal plant model organism (Koornneef and Meinke, 2010). The genome of its many genotypes is sequenced and there is numerous knowledge about their functional genetics and genomics that provides a versatile of tools to deepen knowledge of plant biology (Ehrhardt and Frommer, 2012). Moreover, *A. thaliana* is an important annual species that spread everywhere in the world and can present in various anthropic and wild habitats (Hoffmann, 2002). Thus, one of the most recently used model plants to study plant-endophyte interaction is the *A. thaliana*. This plant has been studied for its interactions with endophytes, such as enterobacteria and *Azorhizobium caulinodans*. Furthermore, the most important factor to use *Arabidopsis* as a model plant is the existence of mutants that can be used for analyzing endophytic colonization. For example, Iniguez et al. (2005) presented that *A. thaliana* and *M. truncatula* mutants can be useful for determining the importance of endophytic bacteria in plant response mechanism. According to this study, plants develop ISR against ethylene and alleviate *Klebsiella* sp. strain Kp342 and *S. typhimurium* colonization. When the ethylene precursor ACC was added to the wild type *M. truncatula* and wheat, the amount of endophytic bacteria decreases, however, *Klebsiella* sp. Kp342 promotes hypercolonization of an ethylene-insensitive *M. trucatula* mutant. Moreover, when wild type plants were treated with the ethylene inhibitor 1-methylcyclopropane, the colonization of

TABLE 2.1 Plant Growth Promoting Bacteria (PGPBs) and Their Beneficial Influences on Plants

Mechanism	Bacteria	Plant	Beneficial Influence*	References
PHYTOSTIMULATION	Herbaspirillum frisingense GSF30^T	Japanese silver grass	IAA	Rothballer et al., 2008
	Pseudomonas spp.	Maize, Sunflower		Li and Ramakrishna, 2011
	Bacillus amyloliquefaciens FZB42	Lemna minor ST		Idris et al., 2007
	Azospirillum spp.	Rice		Steenhoudt and Vanderleyden, 2000
	Azospirillum sp. B510			Isawa et al., 2010
	Pseudomonas spp.	Wheat		Iqbal and Hasnain, 2013
	Actinobacteria spp.	Winter rye		Merzaeva and Shirokikh, 2010
	Streptomyces sp. PT$_2$	Tomato cv. Marmande		Goudjal et al., 2013
	Pseudomonas putida, Rhodococcus spp.	Pea	ACC deaminase	Belimov et al., 2001
	Burkholderia spp. KJ006	Rice		Kwak et al., 2012
	Brevibacterium iodinum, Bacillus licheniformis, Zhihengliuela alba	Red pepper		Siddikee et al., 2011
	Artrobacter spp., Bacillus spp.			Sizderics et al., 2007
	Variovorax paradoxus 5C-2	Pea		Belimov et al., 2009
	Serratia proteamaculans 568	Soybean		Taghavi et al., 2009
	Sinorhizobium meliloti CCNWSX0020	Medicago lupulina		Kong et al., 2015
	Stenotrophomonas maltophilia R551-3	Poplar	IAA	Taghavi et al., 2009
	Pseudomonas putida W619		ACC deaminase	
	Serratia proteamaculans 568	Soybean		
	Burkholderia phytofirmans PsJN	Potato, tomato, maize, barley, onion, canola, grapevine		Weilharter et al., 2011
	Bacillus sp.	Canola and rice		Islam et al., 2009
	Bacillus amyloliquefaciens RWL-1	Rice	GAs	Shahzad et al., 2016
	Bacillus cereus MJ-1	Red pepper		Joo et al., 2005
	Azospirillum sp., Bacillus sp.	Wheat		Kucey, 1988
	Herbaspirillum frisingense GSF30T	Sunflower	JA	Forchetti et al., 2007
		Grass		Straub et al., 2013

TABLE 2.1   *(Continued)*

Mechanism	Bacteria	Plant	Beneficial Influence*	References
	Pseudomonas aeruginosa 7NSK2	Sunflower	SA	De Meyer et al., 1999
	Klepsiella sp. Kp342, *Salmonella typhimurium*	*Medicago truncatula*, *Arabidopsis thaliana*	ISR	Iniquez et al., 2005
	Pseudomonas putida, *Rhodococcus* spp.	Pea	Ethylene	Belimov et al., 2001
	Agrobacterium rhizogenes ATCC-15834	Basil	RA	Bais et al., 2002
	Achromobacter xylosoxidans SF2	Sunflower	JA ABA	Forchetti et al., 2007
	Bacillus licheniformis Ps14	Argentine screwbean	ABA, GA₃, IAA	Sgroy et al., 2009
	Azospirillum lipoferum	Corn	ABA	Cohen et al., 2009
	Azospirillumlipoferum USA 59b strain	Maize		
	Bacillus, Rhizobium, Artrobacter, Azotobacter, Azotospirillum Pseudomonas spp.	Wheat	IAA	Kudoyarova et al., 2014
	Bacillus, Rhizobium spp.			Arkhipova et al., 2007
PHYTOREMEDIATION	*Pseudomonas putida* W619-TCE	Poplar	TCE deg.	Stenuit et al., 2010
	Enterobacter sp. strain PDN3	Poplar (hybrid)		Kang et al., 2012
	Burkholderia cepacia VM1468	Yellow lupine		Weyens et al., 2009a, b
	Pseudomonas putida W619-TCE	Poplar		Weyens et al., 2013
	Enterobacter sp. strain PDN3	Poplar (hybrid)		Kang et al., 2012
	Herbaspirillum sp. K1	Wheat	TCP, PCB	Männistö et al., 2001
	Methylobacterium spp.	*Populus deltoids × nigra*	TNT deg.	Van Aken et al., 2004
	Methylobacterium populi BJ001	Poplar tissues (*Populus deltoidesnigra* DN34)	Methane, TNT, RDX, HMX	
	Bacillus sp. SLS18	Sorghum	Mn and Cd	Luo et al., 2012
	Bacillus pumilus E2S2	*Sedum plumbizincicola*	↑Cd uptake	Ma et al., 2015
	Arhrobacter myosorens 7, *Flavobacterium* sp. L30	Barley	↓Cd	Hu et al., 2007
	Agrobacterium radiobacter 10	Barley		Belimov and Dietz, 2000

TABLE 2.1 *(Continued)*

Mechanism	Bacteria	Plant	Beneficial Influence*	References
PHYTOREMEDIATION	*Variovorax paradoxus, Rhodococcus* sp., *Flavobacterium* sp.	Indian mustard		Belimov et al., 2005
	Pseudomonas sp. ITRI53, *Pseudomonas* sp. MixRI75	Italian ryegrass *L. multiflorum* var. Taurus	Hydrocarbon deg.	Afzal et al., 2011, 2012
	Sphingomonas spp., *Microbacterium* spp.	*Alyssum murale*	Ni-resistant	Belimov et al., 2011
	B. cepacia Bu61pTOM–Bu61	Poplar	Toluene deg.	Taghavi et al., 2005
	Burkholderia cepacia VM1468			
	Burkholderia cepacia G4	European yellow lupine		Baraci et al., 2004
	P. putida VM1450	Pea	2,4-D deg.	Germaine et al., 2006
	Pseudomonas sp.	Populus cv. Hazendans cv. Hoogvorst	MTBE, BTEX, TCE deg.	Germaine et al., 2004
	Burkholderia cepacia VM1468	Poplar	BTEX, TCE deg.	Taghavi et al., 2011
	Achromobacter xylosoxidans	*Phragmites australis, Ipomoea aquatica, Vetiveria, Phragmites australis, Vetiveria zizanioides*	Catechol deg. Phenol deg.	Ho et al., 2009
BIOFERTILIZATION	*Pseudomonas fluorescens*	Pea	Phosphate solubilization	Otieno et al., 2015
	Pseudomonas fluorescens CHA0	Wheat		De Werra et al., 2009
	Achromobacter xiloxidans, Bacillus pumilus	Sunflower		Forchetti et al., 2007
	Rhizobacterium spp.	Corn		Yazdani and Bahmanyar, 2009
	Pantoea agglomerans	Deep water rice	Nitrogen fixation	Verma et al., 2001
	Azospirillum brasilensis (Sp245 strain) *Agrobacterium tumefaciens* (A281 strain) *Leifsoniaxyli* subsp. *xyli* (CTC B07 strain)	Sugarcane		Vinagre et al., 2006
	Azoarcus spp.	Kallar grass		Hurek et al., 2002
	Azoarcus sp. strain BH72	Rice		Egener et al., 1999

TABLE 2.1   (Continued)

Mechanism	Bacteria	Plant	Beneficial Influence*	References
BIOFERTILIZATION	Herbaspirillum seropedicae Z67	Rice		Gyaneshwar et al., 2002
	Azospirillum sp., Azotobacter sp.	Maize		Roesch et al., 2008
	Achoromobacter xiloxidans, Alcaligenes sp., Bacillus pumilus	Sunflower		Forchetti et al., 2007
	Bacillus spp., Klebsiella sp., Acinetobacter sp., Pseudomonas sp., Satphylococcus sp.	Cactus		Puente et al., 2009
	Bacillus cereus, Bacillus marisflavi, Bacillus megaterium, Paenibacillus polymyxa, Paenibacillus massiliensis	Wheat, maize, ryegrass, and willow		Ding et al., 2005
	Burkholderia mimosarum PAS44 Cupriavidus taiwanensis LMG19424	Mimosa nodule		Elliott et al., 2009
	Burkholderia sp., Acinetobacter sp., Rahnella sp., Sphingomonas sp.	Sorghum, mimosa		Doty et al., 2009
BIOCONTROL AND BACTE- RIAL SIGNAL MOLECULES	Burkholderia cepacia G4	European yellow lupine	VOCs	Barac et al., 2004
	Pseudomonas corrugata	Wheat		Trivedi et al., 2008
	B. subtilis GB03	Arabidopsis thaliana		Ryu et al., 2003 Zhang et al., 2007
	Burkholderia cepacia G4	European yellow lupine		Barac et al., 2004
	Pseudomonas aeruginosa, Sinorhizobium meliloti	Medicago truncatula	QS	Mathesius et al., 2003
	Gluconacetobacter diazotrophicus (PAL5 strain)	Sugarcane		Rocha et al., 2007
	Herbaspirillum seropedicae (HRC54 strain)			
	H. rubrisubalbicans (HCC103 strain)			
	H. seropedicae			

TABLE 2.1 (Continued)

Mechanism	Bacteria	Plant	Beneficial Influence[*]	References
BIOCONTROL AND BACTERIAL SIGNAL MOLECULES	H. rubrisubalbicans strain HCC103		QS	Newman et al., 2008
	Xanthomonas campestris pv. vesicatoria	Cabbage, broccoli		Mathesius et al., 2003
	X. campestris pv. campestris	Black rot		
	Sinorhizobium meliloti	M. truncatula genotype A17		Han et al., 2011
	Variovorax paradoxus S110	Potato		Newman et al., 2008
	Xanthomonas campestris pv. vesicatoria	Cabbage, broccoli		Mathesius et al., 2003
	X. campestris pv. campestris	Black rot		
	Sinorhizobium meliloti	M. truncatula genotype A17		Han et al., 2011
	Variovorax paradoxus S110	Potato		Mathesius et al., 2003
	Pseudomonas aeruginosa	Medicago truncatula		
	Sinorhizobium meliloti			
	Gluconacetobacter diazotrophicus (PAL5 strain)	Sugarcane		Rocha et al., 2007
	Herbaspirillum seropedicae (HRC54 strain)			
	H. rubrisubalbicans (HCC103 strain)			
	H. seropedicae			
	G. diazotrophicus strain PAL5 (AD1)			Cavalcante et al., 2007
	H. rubrisubalbicans strain HCC103			
	Gluconacetobacter diazotrophicus Pal5	Sugarcane	Antibiosis	Bertalan et al., 2009
	Bacillus sp. CY22	Platycodon grandiflorum		Cho et al., 2003
	Pseudomonas fluorescens	Apple plantlets		Downing and Thomson, 2000
	Pseudomonas sp.	Egg plant		Ramesh et al., 2012
	Bacillus spp.	Maize		Gond et al., 2015
	Pseudomonas viridiflava	Grass	Antimicrobial	Miller et al., 1998
	Streptomyces griseus	Kandelia candel		Guan et al., 2005

TABLE 2.1   *(Continued)*

Mechanism	Bacteria	Plant	Beneficial Influence*	References
BIOCONTROL AND BACTERIAL SIGNAL MOLECULES	Streptomyces NRRL 30562	Kennedia nigriscans	Antibiotic	Castillo et al., 2002
	Paenibacillus polymyxa	Wheat	Antifungal	Beck et al., 2003
	Pseudomonas fluorescens	Arabidopsis thaliana	SAR	Ton et al., 2002
	Rhizobium sp.	Arabidopsis thaliana	ISR	Van Loon et al., 2007
				Van Oosten et al., 2008
	PGRB's	Arabidopsis thaliana		Van der Ent et al., 2009
	Bacillus spp.	Beans		
	Bacillus pumilus INR7	Pepper		Yi et al., 2013
	Streptomyces sp. strain EN27	Arabidopsis thaliana		Conn et al., 2008
	Paenibacillus alvei K165			Tjamos et al., 2005
	Streptomyces sp.	Cucumber	Biocontrol of Pythium aphanidermatum	Costa et al., 2013
	Pseudomonas mallei (RBG4, ET17)	Eggplant	Biocontrol of Ralstoniaso-lanacearum	Ramesh and Phadke, 2012
	Bacillus spp. (RCh6)			
	Streptomyces sp.	Rice	Siderophore	Gangwar et al., 2012
	Enterobacter sp. 638	Poplar		Taghavi et al., 2009, 2010
	Enterobacter sp. 638	Poplar		

*2,4-D, 2,4-dichlorophenoxyacetic acid; ABA, abscisic acid; ACC deaminase, 1-aminocyclopropane-1-carboxylate deaminase; BTEX, benzene, toluene, ethylbenzene, and xylene; Cd, cadmium; GAs, gibberellins; IAA, indole-3-acetic acid; JA, jasmonic acid; Mn, manganese; Ni, nickel; PCB, polychlorinated biphenyl; RDX, hexahydro-1,3,5-trinitro-1,3,5-triazene; SA, salicylic acid; TCP, 2,3,4,6-tetrachlorophenol; TNT, 2,4,6-trinitrotoluene; VOCs, volatile organic compounds; RA, rosmarinic acid; ISR, induced systemic resistance; DET, decrease ethylene; TCE, trichloroethene; deg, degradation; LP, lipopeptide; MTBE, methyl tert-butyl ether; ISR, induced systemic resistance; SAR, systemic acquired resistance; QS, quarum sensing; ↓, decreasing; ↑, increasing.

(Modified from Ryan et al., 2008).

bacteria was recovered. The presence of type III secretion systems (TTSS-SPI1) and flagella of *Salmonella* pathogenicity island 1, was shown to reduce endophytic colonization and mutants of *S. typhimurium* that do not contain these components showed more endophytic colonization in wheat seedlings and *Medicago sativa* (Iniquez et al., 2005). While *A. thaliana* mutants exhibited only a SA-independent defense response to inhibit the colonization of *Klepsiella* sp. strain Kp42, a colonization by *S. typhimurium* was restricted both SA-dependent and SA-independent pathways. The study with *S. typhimurium* flagella mutants proposes that flagella are recognized by the SA independent response; however, TTSS-SPI1 is recognized by the SA-dependent response required for induction of the PR1 promoter, a gene involved in SA-dependent pathogenesis. Since *Klebsiella* sp. Kp342 is deficient in flagella and TTSS-SPI1 (Dong et al., 2001), it cannot stimulate the SA-dependent responses when interacting with plant, and may result in higher number of colonization in plant (Rosenblueth and Martinez-Romero, 2006). Other bacteria which have flagella such as *Rhizobium* and *Agrobacterium* spp. may also associate closely with plants; except that these are not stimulants of the plant defense mechanisms (Felix et al., 1999). In all *A. thaliana* plants, flagellin plays a role as an inducer for oxidative burst, callus formation, and ethylene production resulting in the stimulation of genes involved in the activation of defense-related genes. Chemoperception systems allow plants to define the existence of molecules from bacteria (Boller, 1995; Rosenblueth and Martinez-Romero, 2006). Therefore, it could be proposed that *A. thaliana* and its endophytes as a significant model system for an overview of the principles orchestrating the endophytic lifestyle by utilizing the abundant knowledge available from the molecular tools and the host plant.

2.7 ROLES AND BIOTECHNOLOGICAL APPLICATIONS OF BACTERIAL ENDOPHYTES

There is a symbiotic relationship between the bacterial endophytes and plants as well as bacterial genes that are expressed with plant existence. These genes are necessary for various aims such as entering and colonizing plants, living in plant tissues and inducing plant growth, competing with pathogens and suppressing them or producing different substances that can be differently expressed by endophytes (Lodewyckx et al., 2002). Endophytic bacteria render important and rich models for studying the genetic expression of bacteria in a variety of natural niches or habitants that are much more different and richer from the culture medium under controlled laboratory conditions. However, there are few studies conducted on this issue (Lugtenberg et al., 2002).

The best example for the approach of biotechnological practice on plant production is the usage of *Agrobacterium tumefaciens* and *Agrobacterium rhizogenes*. These are soil bacteria and pathogenic for some plant varieties. They have genetic tools allowing colonization on plant. They have been used as vectors for transferring covetable genetic information to plant for about 40 years (Păcurar et al., 2011). Disarmed forms of bacteria can insert desired genetic information within a plant genetic construction. If genetic information is inserted on plant DNA, it can be transmitted to daughter cells. Therefore, genetically modified enriched plant cell provides us new genetic variations that can be used in plant breeding (Ziemienowicz, 2014).

Also, previous genomic studies are being made with various endophytic bacteria, including *Azoarcus* spp. (Battistoni et al., 2005), *Klebsiella* spp., *Gluconacetobacter diazotrophicus*, and *Herbaspirillum* sp., which could assist to reveal out the molecular interactions of endophytic bacteria with plants. Besides, *in vivo* expression technology that enables to search gene expression in various niches, including the rhizospheric area (Ramos-González et al., 2005), can also be useful to study gene expression during life cycle of endophytic. Vermeiren et al. (1998), showed the expression of *nifH* gusA fusions of *Azospirillum irakense* and *Pseudomonas stutzeri* in rice roots. The expression of *Azoarcus* spp. nitrogenase genes inside the roots of field-grown Kallar grass was also demonstrated by *in situ* hybridization studies (Hurek et al., 1997). According to results, it seems that grass tissues assure a suitable environment for nitrogen-fixation gene expression; however, N_2 fixation in plant appears to be carbon-limited (Christiansen-Weniger et al., 1992). In acetylene reduction assays, members of genera, *Azospirillum, Herbaspirillum, Serratia,* and *Klebsiella* that were inoculated with different grasses produced ethylene with only a carbon source presence (Egener et al., 1999; Gyaneshwar et al., 2002).

In recent years, production of PGPB or endophytes-based bio-pesticides, which are increasingly market share as plant protection products, is increasing in agricultural production. Because of being natural products, they are environmentally safe and have more advantage than chemical ones. According to targeted material, they have been improved and have used as fungicides, insecticide, herbicide together with various plant growth regulators (Yi et al., 2013). At this point genetic engineering has provided us a new approach to develop novel biological products by inserting the genes responsible for resistance to agricultural crops genome with using vector microorganism. Therefore, crops are modified by resistance protein-encoding genes, and resistance is acquired against biotic and abiotic stresses.

It seems like these bacteria which have integrative plasmids and colonize plant tissue epiphytically and endophytically are a good mediator for biological control. For example, *Clavibacter xyli* containing the Cry-toxins genes (*cry1Ac*) on integrative plasmid were used as biological control agent against corn borers (Thomasino et al., 1995). Thus, *Azospirillum* spp., *R. leguminosarum*, *P. capecia*, and *P. fluorescens*, which are plant colonizing bacteria, were modified by *cry*-genes. *Herbaspirillum seropedica*, an endophytic bacterium that inhabits sugarcane tissue, was transformed by the gene *cry1Ac7* stem from *B. thuringiensis* 234 to suppress growth of *Eldana saccharina* larvae (Downing et al., 2000). Moreover, the Bt-toxin gene products can prevent and decrease both pest and insect vectors that are human disease-causing agents. Likewise, when a nitrogen-fixing cyanobacteria *Anabaena* sp. PCC 7120 is transformed by multiple genes (*cry44*, *cry11A*, and *p20*), it becomes toxic against larvae of yellow fever mosquitoes (Wu et al., 1997).

On the other hand, one of the promising areas about endophytes studies includes their usage of in genetic studies for phytoremediation, such as degradation of organic pollutants from oil products and hazardous chemicals (Afzal et al., 2014). The endophytic bacteria associated with plant tissue may cleaned up soil environment contaminated with not only heavy metals but also complex chemical compounds. In a research regarding this issue, following genetical modification of endophytes with the genes encoding enzyme that can degrade organic pollutants, their ability of degradation was enhanced (Maksimov et al., 2015).

Another important example of the use of endophytic bacteria as a biotechnological tool is the degradation of mycotoxins phenomena. Therefore, it would be possible to gain more safe animal feed in terms of mycotoxins. It was found in a study that a bacterium named E3–39 strain had the ability of degrading the mycotoxin deoxynivalenol. The same bacterium can also reduce fusariotoxin. This characteristic was verified *in vitro* using mass spectrometry. The endophytic bacterium was identified to a group of *Rhizobium-Agrobacterium* (Maksimov et al., 2015).

Actually, acclimation of plantlets is a critical point of work where the beneficial bacteria have an important role. Adaptation of plantlets to acclimation conditions is difficult due to non-functional stoma, not fully efficient roots and photosynthesis organelles. If plantlets are inoculated with beneficial endosymbionts at the final stage of propagation, they can cope with this difficulty (Panigrahi et al., 2015). *Azorhizobium, Azospirillum, Azotobacter, Bacillus, Burkholderia, Curtobacterium, Enterobacter, Halomonas, Methylobacterium, Microbacterium, Methilophylus, Paenibacillus, Pseudomonas, Ralstonia, Rhodococcus, Rhodopseudomonas,* and *Sphingopyxis* belonging to 13 different genera trigger beneficial effect on micropropagation works. Some positive effects of these genera on plantlets were studied by shoot weight, leaf number, axillary shoot growth, rooting, rooted shoots, number, and length of roots, and acclimation of plantlets to *ex vitro* conditions (Orlikowska et al., 2017).

2.8 ENDOPHYTES AS BIOFERTILIZERS: FORESIGHT AND COMPLICATION

Recently, many studies on endophytic bacterial relationships with plants are carried out in restricted conditions. Actually, this approach does not reflect the real plant-endophyte interactions. For a realistic approach, beneficial effect of the bacteria should also be observed in the field (Riggs et al., 2001; Gyaneshwar et al., 2002). Plant growth activity may be effected by conditions of plantation sites such as nitrogen content of soil (Muthukumarasamy et al., 1999, 2002), soil type (De Oliveira et al., 2006), as well as age and variety of host plant (De Oliveira et al., 2006). Besides, many studies showed that use of nitrogen fixing bacteria and N-fertilizer combination decreased the requirement of extra fertilization on field (Saleh et al., 2001). At this point, actual challenge is the optimization of fertilizer amount and nitrogen fixing endophytic bacteria survival rate in rhizospheric soil. Sometimes high nitrogen fertilized soil cause reduction of bacterial colonization like sugarcane and their host bacteria *G. diazothrophicus* and *H. seropedicae* (Bueno dos Reis Junior et al., 2000; Muthukumarasamy et al., 2002). In some events, like *Azospirillum* sp. adsorption on wheat root surface have negative effect if the concentration of Ca^{2+} and $(PO_4)^{3-}$ in the media is over 50 mM (Pinherio et al., 2002). Therefore, indicative low soil fertility (Alfisol soil type) supports the endophytic inoculation and shows a better performance (De Oliviera et al., 2006). On the other hand, the survival rate of endophytic bacteria depends on concentration of nitrogen source especially ammonia in the media (25 mM NH_4NO_3). High concentration of ammonia in the medium causes morphological changes in bacteria and may play a hazardous role in their survival. The use of a compost as a nitrogen source increases the number of bacterial colonization as it decreases the reducing effect of nitrogen fertilizer (Muthukumaraswamy et al., 2007). The genotype and age of the selected plant affect the performance of the bacteria improving the growth of the host plant (Muñoz-Rojas and Caballero-Mellado, 2003; De Oliveira et al., 2006). It was reported that a significant reduction in the *G. diazotrophicus* population due to the age and genotype of the plant (Muñoz-Rojas and Caballero-Mellado, 2003). In some sugarcane varieties, a large number of endophytes are known for a long time. In addition, it causes differences in the number of diazotrophic bacteria in environmental factors such as hydric stress and climatic changes in the soil (Bueno dos Reis Junior et al., 2000). However, additional field trials should be made for the optimization of parameters including time, method of application of the inoculant, and environmental factors.

2.9   CONCLUDING REMARKS AND FUTURE PROSPECTS

World's population is expected to be close to 10 million in the next following 50 years. However, it is not an easy task to supply adequate food for the ever-increasing world population; therefore, various strategies and innovative approaches should be designed in order to achieve this purpose. The utilization of more agricultural land, increased use of herbicides, pesticides, and fertilizers, farm mechanization, transgenic crops and microorganisms to promote plant growth are among the strategies to be followed for more food production. However, many of these solutions are not sustainable, but seem to be beneficial in the short-term. Living with limited resources, forces humans to provide effective, long-term, sustainable, and eco-friendly solutions to provide enough supplements for the world. Hence, the versatile use of PGPBs in agriculture is a complementary approach to solve this problem. In the last 15–20 years, our knowledge about the mechanisms performed by the PGPBs has dramatically increased. Further understanding of the key mechanisms used by endophytic bacteria will enhance the effective employ of these organisms in agricultural applications. Bacterial endophytes with their inherent characteristics, genes, and metabolic pathways are increasingly attractive. Currently, synthetic biology has provided a genuine opening in the design of novel enzymes and microorganisms for biomass fermentation into fuels and different products. Hereafter, plants and their beneficial symbionts could also be altered to elevate the biomass for a rising population in respect of climate change. For this reason, efforts to increase the extensive use of PGPBs should focus primarily on clarification of how these bacteria elicit plant growth.

KEYWORDS

- bacterial endophytes
- biocontrol
- biofertilization
- endophytic bacteria
- phytodegradation
- phytoremediation
- phytostimulation
- plant beneficial microbes
- plant growth-promoting bacteria
- plant interactions with endophytic bacteria
- plant microbiome

REFERENCES

Abhilash, P. C., Powell, J. R., Singh, H. B., & Singh, B. K. Plant–microbe interactions: Novel applications for exploitation in multipurpose remediation technologies. *Trends in Biotechnology*, **2012**, *30*(8), 416–420.

Abreu-Tarazi, M. F., Navarrete, A. A., Andreote, F. D., Almeida, C. V., Tsai, S. M., & Almeida, M. Endophytic bacteria in long-term *in vitro* cultivated "axenic" pineapple microplants revealed by PCR–DGGE. *World J. Microbiol. Biotechnol.*, **2010**, *26*, 555–560.

Afzal, M., Khan, Q. M., & Sessitsch, A. Endophytic bacteria: Prospects and applications for the phytoremediation of organic pollutants. *Chemosphere*, **2014**, *117*, 232–242.

Afzal, M., Yousaf, S., Reichenauer, T. G., & Sessitsch, A. The inoculation method affects colonization and performance of bacterial inoculant strains in the phytoremediation of soil contaminated with diesel oil. *Int. J. Phytoremediat.*, **2012**, *14*, 35–47.

Afzal, M., Yousaf, S., Reichenauer, T. G., Kuffner, M., & Sessitsch, A. Soil type affects plant colonization, activity and catabolic gene expression of inoculated bacterial strains during phytoremediation of diesel. *J. Hazard. Mater.*, **2011**, *186*, 1568–1575.

Andressen, D., Manoochehri, I., Carletti, S., Llorente, B., Tacoronte, M., & Vielma, M. Optimization of the *in vitro* proliferation of jojoba [*Simmondsia chinensis* (Link) Schn.] by using rotable central composite design and inoculation with rhizobacteria. *Bioagro.*, **2009**, *21*, 41–48.

Araújo, W. L., Marcon, J., Maccheroni, W., Van Elsas, J. D., Van Vuurde, J. W. L., & Azevedo, J. L. Diversity of endophytic bacterial populations and their interaction with *Xylella fastidiosa* in citrus plants. *Applied and Environmental Microbiology*, **2002**, *68*(10), 4906–4914.

Arkhipova, T. N., Prinsen, E., Veselov, S. U., Martinenko, E. V., Melentiev, A. I., & Kudoyarova, G. R. Cytokinin producing bacteria enhance plant growth in drying soil. *Plant and Soil*, **2007**, *292*(1/2), 305–315.

Arshad, M., & Frankenberger, W. T. Microbial production of plant hormones. *Plant and Soil*, **1991**, *133*(1), 1–8.

Azevedo, J. L., Maccheroni, W., Pereira, J. O., & De Araújo, W. L. Endophytic microorganisms: A review on insect control and recent advances on tropical plants. *Electronic Journal of Biotechnology*, **2000**, *3*(1), 15, 16.

Bacon, C. W., & White, J. F. *Microbial Endophytes* (pp. 1–129). CRC Press, Florida, USA, **2000**.

Bais, H. P., Walker, T. S., Stermitz, F. R., Hufbauer, R. A., & Vivanco, J. M. Enantiomeric-dependent phytotoxic and antimicrobial activity of (±)-catechin, a rhizosecreted racemic mixture from spotted knapweed. *Plant Physiology*, **2002**, *128*(4), 1173–1179.

Bakker, P. A., Pieterse, C. M., & Van Loon, L. C. Induced systemic resistance by fluorescent *Pseudomonas* spp. *Phytopathology*, **2007**, *97*(2), 239–243.

Barac, T., Taghavi, S., Borremans, B., Provoost, A., Oeyen, L., Colpaert, J. V., Vanqronveld, J., & Van Der Lelie, D. Engineered endophytic bacteria improve phytoremediation of water-soluble, volatile, organic pollutants. *Nature Biotechnology*, **2004**, *22*(5), 583–588.

Barraquio, W. L., Revilla, L., & Ladha, J. K. Isolation of endophytic diazotrophic bacteria from wetland rice. *Plant Soil*, **1997**, *194*, 15–24.

Barry, S. M., & Challis, G. L. Recent advances in siderophore biosynthesis. *Curr. Opin. Chem. Biol.*, **2009**, *13*, 205–215.

Bashan, Y., & Gina, H. Proposal for the division of plant growth-promoting rhizobacteria into two classifications: Biocontrol-PGPB (plant growth-promoting bacteria) and PGPB. *Soil Biology and Biochemistry*, **1998**, *30*(8), 1225–1228.

Battistoni, F., Bartels, D., Kaiser, O., Reamon-Büttner, S. M., Hurek, T., & Reinhold-Hurek, B. Physical map of the *Azoarcus* sp. strain BH72 genome based on a bacterial artificial chromosome library as a platform for genome sequencing and functional analysis. *FEMS Microbiol. Lett.*, **2005**, *249*, 233–240.

Beattie, G. A. Plant-associated bacteria: Survey, molecular phylogeny, genomics and recent advances. In: Gnanamanickam, S. S., (ed.), *Plant-Associated Bacteria* (pp. 1–56). Springer, Dordrecht, **2007**.

Beck, H. C., Hansen, A. M., & Lauritsen, F. R. Novel pyrazine metabolites found in polymyxin biosynthesis by *Paenibacillus polymyxa*. *FEMS Microbiol. Lett.*, **2003**, *220*, 67–73.

Bela, J. K. U., & Kalidas, S. Control of hyperhydricity in anise (*Pimpinella anisum*) tissue culture by *Pseudomonas* spp. *Journal of Herbs, Spices and Medicinal Plants*, **1998**, *6*(1), 57–67.

Belimov, A. A., & Dietz, K. J. Effect of associative bacteria on element composition of barley seedlings grown in solution culture at toxic cadmium concentrations. *Microbiological Research*, **2000**, *155*(2), 113–121.

Belimov, A. A., & Tikhonovich, I. A. Microbiological aspects of sustainability and accumulation of heavy metals in plants. *Skh. Biol.*, **2011**, *3*, 10–15.

Belimov, A. A., Dodd, I. C., Hontzeas, N., Theobald, J. C., Safronova, V. I., & Davies, W. J. Rhizosphere bacteria containing 1-aminocyclopropane-1-carboxylate deaminase increase yield of plants grown in drying soil via both local and systemic hormone signaling. *New Phytologist*, **2009**, *181*(2), 413–423.

Belimov, A. A., Dodd, I. C., Safronova, V. I., Dumova, V. A., Shaposhnikov, A. I., Ladatko, A. G., & Davies, W. J. Abscisic acid metabolizing rhizobacteria decrease ABA concentrations *in planta* and alter plant growth. *Plant Physiology and Biochemistry*, **2014**, *74*, 84–91.

Belimov, A. A., Hontzeasb, N., Safronovaa, V. I., Demchinskayaa, S. V., Piluzzac, G., Bullittac, S., & Glick, B. R. Cadmium-tolerant plant growth-promoting bacteria associated with the roots of Indian mustard (*Brassica juncea,* L. Czern.). *Soil Biology and Biochemistry*, **2005**, *37*(2), 241–250.

Belimov, A. A., Safronova, V. I., Sergeyeva, T. A., Egorova, T. N., Matveyeva, V. A., Tsyganov, V. E., Borisov, A. Y., Tikhonovich, I. A., Kluge, C., Preisfeld, A., Dietz, K. J., & Stepanok, V. V. Characterization of plant growth promoting rhizobacteria isolated from polluted soils and containing 1-aminocyclopropane-1-carboxylate deaminase. *Canadian Journal of Microbiology*, **2001**, *47*(7), 642–652.

Beneduzi, A., Ambrosini, A., & Passaglia, L. M. P. Plant growth-promoting rhizobacteria (PGPR): Their potential as antagonists and biocontrol agents. *Genetics and Molecular Biology*, **2012**, *35*(4), 1044–1051.

Benhamou, N., Kloepper, J. W., & Tuzun, S. Induction of resistance against Fusarium wilt of tomato by combination of chitosan with an endophytic bacterial strain: Ultrastructure and cytochemistry of the host response. *Planta*, **1998**, *204*(2), 153–168.

Benhizia, Y., Benhizia, H., Benguedouar, A., Muresu, R., Giacomini, A., & Squartini, A. Gamma proteobacteria can nodulate legumes of the genus *Hedysarum*. *Syst. Appl. Microbiol.*, **2004**, *27*, 462–468.

Berde, C. V., Bhosale, P. P., & Chaphalkar, S. R. Plasmids of endophytic bacteria as vectors for transformation in plants. *International Journal of Integrative Biology*, **2010**, *9*(3), 113–118.

Berg, G., & Hallmann, J. Control of plant pathogenic fungi with bacterial endophytes. In: Schulz, B. J. E., Boyle, C. J. C., & Sieber, T. N., (eds.), *Microbial Root Endophytes* (pp. 53–69). Springer, Berlin, Heidelberg, **2006**.

Bertalan, M., Albano, R., De Pádua, V., Rouws, L., Rojas, C., Hemerly, A., et al. Complete genome sequence of the sugarcane nitrogen-fixing endophyte *Gluconacetobacter diazotrophicus* Pal5. *BMC Genomics*, **2009**, *10*(1), 450, doi: 10.1186/1471-2164-10-450.

Bloemberg, G. V., & Lugtenberg, B. J. Molecular basis of plant growth promotion and biocontrol by rhizobacteria. *Current Opinion in Plant Biology*, **2001**, *4*(4), 343–350.

Boller, T. Chemoperception of microbial signals in plant cells. *Annu. Rev. Plant Physiol. Plant Mol. Biol.*, **1995**, *46*, 189–214.

Bottini, R., Cassán, F., & Piccoli, P. Gibberellin production by bacteria and its involvement in plant growth promotion and yield increase. *Applied Microbiology and Biotechnology*, **2004**, *65*(5), 497–503.

Bragina, A., Berg, C., Cardinale, M., Shcherbakov, A., Chebotar, V., & Berg, G. *Sphagnum* mosses harbor highly specific bacterial diversity during their whole lifecycle. *The ISME Journal*, **2012**, *6*(4), 802–813.

Buckley, P. M., De Wilde, T. N., & Reed, B. M. Characterization and identification of bacteria isolated from micropropagated mint plants. *In Vitro Cell Dev. Biol.*, **1995**, *31*, 58–64.

Bueno, D. R. J., F., Massena, R. V., Urquiaga, S., & Döbereiner, J. Influence of nitrogen fertilization on the population of diazotrophic bacteria *Herbaspirillum* spp. and *Acetobacter diazotrophicus* in sugar cane *Saccharum* spp. *Plant Soil*, **2000**, *219*, 153–159.

Burg, S. P. Ethylene in plant growth. *Proc. Natl. Acad. Sci. USA*, **1973**, *70*, 591–597.

Cassells, A. C., & Tahmatsidou, V. The influence of local plant growth conditions on non-fastidious bacterial contamination of meristem-tips of *Hydrangea* cultured *in vitro*. *Plant Cell Tissue Organ Cult.*, **1996**, *47*, 15–26.

Castillo, U. F., Strobel, G. A., Ford, E. J., Hess, W. M., Porter, H., Jensen, J. B., Albert, H., Robison, R., Condron, M. A., Teplow, D. B., Stevens, D., & Yaver, D. Munumbicins, wide-spectrum antibiotics produced by *Streptomyces* NRRL, **30562**, endophytic on *Kennedia nigriscans*. *Microbiology*, **2002**, *148*, 2675–2685.

Cavalcante, J. J. V., Vargas, C., Nogueira, E. M., Vinagre, F., Schwarcz, K., Baldani, J. I., Ferreira, P. C. G., & Hemerly, A. S. Members of the ethylene signaling pathway are regulated in sugarcane during the association with nitrogen-fixing endophytic bacteria. *Journal of Experimental Botany*, **2007**, *58*(3), 673–686.

Chalfie, M., Tu, Y., Euskirchen, G., Ward, W. W., & Prasher, D. C. Green fluorescent protein as a marker for gene expression. *Science*, **1994**, *263*, 802–805.

Chandra, S., Bandopadhyay, R., Kumar, V., & Chandra, R. Acclimatization of tissue cultured plantlets: From laboratory to land. *Biotechnol. Lett.*, **2010**, *32*, 1199–1205.

Charkowski, A. O., Barak, J. D., Sarreal, C. Z., & Mandrell, R. E. Differences in growth of *Salmonella enterica* and *Escherichia coli* O157:H7 on alfalfa sprouts. *Appl. Environ. Microbiol.*, **2002**, *68*, 3114–3120.

Chelius, M. K., & Triplett, E. W. Diazotrophic endophytes associated with maize. In: Wymondham, U. K., & Triplett, E. W., (eds.), *Prokaryotic Nitrogen Fixation: A Model System for Analysis of a Biological Process* (pp. 779–791). Horizon Scientific Press, Wymondham, UK, **2000**.

Chen, C., Bauske, E. M., Musson, G., Rodriguezkabana, R., & Kloepper, J. W. Biological control of Fusarium wilt on cotton by use of endophytic bacteria. *Biological Control*, **1995**, *5*(1), 83–91.

Chi, F., Shen, S. H., Cheng, H. P., Jing, Y. X., Yanni, Y. G., & Dazzo, F. B. Ascending migration of endophytic rhizobia, from roots to leaves, inside rice plants and assessment of benefits to rice growth physiology. *Applied and Environmental Microbiology*, **2005**, *71*(11), 7271–7278.

Cho, S. J., Hong, S. Y., Kim, J. Y., Park, S. R., Kim, M. K., Lim, W. J., Shin, E. J., Kim, E. J., Cho, Y. U., & Yun, H. D. Endophytic *Bacillus* sp. CY22 from a balloon flower (*Platycodon grandiflorum*) produces surfactin isoforms. *Journal of Microbiology and Biotechnology*, **2003**, *13*(6), 859–865.

Christiansen-Weniger, C., Groneman, A. F., & Van Veen, J. A. Associative N_2 fixation and root exudation of organic acids from wheat cultivars of different aluminum tolerance. *Plant and Soil*, **1992**, *139*, 167–174.

Christoserdova, L., Chen, S. W., Lapidus, A., & Lindstrom, M. E. Methylotrophy in *Methylobacterium extorquens* AM1 from a genomic point of view. *J. Bacteriol.*, **2003**, *185*, 2980–2987.

Cohen, A. C., Travaglia, C. N., Bottini, R., & Piccoli, P. N. Participation of abscisic acid and gibberellins produced by endophytic *Azospirillum* in the alleviation of drought effects in maize. *Botany*, **2009**, *87*(5), 455–462.

Compant, S., Clément, C., & Sessitsch, A. Plant growth-promoting bacteria in the rhizo- and endosphere of plants: Their role, colonization, mechanisms involved and prospects for utilization. *Soil Biology and Biochemistry*, **2010**, *42*(5), 669–678.

Compant, S., Reiter, B., Sessitsch, A., Nowak, J., Clement, C., & Ait, B. E. Endophytic colonization of *Vitis vinifera* L. by plant growth-promoting bacterium *Burkholderia* sp. strain PsJN. *Appl. Environ. Microbiol.*, **2005**, *71*, 1685–1693.

Conn, V. M., Walker, A. R., & Franco, C. M. M. Endophytic actinobacteria induce defense pathways in *Arabidopsis thaliana*. *Mol. Plant-Microbe Interact.*, **2008**, *21*(2), 208–218.

Conrath, U., Beckers, G. J., Flors, V., García-Agustín, P., Jakab, G., Mauch, F., et al. Priming: getting ready for battle. *Mol. Plant-Microbe Interact.*, **2006**, *19*(10), 1062–1071.

Costa, F. G., Zucchi, T. D., & Melo, I. S. D. Biological control of phytopathogenic fungi by endophytic actinomycetes isolated from maize (*Zea mays* L.). *Brazilian Archives of Biology and Technology*, **2013**, *56*(6), 948–955.

De Almeida, C. V., Andreote, F. D., Yara, R., Tanaka, F. A. O., Azevedo, J. L., & De Almeida, M. Bacteriosomes in axenic plants: Endophytes as stable endosymbionts. *World J. Microbiol. Biotechnol.*, **2009**, *25*, 1757–1764.

De Bary, A. Morphologie und physiologie der pilze, flechten und myxomyceten. In: Abt, E. W., (ed.), *Handbuch der Physiologischen Botanik, 2 Bd., 1.* Verlag, Leipzig, **1866**, doi.org/10.5962/bhl.title.120970.

De Meyer, G., Audenaert, K., & Höfte, M. *Pseudomonas aeruginosa* 7NSK2-induced systemic resistance in tobacco depends on *in planta* salicylic acid accumulation but is not associated with PR1a expression. *European Journal of Plant Pathology*, **1999**, *105*(5), 513–517.

De Oliveira, A. L. M., De Canuto, E. L., Urquiaga, S., Reis, V. M., & Baldani, J. I. Yield of micropropagated sugarcane varieties in different soil types following inoculation with diazotrophic bacteria. *Plant and Soil*, **2006**, *284*, 23–32.

De Werra, P., Péchy-Tarr, M., Keel, C., & Maurhofer, M. Role of gluconic acid production in the regulation of biocontrol traits of *Pseudomonas fluorescens* CHA0. *Applied and Environmental Microbiology*, **2009**, *75*(12), 4162–4174.

Deng, Y., Zhu, Y., Wang, P., Zhu, L., Zheng, J., Li, R., Ruan, L., Peng, D., & Sun, M. Complete genome sequence of *Bacillus subtilis* BSn5, an endophytic bacterium of *Amorphophallus konjac* with antimicrobial activity for the plant pathogen *Erwinia carotovora* subsp. *carotovora*. *Journal of Bacteriology*, **2011**, *193*(8), 2070–2071.

Dias, A. C. F., Costa, F. E. C., Andreote, F. D., Lacava, P. T., Teixeira, M. A., Assumpção, L. C., Araújo, W. L., Azevedo, J. L., & Melo, I. S. Isolation of micropropagated strawberry endophytic bacteria and assessment of their potential for plant growth promotion. *World J. Microbiol. Biotechnol.*, **2009**, *25*, 189–195.

Digat, B., Brochard, P., Hermelin, V., & Touzet, M. Interest of bacterized synthetics substrates MILCAP® for *in vitro* culture. *Acta Hort.*, **1987**, *212*, 375–378.

Dimock, M. B., & Tingey, W. M. Host acceptance behavior of Colorado potato beetle larvae influenced by potato glandular trichomes. *Physiological Entomology*, **1988**, *13*(4), 399–406.

Ding, S. M., Liang, T., Zhang, C. S., Yan, J. C., & Zhang, Z. L. Accumulation and fractionation of rare earth elements (REEs) in wheat: Controlled by phosphate precipitation, cell wall absorption and solution complexation. *Journal of Experimental Botany*, **2005**, *56*(420), 2765–2775.

Dobbelaere, S., Vanderleyden, J., & Okon, Y. Plant growth-promoting effects of diazotrophs in the rhizosphere. *Critical Reviews in Plant Sciences*, **2003**, *22*(2), 107–149.

Dodd, I. C., Zinovkina, N. Y., Safronova, V. I., & Belimov, A. A. Rhizobacterial mediation of plant hormone status. *Annals of Applied Biology*, **2010**, *157*(3), 361–379.

Dong, Y. H., Zhang, X. F., Xu, J. L., & Zhang, L. H. Insecticidal *Bacillus thuringiensis* silences *Erwinia carotovora* virulence by a new form of microbial antagonism, signal interference. *Appl. Environ. Microbiol.*, **2004**, *70*(2), 954–960.

Dong, Y., Glasner, J. D., Blattner, F. R., & Triplett, E. W. Genomic interspecies microarray hybridization: Rapid discovery of three thousand genes in the maize endophyte, *Klebsiella pneumoniae* 342, by microarray hybridization with *Escherichia coli* K-12 open reading frames. *Appl. Environ. Microbiol.*, **2001**, *67*, 1911–1921.

Doty, S. L., Oakley, B., Xin, G., Kang, J. W., Singleton, G., Khan, Z., Vajzovic, A., & Staley, J. T. Diazotrophic endophytes of native black cottonwood and willow. *Symbiosis*, **2009**, *47*(1), 23–33.

Downing, K. J., & Thomson, J. A. Introduction of the *Serratia marcescens* chiA gene into an endophytic *Pseudomonas fluorescens* for the biocontrol of phytopathogenic fungi. *Canadian Journal of Microbiology*, **2000**, *46*(4), 363–369.

Duffy, B. K., & Défago, G. Environmental factors modulating antibiotic and siderophore biosynthesis by *Pseudomonas fluorescens* biocontrol strains. *Applied and Environmental Microbiology*, **1999**, *65*(6), 2429–2438.

Duffy, B., Schouten, A., & Raaijmakers, J. M. Pathogen self-defense: Mechanisms to counteract microbial antagonism. *Annual Review of Phytopathology*, **2003**, *41*(1), 501–538.

Dunaeva, S. E., & Osledkin, Y. S. Bacterial microorganisms associated with the plant tissue culture: Identification and possible role. *Selskokhozyaistvennaya Biologia*, **2015**, doi: 10.15389/agrobiology.2015.1.3eng.

Durbak, A., Yao, H., & Mc Steen, P. Hormone signaling in plant development. *Current Opinion in Plant Biology*, **2012**, *15*(1), 92–96.

Egener, T., Hurek, T., & Reinhold-Hurek, B. Endophytic expression of *nif* genes of *Azoarcus* sp. strain BH72 in rice roots. *Mol. Plant-Microbe İnteract.*, **1999**, *12*(9), 813–819.

Egorshina, A. A., Khairullin, R. M., Sakhabutdinova, A. R., & Lukyantsev, M. A. Involvement of phytohormones in the development of interaction between wheat seedlings and endophytic *Bacillus subtilis* strain 11BM. *Russian Journal of Plant Physiology*, **2012**, *59*(1), 134–140.

Ehrhardt, D., & Frommer, W. New technologies for 21st century plant science. *Plant Cell*, **2012**, *24*, 374–394.

Elliott, G. N., Chou, J. H., Chen, W. M., Bloemberg, G. V., Bontemps, C., Martínez-Romero, E., Velázquez, E., Young, J. P., Sprent, J. I., & James, E. K. *Burkholderia* spp. are the most competitive symbionts of *Mimosa*, particularly under N-limited conditions. *Environmental Microbiology*, **2009**, *11*(4), 762–778.

Fang, J. Y., & Hsu, Y. R. Molecular identification and antibiotic control of endophytic bacterial contaminants from micropropagated *Aglaonema* cultures. *Plant Cell Tissue Organ Cult.*, **2012**, *110*, 53–62.

Felix, G., Duran, J. D., Volko, S., & Boller, T. Plants have a sensitive perception system for the most conserved domain of bacterial flagellin. *Plant J.*, **1999**, *18*, 265–276.

Feller, I. C. Effects of nutrient enrichment on growth and herbivory of dwarf red mangrove (*Rhizophora mangle*). *Ecological Monographs*, **1995**, *65*, 477–505.

Fernandez, O., Theocharis, A., Bordiec, S., Feil, R., Jacquens, L., Clément, C., Fontaine, F., & Barka, E. A. *Burkholderia phytofirmans* PsJN acclimates grapevine to cold by modulating carbohydrate metabolism. *Mol. Plant-Microbe İnteract.*, **2012**, *25*(4), 496–504.

Ferreira, A., Quecine, M. C., Lacava, P. T., Oda, S., Azevedo, J. L., & Araújo, W. L. Diversity of endophytic bacteria from *Eucalyptus* species seeds and colonization of seedlings by *Pantoea agglomerans*. *FEMS Microbiology Letters*, **2008**, *287*(1), 8–14.

Fletcher, J., Leach, J. E., Eversole, K., & Tauxe, R. Human pathogens on plants: Designing a multidisciplinary strategy for research. *Phytopathology*, **2013**, *103*, 306–315.

Forchetti, G., Masciarelli, O., Alemano, S., Alvarez, D., & Abdala, G. Endophytic bacteria in sunflower (*Helianthus annuus* L.): Isolation, characterization, and production of jasmonates and abscisic acid in culture medium. *Applied Microbiology and Biotechnology*, **2007**, *76*(5), 1145–1152.

Fürnkranz, M., Adam, E., Müller, H., Grube, M., Huss, H., Winkler, J., & Berg, G. Promotion of growth, health and stress tolerance of Styrian oil pumpkins by bacterial endophytes. *European Journal of Plant Pathology*, **2012**, *134*(3), 509–519.

Gamalero, E., Lingua, G., Berta, G., & Lemanceau, P. Methods for studying root colonization by introduced beneficial bacteria. *Agronomie*, **2003**, *23*(5/6), 407–418.

Gangwar, M., Sheela, R., & Neerja, S. Investigating endophytic actinomycetes diversity from rice for plant growth promoting and antifungal activity. *International Journal of Advanced Life Sciences*, **2012**, *1*, 10–21.

Gao, M., Teplitski, M., Robinson, J. B., & Bauer, W. D. Production of substances by *Medicago truncatula* that affect bacterial quorum sensing. *Mol. Plant-Microbe Interact.*, **2003**, *16*(9), 827–834.

Germaine, K., Keogh, E., Garcia-Cabellos, G., Borremans, B., Van Der Lelie, D., Barac, T., Oeyen, L., Vangronsveld, J., Moore, F. P., Moore, E. R. B., Campbell, C. D., Ryan, D., & Dowling, D. N. Colonization of poplar trees by *gfp* expressing bacterial endophytes. *FEMS Microbiol. Ecol.*, **2004**, *48*, 109–118.

Glick, B. R., Karaturovíc, D., & Newell, P. A novel procedure for rapid isolation of plant growth-promoting rhizobacteria. *Can. J. Microbiol.*, **1995**, *41*, 533–536.

Glick, B. R., Penrose, D. M., & Li, J. A model for the lowering of plant ethylene concentrations by plant growth promoting bacteria. *J. Theor. Biol.*, **1998**, *190*, 63–68.

Gond, S. K., Bergen, M. S., Torres, M. S., & White. J. F. Endophytic *Bacillus* spp. produce antifungal lipopeptides and induce host defense gene expression in maize. *Microbiological Research*, **2015**, *172*, 79–87.

Goudjal, Y., Toumatia, O., Sabaou, N., Barakate, M., Mathieu, F., & Zitouni, A. Endophytic actinomycetes from spontaneous plants of Algerian Sahara: Indole-3-acetic acid production and tomato plants growth promoting activity. *World Journal of Microbiology and Biotechnology*, **2013**, *29*(10), 1821–1829.

Govindarajan, M., Balandreau, J., Kwon, S. W., Weon, H. Y., & Lakshminarasimhan, C. Effects of the inoculation of *Burkholderia vietnamensis* and related endophytic diazotrophic bacteria on grain yield of rice. *Microbial Ecology*, **2008**, *55*(1), 21–37.

Guan, S. H., Sattler, I., Lin, W. H., Guo, D. A., & Grabley, S. *p*-Aminoacetophenonic acids produced by a mangrove endophyte: *Streptomyces griseus* subspecies. *J. Nat. Prod.*, **2005**, *68*, 1198–1200.

Guerin, P. Sur la presence dun champignon dans livraie. *J. Botanique.*, **1898**, *12*, 230–238 (in French).

Guerney, K. A., & Mantle, P. G. Biosynthesis of 1-N-methylalbonoursin by an endophytic *Streptomyces* sp. *J. Nat. Prod.*, **1993**, *56*, 1194–1198.

Gunatilaka, A. L. Natural products from plant-associated microorganisms: Distribution, structural diversity, bioactivity, and implications of their occurrence. *Journal of Natural Products*, **2006**, *69*(3), 509–526.

Gyaneshwar, P., James, E. K., Reddy, P. M., & Ladha, J. K. *Herbaspirillum* colonization increases growth and nitrogen accumulation in aluminum-tolerant rice varieties. *New Phytol.*, **2002**, *154*, 131–145.

Hallmann, J. *Plant Interactions with Endophytic Bacteria* (pp. 87–119). CABI Publishing, New York, **2001**.

Hallmann, J., & Berg, G. Spectrum and population dynamics of bacterial root endophytes. In: *Microbial Root Endophytes* (pp. 15–31). Springer, Berlin, Heidelberg, **2006**.

Hallmann, J., Kloepper, J. W., Rodriguez-Kabana, R., & Sikora, R. A. Endophytic rhizobacteria as antagonists of *Meloidogyne incognita* on cucumber. *Phytopathology*, **1995**, *85*(10), 1136, https://eurekamag.com/research/031/237/031237007.php (Accessed on 5 October 2019).

Hallmann, J., Quadt-Hallmann, A., Mahaffee, W. F., & Kloepper, J. W. Bacterial endophytes in agricultural crops. *Canadian Journal of Microbiology*, **1997**, *43*(10), 895–914.

Han, J. I., Choi, H. K., Lee, S. V., Orwin, P. M., Kim, J., La Roe, S. L., Kim, T. O., Neil, J., Leadbetter, J. R., Lee, S. Y., Hur, C. G., Spain, J. C., Ovchinnikova, G., Goodwin, L., & Han, C. Complete genome sequence of the metabolically versatile plant growth-promoting endophyte *Variovorax paradoxus* S110. *Journal of Bacteriology*, **2011**, *193*(5), 1183–1190.

Hardoim, P. R., Van Overbeek, L. S., Berg, G., Pirttilä, A. M., Compant, S., Campisano, A., Döring, M., & Sessitsch, A. The hidden world within plants: Ecological and evolutionary considerations for defining functioning of microbial endophytes. *Microbiology and Molecular Biology Reviews*, **2015**, *79*(3), 293–320.

Harman, G. E., Howell, C. R., Viterbo, A., Chet, I., & Lorito, M. *Trichoderma* species opportunistic, avirulent plant symbionts. *Nature Reviews Microbiology*, **2004**, *2*(1), 43–56.

Hartmann, A., Stoffels, M., Eckert, B., Kirchhof, G., & Schloter, M. Analysis of the presence and diversity of diazotrophic endophytes. In: Triplett, E. W., (ed.), *Prokaryotic Nitrogen Fixation: A Model System for Analysis of a Biological Process* (pp. 727–736). Horizon Scientific Press, Wymondham, UK, **2000**.

Hinton, D. M., & Bacon, C. W. *Enterobacter cloacae* is an endophytic symbiont of corn. *Mycopathologia*, **1995**, *129*, 117–125.

Ho, K. L., Lin, B., Chen, Y. Y., & Lee, D. J. Biodegradation of phenol using *Corynebacterium* sp. DJ1 aerobic granules. *Bioresource Technology*, **2009**, *100*(21), 5051–5055.

Hoffmann, M. Biogeography of *Arabidopsis thaliana* (L.) Heynh. (Brassicaceae). *J. Biogeogr.*, **2002**, *29*, 125–134.

Hong, Y., Pasternak, J. J., & Glick, B. R. Biological consequences of plasmid transformation of the plant growth promoting rhizobacterium *Pseudomonas putida* GR12-2. *Can. J. Microbiol.*, **1991**, *37*, 796–799.

Honma, M., & Tokuji, S. Metabolism of 1-aminocyclopropane-1-carboxylic acid. *Agricultural and Biological Chemistry*, **1978**, *42*(10), 1825–1831.

Hu, N., Luo, Y., Wu, L., & Song, J. A field lysimeter study of heavy metal movement down the profile of soils with multiple metal pollution during chelate-enhanced phytoremediation. *International Journal of Phytoremediation*, **2007**, *9*(4), 257–268.

Hurek, T., Egener, T., & Reinhold-Hurek, B. Divergence in nitrogenases of *Azoarcus* spp., Proteobacteria of the beta subclass. *J. Bacteriol.*, **1997**, *179*, 4172–4178.

Hurek, T., Handley, L. L., Reinhold-Hurek, B., & Piche, Y. *Azoarcus* grass endophytes contribute fixed nitrogen to the plant in an unculturable state. *Mol. Plant-Microbe Interact.*, **2002**, *15*, 233–242.

Idris, E. E., Iglesias, D. J., Talon, M., & Borriss, R. Tryptophan-dependent production of indole-3-acetic acid (IAA) affects level of plant growth promotion by *Bacillus amyloliquefaciens* FZB42. *Mol. Plant-Microbe Interact*, **2007**, *20*(6), 619–626.

Illmer, P., & Schinner, F. Solubilization of inorganic calcium phosphates-solubilization mechanisms. *Soil Biology and Biochemistry*, **1995**, *27*(3), 257–263.

Iniguez, A. L., Dong, Y., Carter, H. D., Ahmer, B. M., Stone, J. M., & Triplett, E. W. Regulation of enteric endophytic bacterial colonization by plant defenses. *Mol. Plant-Microbe Interact*, **2005**, *18*(2), 169–178.

Iqbal, A., & Hasnain, S. Auxin producing *Pseudomonas* strains: Biological candidates to modulate the growth of *Triticum aestivum* beneficially. *American Journal of Plant Sciences*, **2013**, *4*(9), **1693**, doi:10.4236/ajps.2013.49206.

Isawa, T., Yasuda, M., Awazaki, H., Minamisawa, K., Shinozaki, S., & Nakashita, H. *Azospirillum* sp. strain B510 enhances rice growth and yield. *Microbes and Environments*, **2010**, *25*(1), 58–61.

Islam, M. R., Madhaiyan, M., Deka, B. H. P., Yim, W., Lee, G., Saravanan, V. S., Fu, Q., Hu, H., & Sa, T. Characterization of plant growth-promoting traits of free-living diazotrophic bacteria and their inoculation effects on growth and nitrogen uptake of crop plants. *Journal of Microbiology and Biotechnology*, **2009**, *19*(10), 1213–1222.

James, E. K., & Olivares, F. B. Infection and colonization of sugar cane and other graminaceous plants by endophytic diazotrophs. *Crit. Rev. Plant Sci.*, **1997**, *17*, 77–119.

James, E. K., Gyaneshwar, P., Mathan, N., Barraquio, Q. L., Reddy, P. M., Iannetta, P. P., Olivares, F. L., & Ladha, J. K. Infection and colonization of rice seedlings by the plant growth promoting bacterium *Herbaspirillum seropedicae* Z67. *Mol Plant Microbe Interact*, **2002**, *15*, 894–906.

Joo, G. J., Kim, Y. M., Kim, J. T., Rhee, I. K., Kim, J. H., & Lee, I. J. Gibberellins-producing rhizobacteria increase endogenous gibberellins content and promote growth of red peppers. *Journal of Microbiology (Seoul, Korea)* **2005**, *43*(6), 510–515.

Kaluzna, M., Mikiciński, A., Sobiczewski, P., Zawadzka, M., Zenkteler, E., & Orlikowska, T. Detection, isolation, and preliminary characterization of bacteria contaminating plant tissue cultures. *Acta Agrobot.*, **2013**, *66*, 81–92.

Kalyaeva, M. A., Ivanova, E. G., Doronina, N. V., Zakharchenko, N. S., Trotsenko, Y. A., & Buryanov, Y. I. The effect of aerobic methylotrophic bacteria on the *in vitro* morphogenesis of soft wheat (*Triticum aestivum*). *Russian Journal of Plant Physiology*, **2003**, *50*(3), 313–317.

Kamilova, F., Validov, S., Azarova, T., Mulders, I., & Lugtenberg, B. Enrichment for enhanced competitive plant root tip colonizers selects for a new class of biocontrol bacteria. *Environmental Microbiology*, **2005**, *7*(11), 1809–1817.

Kang, J. W., Khan, Z., & Doty, S. L. Biodegradation of trichloroethylene by an endophyte of hybrid poplar. *Appl. Environ. Microbiol.*, **2012**, *78*(9), 3504–3507.

Kaul, S., Sharma, T., & Dhar, M. K. Omics tools for better understanding the plant-endophyte interactions. *Frontiers in Plant Science*, **2016**, *7*, 955, doi: 10.3389/fpls.2016.00955.

Kieran, G. J., Liu, X., Cabellos, G. G., Hogan, J. P., Ryan, D., & Dowling, D. N. Bacterial endophyte-enhanced phytoremediation of the organochlorine herbicide 2,4-dichlorophenoxyacetic acid. *FEMS Microbiology Ecology*, **2006**, *57*(2), 302–310.

Kloepper, J. W., Leong, J., Teintze, M., & Schroth, M. N. *Pseudomonas* siderophores: A mechanism explaining disease-suppressive soils. *Curr. Microbiol.*, **1980**, *4*, 317–320.

Kloepper, J. W., Ryu, C. M., & Zhang, S. Induced systemic resistance and promotion of plant growth by *Bacillus* spp. *Phytopathology*, **2004**, *94*, 1259–1266.

Kloepper, J. W., Tipping, E. M., & Lifshitz, R. Plant growth promotion mediated by bacterial rhizosphere colonizers. In: Keister, D. L., & Cregan, P. B., (eds.), *The Rhizosphere and Plant Growth* (pp. 315–326). Kluwer Academic Publishers, Dordrecht, The Netherlands, **1991**.

Knoth, J. L., Kim, S. H., Ettl, G. J., & Doty, S. L. Biological nitrogen fixation and biomass accumulation within poplar clones as a result of inoculations with diazotrophic endophyte consortia. *New Phytologist*, **2014**, *201*(2), 599–609.

Kobayashi, D. Y., & Palumbo, J. D. Bacterial endophytes and their effects on plants and uses in agriculture. *Microbial Endophytes*, **2000**, *19*, 199–233.

Kong, Z. Y., Glick, B. R., Duan, J., Ding, S., Tian, J., McConkey, B. J., & Wei, G. H. Effects of 1-aminocyclopropane-1-carboxylate (ACC) deaminase-overproducing *Sinorhizobium meliloti* on plant growth and copper tolerance of *Medicago lupulina*. *Plant and Soil*, **2015**, *391*(1/2), doi: 10.1007/s11104–015–2434–4.

Koornneef, M., & Meinke, D. The development of Arabidopsis as a model plant. *Plant Journal*, **2010**, *61*, 909–921.

Koumoutsi, A., Chen, X. H., Henne, A., Liesegang, H., Hitzeroth, G., Franke, P., Vater, J., & Borriss, R. Structural and functional characterization of gene clusters directing nonribosomal synthesis of bioactive cyclic lipopeptides in *Bacillus amyloliquefaciens* strain FZB42. *J. Bacteriol.*, **2004**, *186*, 1084–1096.

Kpomblekou, A. K., & Tabatabai, M. A. Effect of low-molecular weight organic acids on phosphorus release and phytoavailabilty of phosphorus in phosphate rocks added to soils. *Agr. Ecosyst. Environ.*, **2003**, *100*, 275–284.

Krause, A., Ramakumar, A., Bartels, D., Battistoni, F., Bekel, T., Boch, J., et al. Complete genome of the mutualistic, N_2-fixing grass endophyte *Azoarcus* sp. strain BH72. *Nature Biotechnology*, **2006**, *24*(11), 1385–1391.

Kucey, R. M. N. Plant growth-altering effects of *Azospirillum brasilense* and *Bacillus* C-1 1–25 on two wheat cultivars. *J. Appl. Bacteriol.*, **1988**, *64*, 187–196.

Kudoyarova, G. R., Melentiev, A. I., Martynenko, E. V., Timergalina, L. N., Arkhipova, T. N., Shendel, G. V., Kuzmina, L. Y., Dodd, I. C., & Veselov, S. Y. Cytokinin producing bacteria stimulate amino acid deposition by wheat roots. *Plant Physiol. Biochem.*, **2014**, *83*, 285e291, doi: 10.1016/j.plaphy.2014.08.015.

Kusari, P., Spiteller, M., Kayser, O., & Kusari, S. Recent advances in research on *Cannabis sativa* L. endophytes and their prospect for the pharmaceutical industry. In: Kharwar, R.

N., Upadhyay, R. S., Dubey, N. K., & Raghuwanshi, R., (eds.), *Microbial Diversity and Biotechnology in Food Security* (pp. 3–15). Springer, New Delhi, **2014**.

Li, K., & Ramakrishna, W. Effect of multiple metal resistant bacteria from contaminated lake sediments on metal accumulation and plant growth. *J. Hazard. Mater.*, **2011**, *189*, 531–539.

Liu, L., Kloepper, J. W., & Tuzun, S. Induction of systemic resistance in cucumber against bacterial angular leaf spot by plant growth-promoting rhizobacteria. *Phytopathology*, **1995**, *85*(8), 843–847.

Lodewyckx, C., Vangronsveld, J., Porteous, F., Moore, E. R., Taghavi, S., Mezgeay, M., & Der Lelie, D. V. Endophytic bacteria and their potential applications. *Critical Reviews in Plant Sciences*, **2002**, *21*(6), 583–606.

Lorentz, R. H., Artico, S., Da Silveira, A. B., Einsfeld, A., & Corção, G. Evaluation of antimicrobial activity in *Paenibacillus* spp. strains isolated from natural environment. *Let. Appl. Microbiol.*, **2006**, *43*, 541–547.

Loy, A., Horn, M., & Wagner, M. Probe base: An online resource for rRNA-targeted oligonucleotide probes. *Nucleic Acids Res*, **2003**, *31*(1), 514–516.

Lucero, M. E., Unc, A., Cooke, P., Dowd, S., & Sun, S. Endophyte microbiome diversity in micropropagated *Atriplex canescens* and *Atriplex torreyi* var *griffithsii. PLoS One*, **2011**, *6*(3), e17693, https://doi.org/10.1371/journal.pone.0017693 (Accessed on 5 October 2019).

Lugtenberg, B., & Kamilova, F. Plant-growth-promoting rhizobacteria. *Annual Review of Microbiology*, **2009**, *63*, 541–556.

Luo, S., Xu, T., Chen, L., Chen, J., Rao, C., Xiao, X., Wan, Y., Zeng, G., Long, F., Liu, C., & Liu, Y. Endophyte-assisted promotion of biomass production and metal-uptake of energy crop sweet sorghum by plant-growth-promoting endophyte *Bacillus* sp. SLS18. *Applied Microbiology and Biotechnology*, **2012**, *93*(4), 1745–1753.

Ma, R. S., Oliveira, F., Nai, M., Rajkumar, Y., Luo, I., Rocha, H., & Freitas, X. X. The hyperaccumulator *Sedum plumbizincicola* harbors metal-resistant endophytic bacteria that improve its phytoextraction capacity in multi-metal contaminated soil. *J. Environ. Manage.*, **2015**, *156*, 62–69.

Madhaiyani, M., Chauhan, P. S., Yim, W. J., Boruah, H. P. D., & Sa, T. M. Diversity and beneficial interactions among Methylobacterium and plants. In: Maheshwari, D. K., (ed.), *Bacteria in Agrobiology: Plant Growth Responses* (pp. 259–284). Springer-Verlag, Berlin, Heidelberg, **2011**.

Madmony, A., Chernin, L., Pleban, S., Peleg, E., & Riov, J. *Enterobacter cloacae*, an obligatory endophyte of pollen grains of Mediterranean pines. *Folia Microbiologica*, **2005**, *50*(3), 209–216.

Maksimov, I. V., Abizgil, D. R. R., & Pusenkova, L. I. Plant growth promoting rhizobacteria as alternative to chemical crop protectors from pathogens. *Appl. Biochem. Microbiol.*, **2011**, *47*(4), 333–345.

Maksimov, I. V., Veselova, S. V., Nuzhnaya, T. V., Sarvarova, E. R., & Khairullin, R. M. Plant growth-promoting bacteria in regulation of plant resistance to stress factors. *Russian Journal of Plant Physiology*, **2015**, *62*(6), 715–726.

Männistö, M. K., Marja, A. T., & Jaakko, A. P. Degradation of 2,3,4,6-tetrachlorophenol at low temperature and low dioxygen concentrations by phylogenetically different groundwater and bioreactor bacteria. *Biodegradation*, **2001**, *12*(5), 291–301.

Marchand, L., Mench, M., Jacob, D. L., & Otte, M. L. Metal and metalloid removal in constructed wetlands, with emphasis on the importance of plants and standardized measurements: A review. *Environmental Pollution*, **2010**, *158*(12), 3447–3461.

Marino, G., Altan, A. D., & Biavati, B. The effect of bacterial contamination on the growth and gas evolution of *in vitro* cultured apricot shoots *In Vitro Cell. Dev. Biol.*, **1996**, *32*, 51–56.

Marler, M. J., Zabinski, C. A., & Callaway, R. M. Mychorrhizae indirectly enhance competitive effects of an invasive forb on a native bunchgrass. *Ecology*, **1999**, *80*, 1180–1186.

Martínez-Aguilar, L., Díaz, R., Peña-Cabriales, J. J., Santos, P. E., Dunn, M. F., & Caballero-Mellado, J. Multichromosomal genome structure and confirmation of diazotrophy in novel plant-associated *Burkholderia* species. *Applied and Environmental Microbiology*, **2008**, *74*(14), 4574–4579.

Mathesius, U., Mulders, S., Gao, M., Teplitski, M., Caetano-Anollés, G., Rolfe, B. G., & Bauer, W. D. Extensive and specific responses of a eukaryote to bacterial quorum-sensing signals. *Proc. Natl. Acad. Sci. USA*, **2003**, *100*(3), 1444–1449.

Mc Inroy, J. A., & Kloepper, J. W. Survey of indigenous bacterial endophytes from cotton and sweet corn. *Plant and Soil*, **1995**, *173*, 337–342.

Melentev, A. I. *Aerobnye Sporoobrazuyushchie Bakterii Bacillus Cohn. v Agroekosistemakh (Aerobic Spore Forming Bacterium Bacillus Cohn. in Agroecosystems)*. Nauka, Moscow, **2007**.

Mercado-Blanco, J., & Lugtenberg, B. J. J. Biotechnological applications of bacterial endophytes. *Curr. Biotechnol.*, **2014**, *3*, 60–75.

Merzaeva, O. V., & Shirokikh, I. G. The production of auxins by the endophytic bacteria of winter rye. *Applied Biochemistry and Microbiology*, **2010**, *46*(1), 44–50.

Miller, C. M., Miller, R. V., Garton-Kenny, D., Redgrave, B., Sears, J., Condron, M. M., Teplow, D. B., & Strobel, G. A. Ecomycins, unique antimycotics from *Pseudomonas viridiflava*. *J. Appl. Microbiol.*, **1998**, *84*, 937–944.

Mishagi, I. J., & Donndelinger, C. R. Endophytic bacteria in symptom-free cotton plants. *Phytopathology*, **1990**, *9*, 808–811.

Mitter, B., Petric, A., Shin, M. W., Chain, P. S. G., Hauberg-Lotte, L., Reinhold-Hurek, B., Nowak, J., & Sessitsch, A. Comparative genome analysis of *Burkholderia phytofirmans* PsJN reveals a wide spectrum of endophytic lifestyles based on interaction strategies with host plants. *Front. Plant. Sci.*, **2013**, *4*, 120, doi: 10.3389/fpls.2013.00120.

Mukhopadhyay, K., Garrison, N. K., Hinton, D. M., Bacon, C. W., Khush, G. S., Peck, H. D., & Datta, N. Identification and characterization of bacterial endophytes of rice. *Mycopathologia*, **1996**, *134*(3), 151–159.

Mundt, J. O., & Hinkle, N. F. Bacteria within ovules and seeds. *Applied and Environmental Microbiology*, **1976**, *32*(5), 694–698.

Muñoz-Rojas, J., & Caballero-Mellado, J. Population dynamics of *Gluconacetobacter diazotrophicus* in sugarcane cultivars and its effect on plant growth. *Microb. Ecol.*, **2003**, *46*, 454–464.

Murthy, B. N. S., Vettakkorumakankav, N., KrishnaRaj, S., Odumeru, J., & Saxena, P. K. Characterization of somatic embryogenesis in *Pelargonium* × *hortorum* mediated by a bacterium. *Plant Cell Reports*, **1999**, *18*(7), 607–613.

Muthukumarasamy, R., Kang, U. G., Park, K. D., Jeon, W. T., Park, C. Y., Cho, Y. S., Kwon, S. W., Song, J., Roh, D. H., & Revathi, G. Enumeration, isolation and identification of diazotrophs from Korean wetland rice varieties grown with long-term application of N and compost and their short-term inoculation effect on rice plants. *J. Appl. Microbiol.*, **2007**, *102*, 981–991.

Muthukumarasamy, R., Revathi, G., & Loganathan, P. Effect of inorganic N on the population, *in vitro* colonization and morphology of *Acetobacter diazotrophicus* (syn. *Gluconacetobacter diazotrophicus*). *Plant and Soil*, **2002**, *243*, 91–102.

Nahrstedt, A., & Butterweck, V. Biologically active and other chemical constituents of the herb of *Hypericum perforatum* L. *Pharmacopsychiatry*, **1997**, *30*(S2), 129–134.

Newman, K. L., Chatterjee, S., Ho, K. A., & Lindow, S. E. Virulence of plant pathogenic bacteria attenuated by degradation of fatty acid cell-to-cell signaling factors. *Mol. Plant-Microbe Interact.*, **2008**, *21*(3), 326–334.

Newman, M. A., Sundelin, T., Nielsen, J. T., & Erbs, G. MAMP (microbe-associated molecular pattern) triggered immunity in plants. *Frontiers in Plant Science*, **2013**, *4*, 139, doi: 10.3389/fpls.2013.00139.

Nikolic, B., Schwab, H., & Sessitsch, A. Metagenomic analysis of the 1-aminocyclopropane-1-carboxylate deaminase gene (acdS) operon of an uncultured bacterial endophyte colonizing *Solanum tuberosum* L. *Archives of Microbiology*, **2011**, *193*(9), 665–676.

Norman, D. J., & Alvarez, A. M. Latent infections of *in vitro* anthurium caused by *Xanthomonas campestris* pv. *dieffenbachiae*. *Plant Cell, Tissue and Organ Culture*, **1994**, *39*, 55–61.

Nowak, J. Benefits of *in vitro* "biotization" of plant tissue cultures with microbial inoculants. *In Vitro Cellular and Developmental Biology–Plant*, **1998**, *34*(2), 122–130.

Orlikowska, T., Nowak, K., & Reed, B. Bacteria in the plant tissue culture environment. *Plant Cell, Tissue and Organ Culture 128*(3), **2017**, 487–508.

Otieno, M., Sidhu, C. S., Woodcock, B. A., Wilby, A., Vogiatzakis, I. N., Mauchline, A. L., Gikungu, M. W., & Potts, S. G. Local and landscape effects on bee functional guilds in pigeon pea crops in Kenya. *Journal of Insect Conservation*, **2015**, *19*(4), 647–658.

Păcurar, D. I., Thordal-Christensen, H., Păcurar, M. L., Pamil, D., Botez, C., & Bellini, C. *Agrobacterium tumefaciens*: From crown gall tumors to genetic transformation. *Physiol. Mol. Plant Pathol.*, **2011**, *76*, 76–81.

Panicker, B., Thomas, P., Janakiram, T., Venugopalan, R., & Narayanappa, S. B. Influence of cytokinin levels on *in vitro* propagation of shy suckering chrysanthemum 'Arka Swarna' and activation of endophytic bacteria. *In Vitro Cell. Dev. – Pl.*, **2007**, *43*, 614–622.

Panigrahi, S., Aruna, L. K., Venkateshwarulu, Y., & Umesh, N. Biohardening of micropropagated plants with PGPR and endophytic bacteria enhances the protein content. In: Kumar, A., (ed.), *Biotechnology and Bioforensics, Forensic and Medical Bioinformatics* (pp. 51–55). Springer, Netherlands, **2015**.

Pedrosa, F. O., Monteiro, R. A., Wassem, R., Cruz, L. M., Ayub, R. A., Colauto, N. B., Fernandez, M. A., Fungaro, M. H., Grisard, E. C., Hungria, M., & Madeira, H. M. Genome of *Herbaspirillum seropedicae* strain SmR1, a specialized diazotrophic endophyte of tropical grasses. *PLoS Genetics*, **2011**, *7*(5), e1002064, doi: org/10.1371/journal.pgen.1002064.

Phillips, L. A., Germida, J. J., Farrell, R. E., & Greer, C. W. Hydrocarbon degradation potential and activity of endophytic bacteria associated with prairie plants. *Soil Biol. Biochem.*, **2008**, *40*, 3054–3064.

Pieterse, C. M., Van Der Does, D., Zamioudis, C., Leon-Reyes, A., & Van Wees, S. C. Hormonal modulation of plant immunity. *Annual Review of Cell and Developmental Biology*, **2012**, *28*, 489–521.

Pieterse, C. M., Zamioudis, C., Berendsen, R. L., Weller, D. M., Van Wees, S. C., & Bakker, P. A. Induced systemic resistance by beneficial microbes. *Annu. Rev. Phytopathol.*, **2014**, *52*, 347–375.

Pinheiro, R. D., Boddey, L. H., & James, E. K. Sprent, J. I., & Boddey, R. M. Adsorption and anchoring of *Azospirillum* strains to roots of wheat seedlings. *Plant Soil*, **2002**, *246*, 151–166.

Pirttilä, A. M., Laukkanen, H., Pospiech, H., Myllylä, R., & Hohtola, A. Detection of intracellular bacteria in the buds of Scotch pine (*Pinus sylvestris* L.) by *in situ* hybridization. *Applied and Environmental Microbiology*, **2000**, *66*(7), 3073–3077.

Pirttilä, A. M., Podolich, O., Koskimäki, J. J., Hohtola, E., & Hohtola, A. Role of origin and endophyte infection in browning of bud derived tissue cultures of Scots pine *Pinus sylvestris* L. *Plant Cell Tissue Organ Cult.*, **2008**, *95*, 47–55.

Pischke, M. S., Huttlin, E. L., Hegeman, A. D., Sussman, M. R. A transcriptome-based characterization of habituation in plant tissue culture. *Plant Physiol.*, **2006**, *140*, 1255–1278.

Poppenberger, B., Leonhardt, W., & Redl, H. Latent persistence of *Agrobacterium vitis* in micropropagated *Vitis vinifera*. *Vitis*, **2002**, *41*(2), 113–114.

Porteous-Moore, F., Barac, T., Borremans, B., Oeyen, L., Vangronsveld, J., Van Der Lelie, D., Campbell, D., & Moore, E. R. B. Endophytic bacterial diversity in poplar trees growing on a BTEX-contaminated site: The characterization of isolates with potential to enhance phytoremediation. *Sys. App. Microl.*, **2006**, *29*, 539–556.

Puente, M. E., Li, C. Y., & Bashan, Y. Endophytic bacteria in cacti seeds can improve the development of cactus seedlings. *Environmental and Experimental Botany*, **2009**, *66*(3), 402–408.

Qin, S., Xing, K., Jiang, J. H., Xu, L. H., & Li, W. J. Biodiversity, bioactive natural products and biotechnological potential of plant-associated endophytic actinobacteria. *Applied Microbiology and Biotechnology*, **2011**, *89*(3), 457–473.

Quambusch, M., Pirttilä, A. M., Tejesvi, M. V., Winkelmann, T., & Bartsch, M. Endophytic bacteria in plant tissue culture: Differences between easy- and difficult-to-propagate *Prunus avium* genotypes. *Tree Physiology*, **2014**, *34*(5), 524–533.

Quoirin, M., & Lepoivre, P. Etude de milieux adaptés aux cultures *in vitro* de *Prunus* sp. *Acta Hortic.*, **1977**, *78*, 437–442.

Rai, M. K., Akhtar, N., & Jaiswal, V. S. Somatic embryogenesis and plant regeneration in *Psidium guajava* L. cv. Banarasi local. *Scientia Horticulturae*, **2007**, *113*(2), 129–133.

Ramesh, R., & Phadke, G. S. Rhizosphere and endophytic bacteria for the suppression of eggplant wilt caused by *Ralstonia solanacearum*. *Crop Protection*, **2012**, *37*, 35–41.

Ramesh, R., Joshi, A. A., & Ghanekar, M. P. Pseudomonads: Major antagonistic endophytic bacteria to suppress bacterial wilt pathogen, *Ralstonia solanacearum* in the eggplant (*Solanum melongena* L.). *World Journal of Microbiology and Biotechnology*, **2009**, *25*(1), 47–55.

Ramos-González, M. I., Campos, M. J., & Ramos, J. L. Analysis of *Pseudomonas putida* KT2440 gene expression in maize rhizosphere: *In vitro* expression technology capture and identification of root-activated promoters. *Journal of Bacteriology*, **2005**, *187*, 4033–4041.

Rashedul, I. M., Madhaiyan, M., Boruah, H. P. D., Yim, W., Lee, G., Saravanan, V. S., Fu, Q., Hu, H., & Sa, T. Characterization of plant growth promoting traits of free-living diazotrophic bacteria and their inoculation effects on growth and nitrogen uptake of crop plants. *J. Microbiol. Biotechnol.*, **2009**, *19*, 1213–1222.

Redman, R. S., Sheehan, K. B., Stout, R. G., Rodriguez, R. J., & Henson, J. M. Thermotolerance generated by plant/fungal symbiosis. *Science*, **2002**, *298*(5598), 1581, doi: 10.1126/science.1072191.

Reed, B. M., & Tanprasert, P. Detection and control of bacterial contaminants of plant tissue cultures: A review of recent literature. *Plant Tissue Cult. Biotechnol.*, **1995**, *1*, 137–142.

Reed, B. M., Buckley, P. M., & De Wilde, T. N. Detection and eradication of endophytic bacteria from micropropagated mint plants. *In Vitro Cell Dev Biol. Plant*, **1995**, *31*, 53–57.

Reinhold-Hurek, B., & Hurek, T. Interactions of gramineous plants with *Azoarcus* spp. and other diazotrophs: Identification, localization, and perspectives to study their function. *Crit. Rev. Plant Sci.*, **1998**, *17*, 29–54.

Riggs, P. J., Chelius, M. K., Iniguez, A. L., Kaeppler, S. M., & Triplett, E. W. Enhanced maize productivity by inoculation with diazotrophic bacteria. *Aust. J. Plant Physiol.*, **2001**, *28*, 829–836.

Rocha, F. R., Papini-Terzi, F. S., Nishiyama, M. Y., Vêncio, R. Z., Vicentini, R., Duarte, R. D., et al. Signal transduction-related responses to phytohormones and environmental challenges in sugarcane. *BMC Genomics*, **2007**, *8*, 71, doi: 10.**1186**/1471–2164–8–71.

Rodriguez, R., & Redman, R. More than 400 million years of evolution and some plants still can't make it on their own: Plant stress tolerance via fungal symbiosis. *Journal of Experimental Botany*, **2008**, *59*(5), 1109–1114.

Roesch, L. F. W., Camargo, F. A., Bento, F. M., & Triplett, E. W. Biodiversity of diazotrophic bacteria within the soil, root and stem of field-grown maize. *Plant and Soil*, **2008**, *302*(1/2), 91–104.

Rosenblueth, M., & Martínez, R. E. *Rhizobium etli* maize populations and their competitiveness for root colonization. *Arch. Microbiol.*, **2004**, *181*, 337–344.

Rosenblueth, M., & Martínez-Romero, E. Bacterial endophytes and their interactions with hosts. *Mol Plant-Microbe Interact*, **2006**, *19*(8), 827–837.

Rothballer, M., Eckert, B., Schmid, M., Fekete, A., Schloter, M., Lehner, A., Pollmann, S., & Hartmann, A. Endophytic root colonization of gramineous plants by *Herbaspirillum frisingense*. *FEMS Microbiol. Ecol.*, **2008**, *66*(1), 85–95.

Russo, A., Carrozza, G. P., Vettori, L., Felici, C., Cinelli, F., & Toffanin, A. Plant beneficial microbes and their application in plant biotechnology. In: Agbo, E. C., (ed.), *Innovations in Biotechnology* (pp. 57–72). InTech, Croatia, **2012**. doi: 10.5772/31466.

Russo, A., Vettori, L., Felici, C., Fiaschi, G., Morini, S., & Toffanin, A. Enhanced micropropagation response and biocontrol effect of *Azospirillum brasiliense* Sp245 on *Prunus cerasifera* L. clone Mr. S 2/5 plants. *J. Biotechnol.*, **2008**, *134*, 312–319.

Ryan, R. P., Germaine, K., Franks, A., Ryan, D. J., & Dowling, D. N. Bacterial endophytes: Recent developments and applications. *FEMS Microbiology Letters*, **2008** *278*(1), 1–9.

Ryan, R. P., Monchy, S., Cardinale, M., Taghavi, S., Crossman, L., Avison, M. B., Berg, G., Van Der Lelie, D., & Dow, J. M. The versatility and adaptation of bacteria from the genus *Stenotrophomonas*. *Nature Reviews Microbiology*, **2009**, *7*(7), 514–525.

Ryu, C. M. Bacterial volatiles as airborne signals for plants and bacteria. In: Lugtenberg, B., (ed.), *Principles of Plant-Microbe Interactions* (pp. 53–64). Springer International Publishing, Switzerland, **2015**.

Ryu, C. M., Farag, M. A., Hu, C. H., Reddy, M. S., Wei, H. X., Paré, P. W., & Kloeppe, J. W. Bacterial volatiles promote growth in *Arabidopsis*. *Proc. Natl. Acad. Sci. USA*, **2003**, *100*(8), 4927–4932.

Saleh, S. A., Mekhemar, G. A. A., El-Soud, A. A. A., Ragab, A. A., & Mikhaeel, F. T. Survival of *Azorhizobium* and *Azospirillum* in different carrier materials: Inoculation of wheat and *Sesbania rostrata*. *Bulletin of Faculty of Agriculture, Cairo University*, **2001**, *52*, 319–338.

Samish, Z., & Dimant, D. Bacterial population in fresh, healthy cucumbers. *Food Manuf.*, **1959**, *34*, 17–20.

Sangeeth, K. P., Bhai, R. S., & Srinivasan, V. *Paenibacillus glucanolyticus*, a promising potassium solubilizing bacterium isolated from black pepper (*Piper nigrum* L.) rhizosphere. *J. Spices Arom. Crops*, **2012**, *21*, 118–124.

Scherling, C., Ulrich, K., Ewald, D., & Weckewerth, W. A metabolic signature of the beneficial interaction of the endophyte *Paenibacillus* sp. isolate and *in vitro*-grown poplar plants revealed by metabolomics. *Mol. Plant-Microbe Interact*, **2009**, *22*, 1032–1037.

Schroth, M. N. *Pseudomonas* siderophores: A mechanism explaining disease-suppressive soils. *Curr. Microbiol.*, **1980**, *4*, 317–320.

Schulz, S., & Dickschat, J. S. Bacterial volatiles: The smell of small organisms. *Natural Product Reports*, **2007**, *24*(4), 814–842.

Scott, R. A., Weil, J., Le, P. T., Williams, P., Fray, R. G., Von Bodman, S. B., & Savka, M. A. Long- and short-chain plant-produced bacterial *N*-acylhomoserine lactones become components of phyllosphere, rhizosphere, and soil. *Mol. Plant–Microbe Interact*, **2006**, *19*(3), 227–239.

Seker, M. G., Sah, I., Kirdok, E., Ekinci, H., Ciftci, Y. O., & Akkaya, O. A hidden plant growth promoting bacterium isolated from *in vitro* cultures of fraser photinia (*Photinia × fraseri*). *Int. J. Agric. Biol.*, **2017**, *19*(6), 1511–1519.

Sessitsch, A., Coenye, T., Sturz, A. V., Vandamme, P., Barka, E. A., Salles, J. F., Van Elsas, J. D., Faure, D., Reiter, B., Glick, B. R., Wang-Pruski, G., & Nowak, J. *Burkholderia phytofirmans* sp. nov., a novel plant-associated bacterium with plant-beneficial properties. *International Journal of Systematic and Evolutionary Microbiology*, **2005**, *55*(3), 1187–1192.

Sessitsch, A., Hardoim, P., Döring, J., Weilharter, A., Krause, A., Woyke, T., et al. Functional characteristics of an endophyte community colonizing rice roots as revealed by metagenomic analysis. *Mol. Plant-Microbe Interact*, **2012**, *25*(1), 28–36.

Sgroy, V., Cassan, F., Masciarelli, O., Del Papa, M. F., Lagares, A., & Luna, V. Isolation and characterization of endophytic plant growth-promoting (PGPB) or stress homeostasis-regulating (PSHB) bacteria associated to the halophyte *Prosopis strombulifera*. *Applied Microbiology and Biotechnology*, **2009**, *85*, 371–381.

Shahzad, R., Waqas, M., Khan, A. L., Asaf, S., Khan, M. A., Kang, S. M., Yun, B. W., & Lee, I. J. Seed-borne endophytic *Bacillus amyloliquefaciens* RWL-1 produces gibberellins and regulates endogenous phytohormones of *Oryza sativa*. *Plant Physiology and Biochemistry*, **2016**, *106*, 236–243.

Shetty, K., Curtis, O. F., Levin, R. E., & Ang, W. W. Prevention of vitrification associated with *in vitro* shoot culture of Oregano (*Origanum vulgare*) by *Pseudomonas* spp. *Journal of Plant Physiology*, **1995**, *147*(3/4), 447–451.

Siciliano, S. D., Fortin, N., Mihoc, A., Wisse, G., Labelle, S., Beaumier, D., Ouellette, D., Roy, R., Whyte, L. G., Banks, M. K., Schwab, P., Lee, K., & Greer, G. W. Selection of specific endophytic bacterial genotypes by plants in response to soil contamination. *Applied and Environmental Microbiology*, **2001**, *67*(6), 2469–2475.

Siddikee, M. D., Ashaduzzaman, P., Chauhan, S., & Tongmin, S. Regulation of ethylene biosynthesis under salt stress in red pepper (*Capsicum annuum* L.) by 1-aminocyclopropane-1-carboxylic acid (ACC) deaminase-producing halotolerant bacteria. *Journal of Plant Growth Regulation*, **2012**, *31*(2), 265–272.

Staniek, A., Woerdenbag, H. J., & Kayser, O. Endophytes: Exploiting biodiversity for the improvement of natural product-based drug discovery. *Journal of Plant Interactions*, **2008**, *3*(2), 75–93.

Steenhoudt, O., & Vanderleyden, J. *Azospirillum*, a free-living nitrogen-fixing bacterium closely associated with grasses: Genetic, biochemical and ecological aspects. *FEMS Microbiology Reviews*, **2000**, *24*(4), 487–506.

Stenuit, B. A., & Agathos, S. N. Microbial 2,4,6-trinitrotoluene degradation: Could we learn from (bio)chemistry for bioremediation and *vice versa*? *Appl. Microbiol. Biotechnol.*, **2010**, *88*, 1043–1064.

Stone, J. K., Bacon, C. W., & White, J. F. An overview of endophytic microbes: endophytism defined. In: Bacon, C. W., & White, J. F., (eds.), *Microbial Endophytes* (pp. 3–30). Marcel Dekker, New York, **2000**.

Straub, D., Yang, H., Liu, Y., Tsap, T., & Ludewig, U. Root ethylene signaling is involved in *Miscanthus sinensis* growth promotion by the bacterial endophyte *Herbaspirillum frisingense* GSF30T. *Journal of Experimental Botany*, **2013**, *64*(14), 4603–4615.

Strobel, G., Daisy, B., Castillo, U., & Harper, J. Natural products from endophytic microorganisms. *Journal of Natural Products*, **2004**, *67*(2), 257–268.

Sunayana, M. R., Sasikala, C., & Ramana, C. V. Rhodestrin: A novel indole terpenoid phytohormone from *Rhodobacter sphaeroides*. *Biotechnol. Lett.*, **2005**, *27*, 1897–1900.

Suzuki, T., Shimizu, M., Meguro, A., Hasegawa, S., Nishimura, T., & Kunoh, H. Visualization of infection of an endophytic actinomycete *Streptomyces galbus* in leaves of tissue-cultured rhododendron. *Actinomycetologica*, **2005**, *19*(1), 7–12.

Sziderics, A. H., Rasche, F., Trognitz, F., Sessitsch, A., & Wilhelm, E. Bacterial endophytes contribute to abiotic stress adaptation in pepper plants (*Capsicum annuum* L.). *Canadian Journal of Microbiology*, **2007**, *53*(11), 1195–1202.

Taghavi, S., Barac, T., Greenberg, B., Borremans, B., Vangronsveld, J., & Van Der Lelie, D. Horizontal gene transfer to endogenous endophytic bacteria from poplar improves phytoremediation of toluene. *Applied and Environmental Microbiology*, **2005**, *71*(12), 8500–8505.

Taghavi, S., Garafola, C., Monchy, S., Newman, L., Hoffman, A., & Weyens, N. Genome survey and characterization of endophytic bacteria exhibiting a beneficial effect on growth and development of poplar trees. *Appl. Environ. Microbiol.*, **2009**, *75*, 748–757.

Taghavi, S., Van Der Lelie, D., Hoffman, A., Zhang, Y. B., Walla, M. D., Vangronsveld, J., Newman, L., & Monchy, S. Genome sequence of the plant growth promoting endophytic bacterium *Enterobacter* sp. 638. *PLoS Genet*, **2010**, *6*(5), e1000943, doi: 10.1371/journal.pgen.1000943.

Taghavi, S., Weyens, N., Vangronsveld, J., Lelie, D., Pirttilä, A. M., & Frank, A. C. Improved phytoremediation of organic contaminants through engineering of bacterial endophytes of trees. In: Pirttila, A. M., & Frank, A. C., (eds.), *Endophytes of Forest Trees* (pp. 205–216). Springer, Netherlands, **2011**.

Tammaro, F., & Xepapadakis, G. Plants used in phytotherapy, cosmetics and dyeing in the Pramanda district (Epirus, North-West Greece). *J. Ethnopharmacol.*, **1986**, *16*, 167–174.

Theocharis, A., Bordiec, S., Fernandez, O., Paquis, S., Dhondt-Cordelier, S., Baillieul, F., Clément, C., & Barka, E. A. *Burkholderia phytofirmans* PsJN primes *Vitis vinifera* L. and confers a better tolerance to low nonfreezing temperatures. *Mol. Plant-Microbe Interact*, **2012**, *25*(2), 241–249.

Thomas, P. Intense association of non-culturable endophytic bacteria with antibiotic-cleansed *in vitro* watermelon and their activation in degenerating cultures. *Plant Cell Rep.*, **2011**, 30, 2313–2325.

Thomas, P. Isolation of *Bacillus pumilus* from *in vitro* grapes as a long-term alcohol-surviving and rhizogenesis inducing covert endophyte. *J. Appl. Microbiol.*, **2004**, *97*, 114–123.

Thomas, P., & Sekhar, A. C. Live cell imaging reveals extensive intracellular cytoplasmic colonization of banana by normally non-cultivable endophytic bacteria. *AoB Plants*, **2014**, *6*, https://doi.org/10.1093/aobpla/plu002 (Accessed on 6 October 2019).

Thomas, P., Kumari, S., Swarna, G. K., & Gowda, T. K. S. Papaya shoot tip associated endophytic bacteria isolated from *in vitro* cultures and host-endophyte interaction *in vitro* and *in vivo*. *Canadian Journal of Microbiology*, **2007**, *53*(3), 380–390.

Timmusk, S., Nicander, B., Granhall, U., & Tillberg, E. Cytokinin production by *Paenibacillus polymyxa*. *Soil Biol. Biochem.*, **1999**, *31*, 1847–1852.

Tjamos, S. E., Flemetakis, E., Paplomatas, E. J., & Katinakis, P. Induction of resistance to *Verticillium dahliae* in *Arabidopsis thaliana* by the biocontrol agent K-165 and pathogenesis-related proteins gene expression. *Mol. Plant Microbe Interact*, **2005**, *18*, 555–561.

Tomasino, S. F., Leister, R. T., Dimock, M. B., Beach, R. M., & Kelly, J. L. Field performance of *Clavibacter xyli* subsp. *cynodontis* expressing the insecticidal protein gene *cryIA* of *Bacillus thuringiensis* against European corn borer in field corn. *Biol. Con.*, **1995**, *5*, 442–448.

Ton, J., Van Pelt, J. A., Van Loon, L. C., & Pieterse, C. M. Differential effectiveness of salic-ylate-dependent and jasmonate/ethylene-dependent induced resistance in Arabidopsis. *Mol. Plant Microbe Interact*, **2002**, *15*(1), 27–34.

Triplett, E. W. Diazotrophic endophytes: Progress and prospects for nitrogen fixation in monocots. *Plant Soil*, **1996**, *186*, 29–38. https://doi.org/10.1007/BF00035052 (Accessed on 6 October 2019).

Trivedi, P., & Tongmin, S. *Pseudomonas corrugata* (NRRL B-**3040**9) mutants increased phosphate solubilization, organic acid production, and plant growth at lower temperatures. *Current Microbiology*, **2008**, *56*(2), 140–144.

Tsavkelova, E. A., Klimova, S. Y., Cherdyntseva, T. A., & Netrusov, A. I. Hormones and hormone-like substances of microorganisms: a review. *Applied Biochemistry and Microbiology*, **2006**, *42*(3), 229–235.

Turner, T. R., James, E. K., & Poole, P. S. The plant microbiome. *Genome Biology*, **2013**, *14*(6), 209, https://doi.org/10.1186/gb-2013–14–6–209 (Accessed on 6 October 2019).

Ueno, K., Cheplick, S., & Kalidas, S. Reduced hyperhydricity and enhanced growth of tissue culture-generated raspberry (*Rubus* sp.) clonal lines by *Pseudomonas* sp. isolated from oregano. *Process Biochemistry*, **1998**, *33*(4), 441–445.

Ulrich, K., Stauber, T., & Ewald, D. *Paenibacillus* a predominant endophytic bacterium colonizing tissue cultures of woody plants. *Plant Cell Tiss. Org.*, **2008**, *93*, 347–351.

Uroz, S., Oger, P. M., Chapelle, E., Adeline, M. T., Faure, D., & Dessaux, Y. A *Rhodococcus* qsdA-encoded enzyme defines a novel class of large-spectrum quorum-quenching lactonases. *Applied and Environmental Microbiology*, **2008**, *74*(5), 1357–1366.

Valdivia, R. H., & Falkow, S. Bacterial genetics by flow cytometry: Rapid isolation of *Salmonella typhimurium* acid-inducible promoters by differential fluorescence induction. *Mol. Microbiol.*, **1996**, *22*, 367–378.

Van Aken, B., Yoon, J. M., & Schnoor, J. L. Biodegradation of nitrosubstituted explosives TNT, RDX, and HMX by a phytosymbiotic *Methylobacterium* sp. associated with poplar tissues (*Populus deltoides* x *nigra* DN34). *Applied and Environmental Microbiology*, **2004**, *70*(1), 508–517.

Van Buren, A. M., Andre, C., & Ishimaru, C. A. Biological control of the bacterial ring rot pathogen by endophytic bacteria isolated from potato. *Phytopathology*, **1993**, *83*, 1406.

Van Der Ent, S., Van Wees, S. C., & Pieterse, C. M. Jasmonate signaling in plant interactions with resistance-inducing beneficial microbes. *Phytochemistry*, **2009**, *70*(13), 1581–1588.

Van Loon, L. C. Plant responses to plant growth-promoting rhizobacteria. *European Journal of Plant Pathology*, **2007**, *119*(3), 243–254.

Van Loon, L. C., Bakker, P. A., & Pieterse, C. M. J. Systemic resistance induced by rhizosphere bacteria. *Ann. Rev. Phytopathol.*, **1998**, *36*, 453–483.

Van Oosten, V. R., Bodenhausen, N., Reymond, P., Van Pelt, J. A., Van Loon, L. C., Dicke, M., & Pieterse, C. M. Differential effectiveness of microbially induced resistance against herbivorous insects in Arabidopsis. *Mol. Plant-Microbe Interact*, **2008**, *21*(7), 919–930.

Van Overbeek, L., & Van Elsas, J. D. Effects of plant genotype and growth stage on the structure of bacterial communities associated with potato (*Solanum tuberosum* L.). *FEMS Microbiol. Ecol.*, **2008**, *64*(2), 283–296.

Van Peer, R., Punte, H. L. M., De Weger, L. A., & Schippers, B. Characterization of root surface and endorhizosphere pseudomonads in relation to their colonization of roots. *Appl. Environ. Microbiol.*, **1990**, *56*, 2462–2470.

Vant Slot, K. A. E., & Knogge W. A dual role for microbial pathogen-derived effector proteins in plant disease and resistance. *Crit. Rev. Plant Dis.*, **2002**, *39*, 471–482.

Verma, S. C., Ladha, J. K., & Tripathi, A. K. Evaluation of plant growth promoting and colonization ability of endophytic diazotrophs from deep water rice. *Journal of Biotechnology*, **2001**, *91*(2), 127–141.

Verma, S. C., Singh, A., Chowdhury, S. P., & Tripathi, A. K. Endophytic colonization ability of two deep-water rice endophytes, *Pantoea* sp. and *Ochrobactrum* sp. using green fluorescent protein reporter. *Biotechnol. Lett.*, **2004**, *26*, 425–429.

Vermeiren, H., Vanderleyden, J., & Hai, W. L. Colonization and *nifH* expression on rice roots by *Alcaligenes faecalis* A15. In: Malik, K. A., Mirza, M. S., & Ladha, J. K., (eds.), *Nitrogen Fixation with Non-Legumes* (pp. 167–177). Kluwer Academic Publishers, London, **1998**.

Vettori, L., Russo, A., Felici, C., Fiaschi, G., Morini, S., & Toffanin, A. Improving micropropagation: Effect of *Azospirillum brasilense* Sp245 on acclimatization of rootstocks of fruit tree. *J. Plant Interact*, **2010**, *5*, 249–259.

Vinagre, F., Vargas, C., Schwarcz, K., Cavalcante, J., Nogueira, E. M., Baldani, J. I., Ferreira, P. C., & Hemerly, A. S. SHR5: A novel plant receptor kinase involved in plant-N2-fixing endophytic bacteria association. *Journal of Experimental Botany*, **2006**, *57*(3), 559–569.

Von Bodman, S. B., Bauer, W. D., & Coplin, D. L. Quorum sensing in plant-pathogenic bacteria. *Annu. Rev. Phytopathol.*, **2003**, *41*, 455–482.

Weilharter, A., Mitter, B., Shin, M. V., Chain, P. S. G., Nowak, J., & Sessitsch, A. Complete genome sequence of the plant growth-promoting endophyte *Burkholderia phytoirmans* strain PsJN. *Bacteriol.*, **2011**, *193*, 3383–3384.

Weyens, N., Beckers, B., Schellingen, K., Ceulemans, R., Croes, S., Janssen, J., Haenen, S., Witters, N., & Vangronsveld, J. Potential of willow and its genetically engineered associated bacteria to remediate mixed Cd and toluene contamination. *Journal of Soils and Sediments*, **2013**, *13*(1), 176–188.

Weyens, N., Croes, S., Dupae, J., Newman, L., Van Der Lelie, D., Carleer, R., & Vangronsveld, J. Endophytic bacteria improve phytoremediation of Ni and TCE co-contamination. *Environ. Pollut.*, **2010b**, *158*, 2422–2427.

Weyens, N., Truyens, S., Dupae, J., Newman, L., Taghavi, S., Van Der Lelie, D., Carleer, R., & Vangronsveld, J. Potential of the TCE-degrading endophyte *Pseudomonas putida* W619-TCE to improve plant growth and reduce TCE phytotoxicity and evapotranspiration in poplar cuttings. *Environ. Pollut.*, **2010a**, *158*, 2915–2919.

Weyens, N., Van Der Lelie, D., Taghavi, S., & Vangronsveld, J. Phytoremediation: Plant-endophyte partnerships take the challenge. *Curr. Opin. Biotechnol.*, **2009b**, *20*, 248–254.

Weyens, N., Van Der Lelie, D., Taghavi, S., Newman, L., & Vangronsveld, J. Exploiting plant-microbe partnerships to improve biomass production and remediation. *Trends Biotechnol.*, **2009a**, *27*, 591–598.

White, J. F., Torres, M. S., Somu, M. P., Johnson, H., Irizarry, I., Chen, Q., & Bergen, M. Hydrogen peroxide staining to visualize intracellular bacterial infections of seedling root cells. *Microscopy Research and Technique*, **2014a**, *77*(8), 566–573.

White, J. F., Torres, M. S., Sullivan, R. F., Jabbour, R. E., Chen, Q., Tadych, M., Irizarry, I., Bergen, M. S., Havkin-Frenkel, D., & Belanger, F. C. Occurrence of *Bacillus amyloliquefaciens* as a systemic endophyte of vanilla orchids. *Microscopy Research and Technique*, **2014b**, *77*(11), 874–885.

Wong-Villarreal, A., & Caballero-Mellado, J. Rapid identification of nitrogen-fixing and legume-nodulating *Burkholderia* species based on PCR 16S rRNA species-specific oligonucleotides. *Systematic and Applied Microbiology*, **2010**, *33*(1), 35–43.

Wu, X., Vennison, S. J., Liu, H., Ben Dov, E., Zaritsky, A., & Boussiba, S. Mosquito larvicidal activity of transgenic *Anabaena* strain PCC, **7120** expressing combinations of genes from *Bacillus thuringiensis* subsp. *israelensis*. *Appl. Environ. Microbiol.*, **1997**, *63*, 1533–1537.

Yazdani, M., Bahmanyar, M. A., Pirdashti, H., & Esmaili, M. A. Effect of phosphate solubilization microorganisms PSM and plant growth promoting rhizobacteria PGPR on yield and yield components of corn *Zea mays* L. *World Academy of Science, Engineering and Technology*, **2009**, *49*, 90–92.

Yi, H. S., Yang, J. W., & Ryu, C. M. ISR meets SAR outside: Additive action of the endophyte *Bacillus pumilus* INR7 and the chemical inducer, benzothiadiazole, on induced resistance against bacterial spot in field grown pepper. *Front. Plant Sci.*, **2013**, *4*, 122, doi: 10.3389/fpls.2013.00122.

Youngbae, S., Kim, S., & Park, C. W. A phylogenetic study of *Polygonum* sect. *tovara* (polygonaceae) based on ITS sequences of nuclear ribosomal DNA. *Plant Biology*, **1997**, *40*, 47–52.

Yuan, Z. C., Haudecoeur, E., Faure, D., Kerr, K. F., & Nester, E. W. Comparative transcriptome analysis of *Agrobacterium tumefaciens* in response to plant signal salicylic acid, indole-3-acetic acid and γ-amino butyric acid reveals signaling cross-talk and *Agrobacterium*–plant co-evolution. *Cellular Microbiology*, **2008**, *10*(11), 2339–2354.

Zakharchenko, N. S., Kochetkov, V. V., Buryanov, Y. A. I., & Boronin, A. M. Effect of rhizosphere bacteria *Pseudomonas aureofaciens* on the resistance of micropropagated plants to phytopathogens. *Appl. Biochem. Microbiol.*, **2011**, *47*, 661–666.

Zhang, C., Tanabe, K., Tani, H., Nakajima, H., Mori, M., & Sakuno, E. Biologically active gibberellins and abscisic acid in fruit of two late-maturing Japanese pear cultivars with contrasting fruit size. *Journal of the American Society for Horticultural Science*, **2007**, *132*(4), 452–458.

Zhang, H. W., Song, Y. C., & Tan, R. X. Biology and chemistry of endophytes. *Nat. Prod. Rep.*, **2006**, *23*, 753–771.

Zhang, L. H. Quorum quenching and proactive host defense. *Trends in Plant Science*, **2003**, *8*(5), 238–244.

Zhang, T., Shi, Z. Q., Hu, L. B., Cheng, L. G., & Wang, F. Antifungal compounds from *Bacillus subtilis* B-FS06 inhibiting the growth of *Aspergillus flavus*. *World Journal of Microbiology and Biotechnology*, **2008**, *24*(6), 783, doi: 10.1007/s11274–007–9533–1.

Ziemienowicz, A. Agrobacterium-mediated plant transformation: Factors, applications and recent advances. *Biocatal. Agric. Biotechnol.*, **2014**, *3*, 95–102.

CHAPTER 3

PLANT GROWTH-PROMOTING ENDOPHYTIC BACTERIA AND THEIR POTENTIAL BENEFITS IN ASIAN COUNTRIES

NITTAYA PITIWITTAYAKUL[1] and SOMBOON TANASUPAWAT[2]

[1]*Department of Agricultural Technology and Environment, Faculty of Sciences and Liberal Arts, Rajamangala University of Technology Isan, Nakhon Ratchasima Campus, Nakhon Ratchasima 30000, Thailand*

[2]*Department of Biochemistry and Microbiology, Faculty of Pharmaceutical Sciences, Chulalongkorn University, Bangkok 10330, Thailand*

3.1 INTRODUCTION

Endophytic bacteria are those bacteria that inhabit inside plant tissues and generally do not cause apparent disease or deleterious effect to their host (Wilson, 1995; Ryan et al., 2008). They are able to colonize the spaces that exist between adjacent cells, and they have been found in all components of plant as well as seeds (Posada and Vega, 2005). Generally, endophytic plant growth-promoting bacteria (PGPB) can stimulate plant growth by direct and indirect mechanisms. Direct beneficial mechanisms are either assisting in resource acquisition (nitrogen fixation and enhancing mineral uptake) or production of phytohormones. Indirect mechanisms involve the ability of PGPB to suppress the growth of various plant pathogens and produce biocontrol agents (Glick, 2012; Ahemad and Kibret, 2014). Nowadays, sustainable agriculture is an alternative way of maintaining high productivity, together with maintaining ecosystems and biodiversity. PGPB are used as the potential eco-friendly and sustainable aspect for plant growth promotion. PGPB may replace or reduce the use of chemical fertilizers and capable of improving the growth and productivity of many plant species, increase agronomic efficiency, and ecological significance. Asian countries have rich plant biodiversity, and there are a very large number of plant species. Endophytic PGPB are ubiquitous as they can be isolated in all plant species, residing in a latent state or actively colonizing plant tissues. Moreover, each individual plant can be inhabited with one or more endophytes. The relative between only a few of these plants and their endophytic biology has ever been completely studied. As a result, the opportunity to discover novel and beneficial endophytic microorganisms among the diversity of plants in different ecosystems is considerable. In this chapter, we described the isolation, identification, diversity, plant growth-promoting activities, and enzyme, including other products of endophytic PGPB in Asian countries.

3.2   ISOLATION AND IDENTIFICATION

The isolation of endophyte bacterial species directly involves the surface sterilization by treating the plant material with a strong oxidant or general disinfectant for a period, followed by a sterile rinse. The most generally used as a surface sterilant is household bleach (NaOCl) diluted in water to a concentration of 2–10% or similar agents (Miche and Balandreau, 2001; Zhang et al., 2006). Furthermore, ethanol (EtOH) (70–95%) can be used as a wetting agent combining with surface sterilization for improving the efficacy of surface sterilization. Lodewyckx et al. (2002) reviewed the procedures for isolation and characterization of endophyte from diverse plant species and concluded that no protocol of surface disinfection results in the complete killing of surface bacteria on 100% of samples without penetrating interior tissues and thereby killing internal colonists. Therefore, comparisons between different studies should be carefully evaluated, taking into account the different surface sterilization methods and conditions used. Moreover, total populations of bacteria accompanying within host plants may depend on the growth media used for isolation. There have been relatively few studies that have analyzed the influence of the culture medium type on the diversity of bacterial isolates. According to Okubo et al. (2009), nodulation-dependent communities of culturable bacterial endophytes from stems of field-grown soybeans were analyzed. In this study, three isolation media, R2A, NA, and PDA, were used to examine the impact of medium type on the diversity of bacterial isolates. The results revealed that the R2A medium exhibited the highest diversity indexes, whereas the PDA medium showed the lowest. Some methods and media for endophytic bacteria isolation that have been reported are shown in Table 3.1. However, the most information on endophytic bacterial diversity that was obtained by using culture-dependent approaches has been so far due to the unknown condition for growth requirements of many bacteria and the presence of cells that are in a viable but non-cultivable state (Shen et al., 2010).

Molecular approaches based on the amplification of the 16S rDNA have been performed for bacterial community analysis in host plants to overcome the limitations of classic isolation procedures that are a culture-based method for isolation of bacteria (culture-dependent approaches). Furthermore, molecular approaches based on 16S rDNA such as amplified rDNA restriction analysis (ARDRA), denaturing gradient gel electrophoresis (DGGE), and restriction fragment length polymorphism (RFLP) can be applied to study population composition including the population dynamics of endophytic bacteria. Additionally, the specific amplification of housekeeping genes such as *atpD* and *gln*II, or genes encoding important traits of endophytic bacteria, for example the *nif* genes required for nitrogen fixation; the *nod* gene required for synthesis and secretion Nod factors relating with the root nodule formation of leguminous plants, can be applied to determine the ability of the endophytic population to participate in important processes within host plant. The diversities of endophytic bacteria in *Caragana microphylla* grown in the desert, rice (*Oryza sativa* L.) roots, peanut roots, wild alpine-subnival plant species and narrow leaf cattail (*Typha angustifolia* L.) have been estimated by using 16S rRNA gene sequence analysis together with ARDRA (Sun et al., 2008; Li et al., 2011; Wang et al., 2013; Dai et al., 2014). RFLP based on 16S-23S internal transcribed spacer (ITS) region was used for characterized the bacterial strains that isolated from nodules of eighteen *Vicia* species in China (Lei et al., 2008). Chen et al. (2015) characterized the endophytic bacteria isolated from *Astragalus* species as traditional Chinese medicine source by the PCR-RFLP of 16S rRNA gene and symbiotic genes as well as the phylogenetic analysis while Kang et al. (2016) characterized the communities of bacteria associated with surface-sterilized pepper plants by using PCR-DGGE (culture-independent) together with culture-dependent (plating). Moreover, nested polymerase chain reaction denaturing gradient gel electrophoresis (PCR-DGGE) with novel specific 16S

TABLE 3.1   Isolation and Identification Methods for Endophytic Bacteria Associated With Host Plants

Endophytic Species	Host Plant Species	Isolation Method	Identification Method	References
[1]*Rhizobium lemnae* [1]*Rhizobium paknamense*	Duckweed (*Lemna aequinoctialis*)	Surface-sterilized plants with 10% sodium hypochlorite (NaOCl) and a few drops of Tween 20, were rinsed with sterilized distilled water (DW) five times, and then ground in sterilized DW. The suspension was placed on 1/10 strength tryptic soy agar (TSA) and incubated at 30°C for 7 days.	• 16S rRNA gene sequence • Housekeeping *recA* and *atpD* gene • *nodC* and *nifH* gene	Kittiwongwattana et al., 2013; Kittiwongwattana et al., 2014
[1]*Rhizobium smilacinae*	Traditional Chinese medicinal plant (*Smilacina japonica*)	The surface-sterilized leaf tissue was cut into small fragments and macerated using a sterile pestle and mortar in sterile DW. The diluted macerated samples in sterile DW were spread onto R2A agar supplemented with 50 μg ml^{-1} of imazalil and incubated at 25°C for 2 weeks.	• 16S rRNA gene sequence • Housekeeping *dnaK*, *glnI*, and *recA* gene	Zhang et al., 2014a
[1]*Bradyrhizobium* sp.	Rice (*Oryza sativa*)	Rice roots and paddy soils were collected and immediately transferred to the laboratory in polyethylene boxes at 4°C. The roots and stems were chemically sterilized in 3% NaOCl (5 min) and soaking in 70% EtOH (5 min) and were thoroughly rinsed with sterilized DW (at least 5 times), cut into 4- to 5-cm-long sections and placed on plate count agar (PCA). The 200 μl of water from the final rinse was spread PCA, as a control to check superficial contamination for each plant.	• 16S rRNA gene sequence • Housekeeping *recA* and *atpD* gene • *glnB* gene • BOX-A1R *nodA*, *nodB*, *nodC*, *pufM*, *bchL*	Piromyou et al., 2015a
[1]*Novosphingobium sediminicola* and [1]*Ochrobactrum intermedium*	Sugarcane	The samples were surface sterilized by sequential washing in 50% EtOH for 1 min, 2% NaOCl for 3 min, and 50% ethanol for 30 sec and then rinsed twice with sterile DW. Each sample was ground in	• 16S rRNA gene sequence	Döbereiner et al., 1972; Muangthong et al., 2015

TABLE 3.1 *(Continued)*

Endophytic Species	Host Plant Species	Isolation Method	Identification Method	References
		a sterile mortar, suspended in an N-free broth (0.1 g K_2HPO_4, 0.4 g KH_2PO_4, 0.2 g $MgSO_4$, 0.1 g NaCl, 0.02 g $CaCl_2$, 0.01 g $FeCl_3$, 0.002 g $NaMoO_4$, and 10 g glucose in 1000 ml distilled water) at 30°C for 24 h. Ten-fold serial dilutions of the suspensions (10^{-1} to 10^{-4}) were spread on Nitrogen Free Agar (NFA) in triplicate and incubated for 72 h.		
[1]*Rhizobium*, [2]*Enterobacter, Stenotrophomonas*	*Pteris vittata*	The sterilized roots (0.2 g) were ground in an autoclaved mortar pestle containing 10 ml sterile DW. Serial dilutions of samples were spread on sucrose-minimal salts low-phosphate (SLP) medium (sucrose 1%, $(NH_4)_2SO_4$ 0.1%, K_2HPO_4 0.05%, $MgSO_4$ 0.05%, NaCl 0.01%, yeast extract 0.05%, $CaCO_3$ 0.05%, pH 7.2) along with 10 mg L^{-1} as sodium arsenate and incubated at 28°C for 72 h.	• 16S rRNA gene sequence	Tiwari et al., 2016
[2]*Enterobacter cloacae* Other: *Paenibacillus xylanexedens, Bacillus*	Date palm (*Phoenix dactylifera* L.)	Seedling roots growing in each soil sample were separated from the plants, pooled, surface steril-ized, and the endophytic bacteria were eluted from the tissues using Ringer's solution as described. The eluted Ringer's solutions were diluted, and 10 µl aliquots were spread onto King's B agar (Kg) (20 g proteose peptone 3, 10 ml glycerol, 1.5 g K_2HPO_4, 1.5 g $MgSO_4 \cdot 7H_2O$ and 15 g agar in l l water), Luria agar (LA) and tryptic soy agar (TSA) rich media.	• 16S rRNA gene sequence analysis	Rashid et al., 2012; Yaish et al., 2015

TABLE 3.1 *(Continued)*

Endophytic Species	Host Plant Species	Isolation Method	Identification Method	References
[2]*Klebsiella* four species Others: *Microbacterium*, two species of *Paenibacillus*, three *Bacillus* species	Korean rice cultivars	Plant tissue samples were surface sterilized with 70% EtOH (1 min) and shaken in 1.2% (w/v) NaClO solution (15 min). Samples that washed three times with sterile DW with shaking (15 min each) were ground with sterilized mortar and pestle and inoculated on nitrogen-free semi-solid agar media. After incubation at 30°C for 2 days, the inoculants were transferred to fresh nitrogen-free media and then incubated at 30°C for 2 days.	• *nifH* genes • 16S rRNA gene sequence analysis	Ji et al., 2014
[1]*Agrobacterium tumefaciens* [1]*Sphingopyxis chilensis* [2]*Enhydrobacter aerosaccus*	Pepper plants (*Capsicum annuum* L. cv. Nokkwang)	The root, stem, and leaf parts were aseptically cut into pieces. Surface disinfection was done by stepwise washing in 70% EtOH (1 min), 1% NaOCl (1 min), and 70% EtOH (30 sec), followed by three rinses with sterile DW for 1 min each. The other disinfected pieces (0.5 g) were macerated by grinding in sterile mortars. Ground plant material was transferred into a sterile culture tube with 3 mL 120 mM sodium phosphate buffer (19.9 g $Na_2HPO_4 \cdot H_2O$, 1.27 g $NaH_2PO_4 \cdot 2H_2O$, H_2O 1 l, pH 8.0) and agitated at 120 rpm for 2 h. The suspension was serially diluted 10-fold to 10^{-5} and spread onto 0.3% TSA with 100 µg mL^{-1} cycloheximide and incubated at 28°C for 7 days.	• 16S rRNA gene sequence analysis • DGGE of PCR-amplified 16S rDNAs	Kang et al., 2016
[2]*Pseudomonas* sp., [2]*Enterobacter* sp. Others: *Bacillus* sp.	*Aloe vera*	The leaves, stems, and roots were detached with a sterile knife, washed with sterile DW, and left to drain (10–15 min). Pieces of tissue (2–3 cm) were surface sterilized (Azevedo et al., 2000) with	• 16S rRNA gene sequence analysis	Azevedo et al., 2000; Akinsanya et al., 2015

TABLE 3.1 *(Continued)*

Endophytic Species	Host Plant Species	Isolation Method	Identification Method	References
		modifications, immersed in 90% EtOH (5min), 3% NaOCl solution (2 min), and into 75% EtOH (3 min). The samples were rinsed three times in sterile DW and drained in a laminar flow hood. They were cut longitudinally with a sterile scalpel and laid, with the exposed inner surface facing downwards, on NA plates and incubated at 30°C for 36–48 h to obtain the cultures.		
[1]*Novosphingobium* sp. [1]*Asticccacaulis* sp. [1]*Rhodobacter* sp. [1]*Rhizobium* sp. [3]*Burkholderia vietnamiensis* [3]*B. tropica* [3]*B. kururiensis* [3]*B. ambifaria* [3]*Herbaspirillum* sp.	Nipa Palm (*Nypa fruticans*)	Root samples were first washed with 70% EtOH (30 sec) and then surface-sterilized with diluted commercial bleach containing 0.5% NaOCl (5 min), followed by rinsing three times with sterilized water for 10 min each time. In view of the relatively high concentration of sucrose in the sap of the Nipa palm, surface-sterilized root cut vertically in 10 mm long was inoculated into a gellan gum soft gel medium that contained 0.2% w/v sucrose as sole carbon source.	• 16S rRNA gene sequence analysis	Tang et al., 2010

[1]*Alphaproteobacteria;* [2]*Gammaproteobacteria;* [3]*Betaproteobacteria.*

rDNA-targeted primers were successfully used to characterize the endophytic diversity in *Dendrobium officinale* from three different sources in China. Consequently, the good alternative for investigation of communities and roles of endophytes was the nested PCR-DGGE method based on the novel primers (Yu et al., 2013).

3.3   DIVERSITY OF PGPB

The microbial endophytes have commonly been found in every plants species. Partida-Martínez and Heli (2011) concluded that the existence of microorganisms in and on plants must be considered as the rule, rather than the exception. Endophytic organisms have been isolated from different parts of the plant and from both monocotyledonous and dicotyledonous plants, ranging from woody tree species, such as oak and pear, to herbaceous crop plants such as sugar beet and maize (Ryan et al., 2008). Although under various stresses such as heavy metals and salinity, some endophytic bacteria have been observed (Wani et al., 2007a, b; Sheng et al., 2008; Sun et al., 2010; Zhang et al., 2011; Babu et al., 2015; Li et al., 2016a; Zhao et al., 2016). Many factors, plant genotype, growth stage, and physiological status, type of plant tissue, environmental (soil) conditions, and agricultural practices also define endophytic colonization and endosphere community structures (Hardoim et al., 2008). The relatively few studies that have analyzed the influences of diverse plant growth stage on endophytic diversity have been reported. According to Shi et al. (2014), PCR-based Illumina pyrosequencing was used to examine the endophytic bacterial diversity and space-time dynamics in sugar beet (*Beta vulgaris* L.) in China. They reported that the plant genotype and plant growth stage of sugar beet influenced to the dynamics of endophytic bacteria communities. The great numbers of endophytic bacteria were detected during rosette formation and tuber growth and were reduced during seedling growth and sucrose accumulation, as observed by OTUs (Operational Taxonomic Units). Moreover, Kang et al. (2016) studied the communities' structures of the endophytic bacteria associated with surface-sterilized pepper plants grown in the different field soils by using culture-independent (PCR-DGGE) and culture-dependent (plating) methods. Field soils used in this study were collected from two different geographic locations; Deckso and Gwangyang in Korea. The results showed that in the same plant species, when propagated in diverse soil environments, the endophytic communities were different. Besides the determinants as described above, intrinsic bacterial traits significant for colonization have major roles as factors of endophytic diversity. Therefore, comparisons between various plant endophyte communities are difficult. According to Reinhold-Hurek and Hurek (2011), the majority of the culturable isolated endophytic bacteria species are members of *Proteobacteria*, while *Firmicutes*, *Actinobacteria*, and also *Bacteroides* are little. The presence of bacterial endophyte communities depends on their growth competence on a synthetic medium. Endophytic species are closely related to epiphytic species, and belong mainly to the alpha-, beta-, and gamma-*Proteobacteria* subgroups (Kuklinsky-Sobral et al., 2004). Therefore, the diversity of endophytic bacteria in alpha-, beta-, and gamma-*Proteobacteria* subgroups are reported from different host plant species (Table 3.2).

3.4   PLANT GROWTH-PROMOTING ACTIVITY AND PRODUCTS

PGPB stimulates plant growth in two mechanisms as direct and indirect mechanisms (Bashan and de-Bashan, 2005). The direct mechanism of PGPB is the major step involved to support plant growth in a forward and direct manner that contained nitrogen fixation, phytohormones production, phosphate solubilization, and iron availability. The indirect mechanisms referred to the PGPB that acted as biocontrol, indirectly promote plant growth by producing antagonistic substances or inducing resistance to pathogens in order to prevent the deleterious effects of phytopathogenic microorganisms (bacteria, fungi, and viruses). Endophytic bacteria isolated from host plant species in Asian countries and their plant growth-promoting traits are shown in Table 3.3.

3.4.1   NITROGEN FIXATION

In developing countries like South-East Asia and the developed world, agricultural practices that use high amounts of fertilizer have led to considerable increases in the productivity of food production, but high energy and environmental costs associated with fertilizer use necessitate the explore for alternative approaches of biological soil management. This fixation occurs through symbiotic and non-symbiotic N_2-fixing bacteria possessing the nitrogenase enzyme that converts atmospheric elemental

TABLE 3.2 Diversity of Endophytic Bacteria in Different Host Plant Species in Asia

Country	Species	Isolation Sources	References
Thailand	[1]*Bradyrhizobium* sp. SUTN9-2	Leguminous weed (*Aeschynomene americana* L.)	Piromyou et al., 2015b
	[1]*Bradyrhizobium* sp.	Rice (*Oryza sativa*)	Piromyou et al., 2015a
	[1]*Novosphingobium sediminicola*, [1]*Ochrobactrum intermedium*	Industrial variety wild and chewing sugarcane	Muangthong et al., 2015
	[1]*Novosphingobium nitrogenifigen*, [2]*Pseudomonas*, [2]*Klebsiella*, [2]*Pantoea*	Rice	Rangjaroen et al., 2014
	[1]*Rhizobium lemnae*	Duckweed (*Lemna aequinoctialis*)	Kittiwongwattana et al., 2014
China	[1]*Rhizobium smilacinae*	Traditional Chinese medicinal plant (*Smilacina japonica*)	Zhang et al., 2014a
	[1]*Ochrobactrum anthropi* Mn1 strain	Roots of Jerusalem artichoke	Meng et al., 2011, 2014
	[1]*Sphingomodaceae*, [3]*Burkholderiaceae*	*Caragana microphylla* grown in the desert	Dai et al., 2014
	[2]*Pseudoxanthomonas gei* sp. nov.	Stem of *Geum aleppicum* Jacq.	Zhang et al., 2014b
	[3]*Burkholderia* sp.	*Dendrobium officinale*	Yu et al., 2013
	[1]*Brevundimonas diminuta*, [1]*Methylobacterium* sp., [1]*Sinorhizobium terangae*, [1]*Novosphingobium tardaugens*, [1]*Caulobacter* sp., [1]*Kaistina koreensis*, [2]*Stenotrophomonas maltophilia*, [2]*Enterobacter* sp., [2]*Pantoea* sp., [2]*Stenotrophomonas* sp., [2]*Acinetobacter baumannii*, [2]*Alkanindiges illinoisensis*, [2]*Methylophaga marina*, [2]*Plesiomonas shigelloides*, [2]*Pseudomonas stutzeri*, [3]*Achromobacter xylosoxidans*, [3]*Burkholderia fungirum*, [3]*Burkholderia* sp., [3]*Acidovorax facilis*, [3]*Comamonas testosterone*, [3]*Curibacter gracilis*,	Roots of rice (*Oryza sativa* L.)	Sun et al., 2008

TABLE 3.2 *(Continued)*

Country	Species	Isolation Sources	References
China (continued)	³Delftia acidovorans, ³Delftia tsuruhatensis, ³Herbaspirillum frisingense, ³Hydrogenophaga taeniospiralis, ³Variovorax sp., ³Duganella violaceinigra, ³Methyloversatilis universalis, ³Gallionella ferruginea, ³Sterolibactrium denitrificans		
	¹Agrobacterium	Legume species of Astragalus	Chen et al., 2015
	²Pantoea sp. Sd-1	Rice seeds	Xiong et al., 2014
	³Burkholderia sp. SaZR4, SaMR10, ³Variovorax sp. SaNR1, ¹Sphingomonas sp. SaMR12	Sedum alfredii plants	Zhang et al., 2013
	²Enterobacter, ²Serratia, ²Stenotrophomonas, ²Pseudomonas	Peanut plants	Wang et al., 2013
	²Raoultella sp.	Sugarcane roots	Luo et al., 2016
	²Pseudomonas spp.	Wheat	Yang et al., 2011
	¹Pelomonas, ¹Rhizobium, ²Pseudomonas, ²Aeromonas, ³Rhodoferax, ³Ulginosibacterium	Narrowleaf cattail (Typha angustifolia L.) roots	Li et al., 2011
	Others: Sulfurospirillum, Ilyobacter, and Bacteroides		
	¹Sphingomonas, ²Pseudomonas	Wild alpine-subnival plant species	Sheng et al., 2011
	¹Sphingomonas, ²Pantoea, ²Enterobacter	Roots of elephant grass	Li et al., 2016b
	Others: Bacillus		
	³Variovorax paradoxus	Salicornia europaea grown under extreme salinity	Zhao et al., 2016
	²Enterobacter sp.	Roots of Sorghum sudanense grown in a Cu mine wasteland soils	Li et al., 2016c

TABLE 3.2   *(Continued)*

Country	Species	Isolation Sources	References
China (continued)	[2]*Klebsiella variicola* strain DX120E	Sugarcane crops	Lin et al., 2015
	[1]*Rhizobium oryzicola*	Surface-sterilized rice roots	Zhang et al., 2015
	[1]*Ochrobactrum endophyticum*	Roots of *Glycyrrhiza uralensis* F.	Li et al., 2016a
	[1]*Novosphingobium oryzae*	Rice roots	Zhang et al., 2016
	[1]*Rhizobium populi*	Storage liquid in the stems of *Populus euphratica* trees	Rozahon et al., 2014
	[1]*Paracoccus sphaerophysae*	Root nodules of *Sphaerophysa salsula*	Deng et al., 2011
India	[2]*Pseudomonas* and [2]*Stenotrophomonas*	*Origanum vulgare*	Bafana, 2013
	[2]*Enterobacter* and [2]*Stenotrophomonas*	Roots of Indian ecotype *Pteris vittata*	Tiwari et al., 2016
	[1]*Rhizobium*		
	[2]*Pseudomonas putida*, [2]*Enterobacter cloacae*	Saffron (*Crocus sativus*)	Sharma et al., 2015
	Others:		
	Bacillus licheniformis, *B. subtilis*, *B. cereus*, *B. humi*, *B. pumilus*, *B. safensis*, *Brevibacillus* sp., *Paenibacillus elgii*, *Staphylococcus hominis*		
	[2]*Pantoea* sp. and [2]*Pseudomonas* sp.	Fruit tissue	Jasim et al., 2015
	[3]*Polaromonas* sp. and [3]*Ralstonia* sp.		
	[1]*Paracoccus sanguinis*, [2]*Klebsiella pneumoniae*, [2]*Cedecea davisae*, [2]*Klebsiella oxytoca* and [2]*Erwinia tasmaniensis*	Banana shoot-tips	Thomas and Sekhar, 2017

TABLE 3.2 *(Continued)*

Country	Species	Isolation Sources	References
India (continued)	[2]*Pseudomonas putida* BP25	Root endosphere of Panniyur-5, black pepper	Sheoran et al., 2015
	[2]*Klebsiella pneumonia*	*Piper nigrum* (Black pepper)	Jasim et al., 2014
	[2]*Serratia* sp., [2]*Pseudomonas* sp., [2]*Pantoea* sp. Others: *Lysinibacillus* sp., *Bacillus* sp.	Ethnomedicinal plants	Nongkhlaw and Joshi, 2014
	[2]*Pseudomonas aeruginosa* BP35	Black pepper	Kumar et al., 2013
	[2]*Pseudomonas* spp. [2]*Erwinia* spp. [1]*Rhizobium, Agrobacterium*	Root-nodule of chickpea (*Cicer arietinum* L.) and mothbean (*Vigna aconitifolia* L.)	Sharma et al., 2012
	[3]*Burkholderia tropica, B. unamae* and *B. cepacia*	*Lycopodium cernuum* L.	Ghosh et al., 2016
Korea	[2]*Pseudomonas*	Ginseng plants of varying age	Vendan et al., 2010
	[2]*Pseudomonas* sp. HNR13	Roots of Chinese cabbage	Haque et al., 2016
	[3]*Burkholderia* sp. strain KJ006	Rice	Kwak et al., 2012
	[2]*Serratia* sp. RSC-14	Roots of *Solanum nigrum*	Khan et al., 2015
	[2]*Pseudomonas koreensis* AGB-1	Roots of *Miscanthus sinensis* growing in mine-tailing soil	Babu et al., 2015
	[2]*Klebsiella* four species Others: *Microbacterium*, two species of *Paenibacillus*, three *Bacillus* species	Leaves, stems, and roots of 10 rice cultivars	Ji et al., 2014
	[1]*Agrobacterium tumefaciens*	Pepper plants (*Capsicum annuum* L. cv. Nokkwang)	Kang et al., 2016

TABLE 3.2   *(Continued)*

Country	Species	Isolation Sources	References
Korea (continued)	[1]*Sphingopyxis chilensis*		
	[2]*Enhydrobacter aerosaccus*		
	[1]*Martelella endophytica* sp. nov.	Root sample of a halophyte, *Rosa rugosa*	Bibi et al., 2013
	[1]*Hoeflea suaedae* sp. nov.	Root of a halophyte (*Suaeda maritima*)	Chung et al., 2013
	[1]*Rhizobium halophytocola* sp. nov.	Root of *Rosa rugosa*	Bibi et al., 2012
Japan	[1]*Azospirillum* sp. B510	Stems of rice plants (*Oryza sativa* cv. Nipponbare)	Yasuda et al., 2009; Kaneko et al., 2010
	[2]*Pseudomonas fluorescens* Others: *Bacillus* sp., *Streptomyces luteogriseus*	*Carex kobomugi* roots	Matsuoka et al., 2013
Saudi Arabia	[2]*Enterobacter cloacae*	Non-nodulating roots of *Medicago sativa*	Khalifa et al., 2016
Oman	[2]*Enterobacter cloacae* Others: *Paenibacillus xylanexedens*	Date palm (*Phoenix dactylifera* L.) seedling roots	Yaish et al., 2015
	[1]*Sphingomonas* sp. LK11	Leaves of *Tephrosia apollinea*	Khan et al., 2014
Pakistan	[2]*Acinetobacter* Others: *Bacillus*	Three grasses, *Lolium perenne*, *Leptochloa fusca*, *Brachiaria mutica*, and two trees, *Lecucaena leucocephala* and *Acacia ampliceps*	Fatima et al., 2015

TABLE 3.2 *(Continued)*

Country	Species	Isolation Sources	References
Singapore	[1]*Methylobacterium* sp. strain L2-4	*Jatropha curcas* L.	Madhaiyan et al., 2015, 2018
Iran	[2]*Pseudomonas fluorescens* strain REN1	Roots of rice plants	Etesami et al., 2014
Taiwan	[3]*Achromobacter xylosoxidans* strain F3B	Plants that are predominantly located in a constructed wetland, including reed (*Phragmites australis*) and water spinach (*Ipomoea aquatica*)	Ho et al., 2012, 2013
Malaysia	[1]*Novosphingobium* sp. [1]*Asticcacaulis* sp. [1]*Rhodobacter* sp. [1]*Rhizobium* sp. [3]*Burkholderia vietnamiensis, B. tropica, B. kururiensis,* and *B. ambifaria* [3]*Herbaspirillum* sp. [2]*Pseudomonas* sp. and *Enterobacter* sp. Others: *Bacillus* sp.	Root tissues of Nipa Palm (*Nypa fruticans*) *Aloe vera*	Tang et al., 2010 Akinsanya et al., 2015
Vietnam	[2]*Pseudomonas aeruginosa* 23 (1-1)	Watermelon roots	Nga et al., 2010

[1]*Alphaproteobacteria,* [2]*Gammaproteobacteria,* [3]*Betaproteobacteria.*

TABLE 3.3 Endophytic Bacteria Isolated from Host Plant Species, Their Plant Growth Promoting Traits and Beneficial Interactions

PGPB	Plant Growth Promoting Trait	Beneficial Interactions	References
Ochrobactrum anthropi Mn1	Nitrogen fixation	Root morphological optimization and enhanced nutrient uptake	Meng et al., 2014
Burkholderia (Dominant genus)	Nitrogen fixation	Play potentially important roles in *D. officinale*	Yu et al., 2013
Raoultella sp.	Nitrogen fixation	Increased sugarcane plant biomass, total nitrogen, nitrogen concentration, and chlorophyll, and relieved nitrogen-deficiency symptoms of plants under a nitrogen-limiting condition	Luo et al., 2016
Variovorax paradoxus	Indole-3-acetic acid (IAA) production and phosphate-solubilizing activities	Excellent seed germination at high NaCl concentration and increased in plant shoot length, indicating that PGPB could protect *S. europaea* seedling from injury caused by salt stress	Zhao et al., 2016
Enterobacter sp., *Sphingomonas* sp., *Pantoea* sp., *Bacillus* sp.	ACC deaminase, IAA production, siderophores, and arginine decarboxylase	Four endophytic bacteria from elephant grass (*Pennisetum purpureum* Schumach) promoted plant growth and biomass yield, alleviated the harmful effects of salt stress on Hybrid *Pennisetum*	Li et al., 2016b
Enterobacter cloacae	IAA production	Enhance canola root elongation when grown under normal and saline conditions as demonstrated by a gnotobiotic root elongation assay	Yaish et al., 2015
Pseudomonas fluorescens REN1	ACC deaminase activity and IAA production	ACC deaminase containing *P. fluorescens* REN1 increased *in vitro* root elongation and endophytically colonized the root of rice seedlings	Etesami et al., 2014
Klebsiella, Microbacterium, Bacillus, Paenibacillus	IAA production, Siderophore producing activity, and Phosphate-solubilizing activity	Rice seeds treated with these PGPB showed improved plant growth, increased height and dry weight and antagonistic effects against fungal pathogens	Ji et al., 2014

TABLE 3.3 *(Continued)*

PGPB	Plant Growth Promoting Trait	Beneficial Interactions	References
Sphingomonas SaMR12	IAA production and Zinc phytoremediation	SaMR12 inoculation significantly enhanced the efficiency of zinc phytoextraction by increasing *S. alfredii* biomass, promoting zinc absorption, improving root morphology, and enhancing root exudates	Chen et al., 2014
Sphingomonas sp. LK11	Gibberellins and IAA production	Increased growth attributes of tomato plants (shoot length, chlorophyll contents, shoot, and root dry weights) compared to the control	Khan et al., 2014
Serratia sp. RSC-14	Phosphate solubilization and IAA production	Relived the toxic effects of Cd-induced stress by significantly increasing root/shoot growth, biomass production, and chlorophyll content and decreasing malondialdehyde (MDA) and electrolytes content of *Solanum nigrum* host plant	Khan et al., 2015
Enterobacter cloacae	Phosphate solubility and IAA production	Inoculation of *Pisum sativum* with MSR1 significantly improved the growth parameters (the length and dry weight)	Khalifa et al., 2016
Burkholderia tropica, B. unamae and *B. cepacia*	Phosphorus (Pi) solubilization	Supply soluble phosphate to the host plant to maintain its good health	Ghosh et al., 2016
Pseudomonas fluorescens, Bacillus sp., *Streptomyces luteogriseus*	Siderophore production and inorganic phosphate solubilization	Contribute to the Fe and P uptakes by *C. kobomugi* by increasing availability in the soil	Matsuoka et al., 2013
Pseudomonas aeruginosa strain BP35	Antagonistic effect	Protection to black pepper against infections by *Phytophthora capsici* and *Radopholus similis*	Kumar et al., 2013
Pseudomonas aeruginosa	Antagonistic effect	Antifungal compound, phenazine 1-carboxylic acid against *Pythium myriotylum*	Jasim et al., 2014

TABLE 3.3   *(Continued)*

PGPB	Plant Growth Promoting Trait	Beneficial Interactions	References
Azospirillum sp. B510	Antagonistic effect	Enhanced resistance against diseases caused by the virulent rice blast fungus *Magnaporthe oryzae* and by the virulent *Xanthomonas oryzae*	Yasuda et al., 2009; Isawa et al., 2010; Kaneko et al., 2010
Pseudomonas putida BP25	Antagonistic effect	Volatile substances exhibited anti-pathogens, *Phytophthora capsici, Pythium myriotylum, Giberella moniliformis, Rhizoctonia solani, Athelia rolfsii, Colletotrichum gloeosporioides* and plant-parasitic nematode, *Radopholus similis*	Sheoran et al., 2015

dinitrogen (N$_2$) into ammonia (Shiferaw et al., 2004). The nitrogen-fixing endophytic bacteria widely presented as symbionts are *Rhizobium, Brady-rhizobium, Sinorhizobium,* and *Mesorhizobium* with the leguminous plants while *Frankia* with non-leguminous trees and shrubs (Zahran, 2001). However, there is some report about plant colonization by Bradyrhizobia found not only in leguminous plants but also in non-leguminous species such as rice (Piromyou et al., 2015a). Non-symbiotic nitrogen fixation is carried out by free-living diazotrophs and can enhance non-legume plant growth. The important free-living and associative nitrogen-fixing genera have been reported, are *Azospirillum, Azotobacter, Gluconaceto-bacter, Azoarcus, Achromobacter, Burkholderia, Enterobacter, Herbaspi-rillum, Klebsiella, Mycobacterium, Pseudomonas, Rhodobacter, Serratia, Bacillus, Clostridia,* and *Citrobacter* (Hayat et al., 2012). Nitrogenase (*nif*) genes, for nitrogen fixation are found in both symbiotic and free-living systems (Reed et al., 2011). Muangthong et al. (2015) reported the strains of *Novosphingobium sediminicola* and *Ochrobactrum intermedium* isolated from sugarcane in Thailand showing the nitrogenase activity. In addition, Tang et al. (2010) reported that root-associating bacteria of the nipa palm (*Nypa fruticans*), *Burkholderia vietnamiensis* exhibited the activity of nitrogen fixation. *Raoultella* sp. strain L03 was isolated from surface-sterilized sugarcane root in China. This strain is proficient to fix nitrogen in association with the plant host. Inoculation of this strain increased sugarcane plant biomass, total nitrogen, nitrogen concentration, and chlorophyll, and relieved nitrogen-deficiency signs of plants under a nitrogen-limiting condition (Luo et al., 2016). The *Ochrobactrum anthropi* Mn1 strain stimulated the growth of *Jerusalem artichoke* host plant by plant growth-promoting effect as symbiotic nitrogen fixation, root morphological optimization, and improved nutrient availability and plant uptake (Meng et al., 2014).

3.4.2 PHYTOHORMONES

Phytohormones play key roles as signals and regulators of growth and expansion in plants. The capacity to produce them is frequently considered as a trait of the plant kingdom. While this strategy is probably successful, soil, and plant-associated prokaryotes may also produce or transform phyto-hormones under *in vitro* (Costacurta and Vanderleyden, 1995). Various PGPB are known to produce phytohormone, namely, auxins, cytokinin, and gibberellins (Pliego et al., 2011; Hayat et al., 2012). Tiwari et al. (2016) reported that the arsenic resistance endophytes obtained from the roots of *Pteris vittata* belonged to group *Proteobacteria* as *Rhizobium* sp. strain E1, *Enterobacter* sp. strain E3 and E5 including *Stenotrophomonas* sp. strain E7 showed the capability of indole-3-acetic acid (IAA) production. The *Variovorax paradoxus* belonged to β-*Proteobacteria* and strain of *Bacillus endophyticus*, *Bacillus tequilensis*, *Planococcus rifietoensis*, and *Arthrobacter agilis* could enhance halophyte *Salicornia europaea* plant growth under saline stress environments by IAA production together with the other plant growth-promoting traits. All strains showed excellent seed germination at high NaCl concentration and also showed a significant increase in plant shoot length, indicating that PGPB could protect *S. europaea* seedling from injury caused by salt stress (Zhao et al., 2016). According to Li et al. (2016b), four bacterial genera: *Sphingomonas* (strain pp01), *Pantoea* (strain pp02), *Bacillus* (strain pp04) and *Enterobacter* (strain pp06) were able to produce IAA at the range of 10.50–759.19 mg/L, where *Enterobacter* sp. established the highest value. The four endophytic bacteria isolated from elephant grass (*Pennisetum purpureum* Schumach) significantly enhanced plant growth and biomass yield, alleviated the harmful effects of salt stress on hybrid *Pennisetum*. An endophytic bacteria *Serratia* sp. RSC-14 could tolerate to cadmium and secreted phytohormones such as IAA (54 μg/mL). RSC-14 inoculation relived the toxic effects of Cd-induced stress by significantly increasing root/shoot growth, biomass production, and chlorophyll content and decreasing malondialdehyde (MDA) and electrolytes content of *Solanum nigrum* host plant (Khan et al., 2015). *Enterobacter cloacae* MSR1 isolated from the non-nodulation roots of *Medicago sativa* and *Enterobacter cloacae* PD-P6 isolated from date palm (*Phoenix dactylifera* L.) possessed the multiple plant-growth promoting characteristics, one of them is IAA production (Yaish et al., 2015; Khalifa et al., 2016). Inoculation of *Pisum sativum* with strain MSR1 significantly enhanced the growth parameters (the length and dry weight) of this economically significant grain legume compared to the non-treated plants. PD-P6 (*Enterobacter cloacae*) was able to enhance canola root elongation when grown under normal and saline conditions as demonstrated by a gnotobiotic root elongation assay. *Sphingomonas* sp. LK11 isolated from the leaves of *Tephrosia apollinea* showed significantly increased growth attributes (shoot length, chlorophyll contents, shoot, and root dry weights) of tomato plants by producing two kinds of phytohormones as gibberellins and IAA (Khan et al., 2014). Chen et al. (2014) reported about the endophytic bacterium,

Sphingomonas SaMR12 isolated from *Sedum alfredii* seems to increase plant biomass by IAA production and zinc phytoextraction. In addition, Jasim et al. (2014) observed that the phytochemicals from *Piper nigrum* had a stimulating effect on IAA production by *Klebsiella pneumoniae* isolated from the same plant. Sharma et al. (2012) found the most isolates belonged to genus *Pseudomonas* associated with chickpea (*Cicer arietinum* L.) and moth bean (*Vigna aconitifolia* L.) that could produce IAA. Ten strains of genus *Klebsiella* (Gammaproteobacteria), *Bacillus*, and *Paenibacillus* (Bacilli) and *Microbacterium* (Actinobacteria) isolated from Korean rice cultivars showed the highest IAA production and rice seeds treated with these PGPB could enhance plant growth, increase height and dry weight (Ji et al., 2014).

3.4.3 MINERAL NUTRIENT SOLUBILIZATION

Most of this phosphorus is found in insoluble forms, therefore, not available for plant growth. Mineral phosphate solubilization by endophytic bacteria could aid in the availability of phosphate to host crops during initial colonization and subsequently promote plant growth (Kuklinsky-Sobral et al., 2004). The principal mechanism for mineral phosphate solubilization has been proposed. Organic acids and acid phosphatases produced by microorganisms play a major role in the mineralization of organic phosphorus in soil (Rodríguez and Fraga, 1999). The most proficient PGPB has been described as the potential phosphate solubilization to stimulate growth of host plant together with IAA production belonging to genera as follows: group γ-*Proteobacteria* – *Pseudomonas, Enterobacter, Klebsiella, Serratia, Erwinia*; group α-*Proteobacteria* – *Phyllobacterium, Rhizobium, Agrobacterium*; group β-*Proteobacteria* – *Variovorax* (Sharma et al., 2012; Ji et al., 2014; Khan et al., 2015; Khalifa et al., 2016; Zhao et al., 2016). According to Ghosh et al. (2016), *Burkholderia tropica, Burkholderia unamae*, and *Burkholderia cepacia* associated with *Lycopodium cernuum* L. from phosphate starved red lateritic soil of West Bengal showed the phosphate solubilizing capability. The isolated strains could release of bound phosphates from ferric phosphate ($FePO_4$), aluminum phosphate ($AlPO_4$), and four different complex rock phosphates. The results indicated that their phosphate solubilizing efficacy is very good. Moreover, all strains were effective to supply soluble phosphate to the host plant to maintain its good health.

3.4.4   SIDEROPHORE PRODUCTION

Iron is a vital bioelement for virtually all forms of life. All microorganisms known so far, with the exception of certain lactobacilli, require iron. In aerobic conditions, microorganisms need iron for a variety of functions, including reduction of oxygen for the ATP synthesis, for heme formation, and for other essential purposes. The aerobic atmosphere of the planet has caused the surface iron exists predominantly in its ferric state and reacts to form highly insoluble hydroxides and oxyhydroxides, making it largely unavailable to microorganisms. To acquire adequate iron, bacteria have developed active strategies to solubilize this metal for its proficient Fe uptake. Siderophores produced by bacteria are one of the most commonly found strategies. Bacterial siderophores are low-molecular-mass molecules with high specificity and affinity for chelating or binding Fe^{3+} (Neilands, 1983; Miethke and Marahiel, 2007; Rajkumar et al., 2010). Moreover, the biocontrol activity of PGPB can be achieved by siderophore production. Siderophores play an important role in the suppression of plant-pathogen by chelation of Fe thereby creating competition for iron. Under iron stress conditions confers upon these antagonistic organisms as an added advantage, resulting in the exclusion of pathogens due to iron starvation (O'Sullivan and O'Gara, 1992). Eight isolates of endophytic root-nodule bacteria from chickpea (*Cicer arietinum* L.) and moth bean (*Vigna aconiti-folia* L.) belonging to *Pseudomonas* and *Rhizobium* exhibited positive results for siderophore production (Sharma et al., 2012). Ji et al. (2014) also revealed that six isolates associated with Korean rice cultivars, belonging to genera *Klebsiella* (Gammaproteobacteria), *Bacillus*, and *Paenibacillus* (Bacilli) and *Microbacterium* (Actinobacteria) showed high siderophore producing activity. Endophytic bacteria isolated from elephant grass (*Penni-setum purpureum* Schumach) were identified as *Sphingomonas*, *Bacillus*, and *Enterobacter*. These isolated strains showed siderophore production ability, where *Bacillus* recorded the optimum siderophore 64.06% units (Li et al., 2016b). Matsuoka et al. (2013) reported that *Pseudomonas fluore-scens*, *Bacillus* sp., and *Streptomyces luteogriseus* isolated from roots of *Carex kobomugi* exhibited siderophore production and inorganic phosphate solubilization under Fe or P limited condition. *C. kobomugi* showed higher Fe and P content. Colonization of root tissue by these bacteria contributes to the Fe and P uptakes by *C. kobomugi* by increasing availability in the soil. Tiwari et al. (2016) have detected two isolates positive for sidero-phore production belonging to *Rhizobium* and *Bacillus* that might have a presumptive role in Arsenic (As) uptake, accumulation, and detoxification in association with *P. vittata* ecotype.

3.4.5 ANTAGONISTIC ACTIVITY AND PHYTOPATHOGEN BIOCONTROL

Plant pathogens are a major and chronic problem for crop production and global ecosystem stability. The PGPB prevents the plants from phytopathogens (fungal, bacterial, and viral diseases), including insect and nematode pests, by several mechanisms. PGPB colonization and defensive retention of host plants are enabled by production of siderophores (Miethke and Marahiel, 2007), cell wall lytic enzymes (Backman and Sikora, 2008), antibiotic metabolites (Schouten et al., 2004) and induction of systematic resistance in plants (Kloepper et al., 2004; Van Loon et al., 2004). According to Yasuda et al. (2009), *Azospirillum* sp. B510 isolated from the stem of rice plants had the ability of disease resistance in host plants. Inoculation *Azospirillum* sp. B510 with rice plants showed enhanced resistance against diseases caused by the virulent rice blast fungus *Magnaporthe oryzae* and by the virulent bacterial pathogen *Xanthomonas oryzae*. Moreover, *Pseudomonas* sp. HNR13 including with *Bacillus* sp. and *Microbacterium* sp. isolated from the Chinese cabbage roots also showed strong antagonistic activity against the soil-borne fungal pathogens; *Pythium ultimum*, *Phytophthora capsici*, *Fusarium oxysporum*, and *Rhizoctonia solani* while most of the isolates are members of the genus *Bacillus* that could produce cell wall degrading enzymes, exhibited high antagonism against the tested food-borne pathogenic bacteria (Haque et al., 2016). The *Pseudomonas putida* BP25 inhibited a broad range of pathogens such as *Phytophthora capsici*, *Pythium myriotylum*, *Giberella moniliformis*, *Rhizoctonia solani*, *Athelia rolfsii*, *Colletotrichum gloeosporioides* and plant parasitic nematode, *Radopholus similis* by its volatile substances (Sheoran et al., 2015).

3.4.6 1-AMINOCYCLOPROPANE-1-CARBOXYLASE (ACC) DEAMINASE AND OTHER PRODUCTS

The enzyme ACC deaminase was first recovered by Honma and Shimo-mura (1978). This enzyme catalyzes the cleavage of ACC, the immediate precursor, to ammonia and α-ketobutyrate (KB), therefore reducing the amount of ethylene synthesized by the plant (Gamalero and Glick, 2015). Ethylene, a stress phytohormone, was reduced to allow the plant to be more resistant to a wide variety of environment stresses. It can function as an efficient plant growth regulator at a very low concentration as low as 0.05 μL/L (Abeles et al., 1992). PGPB belonging to *Proteobacteria* subgroup, *Azospirillum, Rhizobium, Agrobacterium, Achromobacter, Burkholderia, Ralstonia, Pseudomonas,* and *Enterobacter* have been reported to possess ACC deaminase activity/genes (Blaha et al., 2006). PGPB act as a sink for ACC with the consequence that lowering plant ACC levels, decreasing the amount of ACC within the plant that can be converted into ethylene (Santoyo et al., 2016). ACC deaminase containing PGPB can significantly decrease the extent of plant growth inhibition that accrues from the stresses such as flooding, high salinity, drought, the presence of fungal and bacterial pathogens, nematode damage, the presence of high levels of metals and organic contaminants including extremes of temperature (Gamalero and Glick, 2015). *Pseudomonas stutzeri* A1501 carries a single gene encoding ACC deaminase, designated *acdS* gene. The *acdS* mutant lacking of ACC deaminase activity was less resistant to NaCl and $NiCl_2$ compared with the wild type. Moreover, the nitrogenase activity of the lack of ACC deaminase-inactivated mutant was greatly impaired under salt stress condition (Han et al., 2015). The four strains tested, *Sphingomonas* (strain pp01), *Pantoea* (strain pp02), *Bacillus* (strain pp04) and *Enterobacter* (strain pp06) demonstrated relatively high levels of ACC deaminase activity, where *Pantoea* sp. presented the highest ACC deaminase activity at 1106.66±78.59 nmol KB/h/mg (Li et al., 2016b). Jasim et al. (2015) isolated and identified the endophytic bacteria from the fruit tissue of *Elettaria cardamomum* by using both culture-based and PCR based methods. PCR-based screening identified the isolates EcB 2 (*Pantoea* sp.), EcB 7 (*Polaromonas* sp.), EcB 9 (*Pseudomonas* sp.), EcB 10 (*Pseudomonas* sp.) and EcB 11 (*Ralstonia* sp.) as positive for ACC deaminase. *Pseudomonas fluorescens* REN1 containing ACC deaminase isolated from roots of rice, increased *in vitro* root elongation, and endophytically colonized the root of rice seedlings significantly, as compared to control under constant flooded conditions (Etesami et al., 2014).

Generally, plant growing in a unique environmental setting having special ethnobotanical uses having extreme age or interesting endemic location possesses novel endophytic microorganisms which can supply new leads. Recently, new endophytic bioactive metabolites, having a wide variety of biological activities as an antibiotic, antiviral, anticancer, anti-inflammatory, and antioxidant have been characterized (Strobel and Daisy, 2003). Pundir et al. (2014) isolated endophytic bacteria from tomato, *Aloe vera*, chilli, radish, cauliflower, cabbage, arjun, pomegranate, grass, carrot, coriander, guava, stevia, mint, garlic, peas, giloy, turmeric, neem and rose that were collected from different areas of Ambala, Haryana. Bioprospecting potentials as antimicrobial activity, antibiotic suscep-tibility pattern, enzyme activity, and dye degradation ability of thirty isolates were observed. Twenty isolates exhibited both antifungal (*Asper-gillus fumigatus, Aspergillus* sp. and *Candida albicans*) and antibacterial activity (*Staphylococcus epidermidis, Bacillus amyloliquefaciens, Esch-erichia coli*, and *Salmonella enterica* ser. *typhi*). Moreover, 33.33% of isolates showed urease activity, 66.66% amylase activity, 50% esterase activity, while 53.33% malachite green degradation. According to El-Deeb et al. (2013), among 28 endophytic bacterial isolates form organs of *Plec-tranthus tenuiflorus* belonging to genus *Pseudomonas, Acinetobacter, Bacillus, Micrococcus, Paenibacillus* showed the stronger activities in extracellular enzymes, for example, amylase, esterase, lipase, protease, pectinase, xylanase, and cellulase. Sharma et al. (2015) reported about the diversity of culturable bacterial endophyte associated with Saffron that is a medicinally important plant in India. Fifty-four bacterial strains were identified as *Bacillus, Paenibacillus, Pseudomonas, Brevibacillus, Entero-bacter*, and *Staphylococcus*, 81% isolates showed lipase activity, 57% cellulase, 48% protease, 38% amylase, 33% chitinase and 29% showed pectinase activity. Moreover, Akinsanya et al. (2015) revealed for the first time the endophytic bacteria communities in *Aloe vera* that were mostly *Pseudomonas* sp., *Enterobacter* sp., and *Bacillus* sp. and these isolates produced bioactive compounds with good antimicrobial and antioxidant activities.

3.5 CONCLUSION

Nowadays, the endophytic bacterial diversity in different host plant species is characterized based on both cultivation-dependent and culti-vation-independent methods, including molecular techniques in order to understand the specific association between plants and endophytes as well as between the communities of various endophytes associated within a single plant. A numerous diversity of endophytic bacteria isolated from a large number of host plants have been studied. The mostly endophytic bacteria were found belonging to α-, β- and α-*Proteobacteria* subgroup. These PGPB have multiple mechanisms to promote plant growth, such as the production of enzymes, bioactive factors, antibiotics, metabolites as well as phytohormones. PGPB are the potential tools for sustainable agriculture and trend for the future. The use of PGPB is an environmental friendly approach. Moreover, endophytic bacteria are the important sources of natural products that are biological control agents (BCA) and other bioactive compounds by their different functional roles for potential use in medicine, agriculture, or industry.

KEYWORDS

- **16S rRNA gene sequence**
- **1-aminocyclopropane-1-carboxylase deaminase**
- **antagonistic activity**
- **endophytic bacteria**
- **mineral nutrient solubilization**
- ***nifH* gene**
- **nitrogen fixation**
- **phytohormones**
- **phytopathogen biocontrol**
- **plant growth-promoting activity**
- **siderophore production**

REFERENCES

Abeles, F. B., Morgan, P. W., & Saltveit, M. E. *Ethylene in Plant Biology.* Academic Press, New York, **1992**.

Ahemad, M., & Kibret, M. Mechanisms and applications of plant growth promoting rhizobacteria: Current perspective. *J. King. Saud. Univ. Sci.,* **2014**, *26*(1), 1–20.

Akinsanya, M. A., Goh, J. K., Lim, S. P., & Ting, A. S. Diversity, antimicrobial and antioxidant activities of culturable bacterial endophyte communities in *Aloe vera. FEMS Microbiol. Lett.,* **2015**, *362*(23), fnv184, doi: 10.1093/femsle/fnv184.

Azevedo, J. L., Maccheroni, W., Pereira, J. O., & Luiz De Araújo, W. Endophytic microorganisms: A review on insect control and recent advances on tropical plants. *Electron J. Biotechnol.*, **2000**, *3*, 15–16.

Babu, A. G., Shea, P. J., Sudhakar, D., Jung, I. B., & Oh, B. T. Potential use of *Pseudomonas koreensis* AGB-1 in association with *Miscanthus sinensis* to remediate heavy metal (loid)-contaminated mining site soil. *J. Environ. Manage.*, **2015**, *151*, 160–166.

Backman, P. A., & Sikora, R. A., Endophytes: An emerging tool for biological control. *Biol. Control.*, **2008**, *26*, 1–3.

Bafana, A. Diversity and metabolic potential of culturable root-associated bacteria from *Origanum vulgare* in sub-Himalayan region. *World J. Microbiol. Biotechnol.*, **2013**, *29*(1), 63–74.

Bashan, Y., & De-Bashan, L. E. Bacteria, plant growth-promoting. In: Hillel, D., (ed.), *Encyclopedia of Soils in the Environment*, (Vol. 1, pp. 103–115). Elsevier, Oxford, UK, **2005**.

Bibi, F., Chung, E. J., Khan, A., Jeon, C. O., & Chung, Y. R. *Martelella endophytica* sp. nov., an antifungal bacterium associated with a halophyte. *Int. J. Syst. Evol. Microbiol.*, **2013**, *63*, 2914–2919.

Bibi, F., Chung, E. J., Khan, A., Jeon, C. O., & Chung, Y. R. *Rhizobium halophytocola* sp. nov., isolated from the root of a coastal dune plant. *Int. J. Syst. Evol. Microbiol.*, **2012**, *62*, 1997–2003.

Blaha, D., Prigent-Combaret, C., Mirza, M. S., & Moënne-Loccoz, Y. Phylogeny of the 1-aminocyclopropane-1-carboxylic acid deaminase-encoding gene *acdS* in phytobeneficial and pathogenic *Proteobacteria* and relation with strain biogeography. *FEMS Microbiol. Ecol.*, **2006**, *56*, 455–470.

Chen, B., Shen, J., Zhang, X., Pan, F., Yang, X., & Feng, Y. The endophytic bacterium, *Sphingomonas* SaMR12, improves the potential for zinc phytoremediation by its host, *Sedum alfredii*. *PLoS One*, **2014**, *9*(9), e106826, https://doi.org/10.1371/journal.pone.0106826 (Accessed on 7 October 2019).

Chen, W., Sun, L., Lu, J., Bi, L., Wang, E., & Wei, G. Diverse nodule bacteria were associated with *Astragalus* species in arid region of northwestern China. *J. Basic Microbiol.*, **2015**, *55*(1), 121–128.

Chung, E. J., Park, J. A., Pramanik, P., Bibi, F., Jeon, C. O., & Chung, Y. R. *Hoeflea suaedae* sp. nov., an endophytic bacterium isolated from the root of the halophyte *Suaeda maritima*. *Int. J. Syst. Evol. Microbiol.*, **2013**, *63*, 2277–2281.

Costacurta, A., & Vanderleyden, J. Synthesis of phytohormones by plant associated bacteria. *Crit. Rev. Microbiol.*, **1995**, *21*, 1–18.

Dai, J. X., Liu, X. M., & Wang, Y. J. Diversity of endophytic bacteria in *Caragana microphylla* grown in the desert grassland of the Ningxia Hui autonomous region of China. *Genet Mol. Res.*, **2014**, *13*(2), 2349–2358.

Deng, Z. S., Zhao, L. F., Xu, L., Kong, Z. Y., Zhao, P., Qin, W., Chang, J. L., & Wei, G. H. *Paracoccus sphaerophysae* sp. nov., a siderophore-producing, endophytic bacterium isolated from root nodules of *Sphaerophysa salsula*. *Int. J. Syst. Evol. Microbiol.*, **2011**, *61*, 665–669.

Döbereiner, J., Day, J. M., & Dart, P. J. Nitrogenase activity in the rhizosphere of sugarcane and other tropical grasses. *Plant Soil*, **1972**, *37*(1), 191–196.

El-Deeb, B., Fayez, K., & Gherbawy, Y. Isolation and characterization of endophytic bacteria from *Plectranthus tenuiflorus* medicinal plant in Saudi Arabia desert and their antimicrobial activities. *J. Plant Interact.*, **2013**, *8*(1), 56–64.

Etesami, H., Mirseyed, H. H., & Alikhani, H. A. Bacterial biosynthesis of 1-aminocy-clopropane-1-caboxylate (ACC) deaminase, a useful trait to elongation and endophytic colonization of the roots of rice under constant flooded conditions. *Physiol. Mol. Biol. Plants*, **2014**, *20*(4), 425–434.

Fatima, K., Afzal, M., Imran, A., & Khan, Q. M. Bacterial rhizosphere and endosphere populations associated with grasses and trees to be used for phytoremediation of crude oil contaminated soil. *Bull. Environ. Contam. Toxicol.*, **2015**, *94*(3), 314–320.

Gamalero, E., & Glick, B. R. Bacterial modulation of plant ethylene levels. *Plant Physiol.*, **2015**, *169*(1), 13–22.

Ghosh, R., Barman, S., Mukherjee, R., & Mandal, N. C. Role of phosphate solubilizing *Burkholderia* spp. for successful colonization and growth promotion of *Lycopodium cernuum* L. (Lycopodiaceae) in lateritic belt of Birbhum district of West Bengal, India. *Microbiol. Res.*, **2016**, *183*, 80–91.

Glick, B. R. Plant growth-promoting bacteria: Mechanisms and applications. *Scientifica*, **2012**, 963401, doi:10.6064/2012/963401.

Han, Y., Wang, R., Yang, Z., Zhan, Y., Ma, Y., Ping, S., Zhang, L., Lin, M., & Yan, Y. 1-Aminocyclopropane-1-carboxylate deaminase from *Pseudomonas stutzeri* A1501 facilitates the growth of rice in the presence of salt or heavy metals. *J. Microbiol. Biotechnol.*, **2015**, *25*(7), 1119–1128.

Haque, M. A., Yun, H. D., & Cho, K. M. Diversity of indigenous endophytic bacteria associated with the roots of Chinese cabbage (*Brassica campestris* L.) cultivars and their antagonism towards pathogens. *J. Microbiol.*, **2016**, *54*(5), 353–363.

Hardoim, P. R., Van Overbeek, L. S., & Elsas, J. D. Properties of bacterial endophytes and their proposed role in plant growth. *Trends Microbiol.*, **2008**, *16*(10), 463–471.

Hayat, R., Ahmed, I., & Sheirdil, R. I. An overview of plant growth promoting rhizobacteria (PGPR) for sustainable agriculture. In: Ashraf, M., Öztürk, M., Ahmad, M. S. A., & Aksoy, A., (eds.), *Crop Production for Agricultural Improvement* (pp. 557–579). Part 3. Springer, Dordrecht, **2012**.

Ho, Y. N., Hsieh, J. L., & Huang, C. C. Construction of a plant-microbe phytoremediation system: Combination of vetiver grass with a functional endophytic bacterium, *Achromobacter xylosoxidans* F3B, for aromatic pollutants removal. *Bioresour. Technol.*, **2013**, *145*, 43–47.

Ho, Y. N., Mathew, D. C., Hsiao, S. C., Shih, C. H., Chien, M. F., Chiang, H. M., & Huang, C. C. Selection and application of endophytic bacterium *Achromobacter xylosoxidans* strain F3B for improving phytoremediation of phenolic pollutants. *J. Hazard Mater.*, **2012**, *219–220*, 43–49.

Honma, M., & Shimomura, T. Metabolism of 1-aminocyclopropane-1-carboxylic acid. *Agric. Biol. Chem.*, **1978**, *42*, 1825–1831.

Isawa, T., Yasuda, M., Awazaki, H., Minamisawa, K., Shinozaki, S., & Nakashita, H. *Azospirillum* sp. strain B510 enhances rice growth and yield. *Microbes Environ.*, **2010**, *25*(1), 58–61.

Jasim, B., Anish, M. C., Shimil, V., Jyothis, M., & Radhakrishnan, E. K. Studies on plant growth promoting properties of fruit-associated bacteria from *Elettaria cardamomum* and molecular analysis of ACC deaminase gene. *Appl. Biochem. Biotechnol.*, **2015**, *177*(1), 175–189.

Jasim, B., Jimtha, J. C., Shimil, V., Jyothis, M., & Radhakrishnan, E. K. Studies on the factors modulating indole-3-acetic acid production in endophytic bacterial isolates from

Piper nigrum and molecular analysis of *ipdc* gene. *J. Appl. Microbiol.*, **2014**, *117*(3), 786–799.

Ji, S. H., Gururani, M. A., & Chun, S. C. Isolation and characterization of plant growth promoting endophytic diazotrophic bacteria from Korean rice cultivars. *Microbiol. Res.*, **2014**, *169*(1), 83–98.

Kaneko, T., Minamisawa, K., Isawa, T., Nakatsukasa, H., Mitsui, H., Kawaharada, Y., et al. Complete genomic structure of the cultivated rice endophyte *Azospirillum* sp. B510. *DNA Res.*, **2010**, *17*(1), 37–50.

Kang, S. A., Han, J. W., & Kim, B. S. Community structures and antagonistic activities of the bacteria associated with surface-sterilized pepper plants grown in different field soils. *Arch. Microbiol.*, **2016**, *198*(10), 1027–1034.

Khalifa, A. Y., Alsyeeh, A. M., Almalki, M. A., & Saleh, F. A. Characterization of the plant growth promoting bacterium, *Enterobacter cloacae* MSR1, isolated from roots of non-nodulating *Medicago sativa*. *Saudi J. Biol. Sci.*, **2016**, *23*(1), 79–86.

Khan, A. L., Waqas, M., Kang, S. M., Al-Harrasi, A., Hussain, J., Al-Rawahi, A., Al-Khiziri, S., Ullah, I., Ali, L., Jung, H. Y., & Lee, I. J. Bacterial endophyte *Sphingomonas* sp. LK11 produces gibberellins and IAA and promotes tomato plant growth. *J. Microbiol.*, **2014**, *52*(8), 689–695.

Khan, A. R., Ullah, I., Khan, A. L., Park, G. S., Waqas, M., Hong, S. J., Jung, B. K., Kwak, Y., Lee, I. J., & Shin, J. H. Improvement in phytoremediation potential of *Solanum nigrum* under cadmium contamination through endophytic-assisted *Serratia* sp. RSC-14 inoculation. *Environ. Sci. Pollut. Res. Int.*, **2015**, *22*(18), 14032–14042.

Kittiwongwattana, C., & Thawai, C. *Rhizobium lemnae* sp. nov., a bacterial endophyte of *Lemna aequinoctialis*. *Int. J. Syst. Evol. Microbiol.*, **2014**, *64*, 2455–2460.

Kittiwongwattana, C., & Thawai, C. *Rhizobium paknamense* sp. nov., isolated from lesser duckweeds (*Lemna aequinoctialis*). *Int. J. Syst. Evol. Microbiol.*, **2013**, *63*, 3823–3828.

Kloepper, J. W., Ryu, C. M., & Zhang, S. Induced systemic resistance and promotion of plant growth by *Bacillus* species. *Phytopathology*, **2004**, *94*, 1259–1266.

Kuklinsky-Sobral, J., Araujo, W., Mendes, R., Geraldi, I., Pizzirani-Kleiner, A., & Azevedo, J. Isolation and characterization of soybean-associated bacteria and their potential for plant growth promotion. *Environ. Microbiol.*, **2004**, *6*, 1244–1251.

Kumar, A., Munder, A., Aravind, R., Eapen, S. J., Tümmler, B., & Raaijmakers, J. M. Friend or foe: Genetic and functional characterization of plant endophytic *Pseudomonas aeruginosa*. *Environ. Microbiol.*, **2013**, *15*(3), 764–779.

Kwak, M. J., Song, J. Y., Kim, S. Y., Jeong, H., Kang, S. G., Kim, B. K., Kwon, S. K., Lee, C. H., Yu, D. S., Park, S. H., & Kim, J. F. Complete genome sequence of the endophytic bacterium *Burkholderia* sp. strain KJ006. *J. Bacteriol.*, **2012**, *194*(16), 4432–4433.

Lei, X., Wang, E. T., Chen, W. F., Sui, X. H., & Chen, W. X. Diverse bacteria isolated from root nodules of wild *Vicia* species grown in temperate region of China. *Arch. Microbiol.*, **2008**, *190*(6), 657–671.

Li, L., Li, Y. Q., Jiang, Z., Gao, R., Nimaichand, S., Duan, Y. Q., Egamberdieva, D., Chen, W., & Li, W. J. *Ochrobactrum endophyticum* sp. nov., isolated from roots of *Glycyrrhiza uralensis*. *Arch. Microbiol.*, **2016a**, *198*(2), 171–179.

Li, X., Geng, X., Xie, R., Fu, L., Jiang, J., Gao, L., & Sun, J. The endophytic bacteria isolated from elephant grass (*Pennisetum purpureum* Schumach) promote plant growth and enhance salt tolerance of Hybrid Pennisetum. *Biotechnol. Biofuels*, **2016b**, *9*(1), 190, doi: 10.1186/s13068–016–0592–0.

Li, Y. H., Liu, Q. F., Liu, Y., Zhu, J. N., & Zhang, Q. Endophytic bacterial diversity in roots of *Typha angustifolia* L. in the constructed Beijing Cuihu Wetland (China). *Res. Microbiol.*, **2011**, *162*(2), 124–131.

Li, Y., Wang, Q., Wang, L., He, L. Y., & Sheng, X. F. Increased growth and root Cu accumulation of *Sorghum sudanense* by endophytic *Enterobacter* sp. K3–2: Implications for *Sorghum sudanense* biomass production and phytostabilization. *Ecotoxicol. Environ. Saf.*, **2016c**, *124*, 163–168.

Lin, L., Wei, C., Chen, M., Wang, H., Li, Y., Li, Y., Yang, L., & An, Q. Complete genome sequence of endophytic nitrogen-fixing *Klebsiella variicola* strain DX120E. *Stand Genomic Sci.*, **2015**, *10*, 22, doi: 10.1186/s40793–015–0004–2.

Lodewyckx, C., Vangronsveld, J., Porteous, F., Moore, E. R. B., Taghavi, S., Mezgeay, M., & Van Der Lelie, D. Endophytic bacteria and their potential applications. *Crit. Rev. Plant Sci.*, **2002**, *21*, 583–606.

Luo, T., Ou-Yang, X. Q., Yang, L. T., Li, Y. R., Song, X. P., Zhang, G. M., Gao, Y. J., Duan, W. X., & An, Q. *Raoultella* sp. strain L03 fixes N$_2$ in association with micropropagated sugarcane plants. *J. Basic Microbiol.*, **2016**, *56*(8), 934–940.

Madhaiyan, M., Alex, T. H., Ngoh, S. T., Prithiviraj, B., & Ji, L. Leaf-residing *Methylobacterium* species fix nitrogen and promote biomass and seed production in *Jatropha curcas*. *Biotechnol. Biofuels*, **2015**, *8*, 222, doi: 10.1186/s13068–015–0404–y.

Madhaiyan, M., Chan, K. L., & Ji, L. Draft genome sequence of *Methylobacterium* sp. strain L2–4, a leaf-associated endophytic N-fixing bacterium isolated from *Jatropha curcas* L. *Genome Announc*, **2018**, *6*(19), doi: 10.1128/genomeA.01306–14.

Matsuoka, H., Akiyama, M., Kobayashi, K., & Yamaji, K. Fe and P solubilization under limiting conditions by bacteria isolated from *Carex kobomugi* roots at the Hasaki coast. *Curr. Microbiol.*, **2013**, *66*(3), 314–321.

Meng, X., Long, X., Kang, J., Wang, X., & Liu, Z. Isolation and identification of endogenic nitrogen-fixing bacteria in the roots of Jerusalem artichoke (*Helianthus tuberosus* L). *Acta Prataculturae Sinica*, **2011**, *20*, 157–163.

Meng, X., Yan, D., Long, X., Wang, C., Liu, Z., & Rengel, Z. Colonization by endophytic *Ochrobactrum anthropi* Mn1 promotes growth of Jerusalem artichoke. *Microb. Biotechnol.*, **2014**, *7*(6), 601–610.

Miche, L., & Balandreau, J. Effects of rice seed surface sterilization with hypochlorite on inoculated *Burkholderia vietnamiensis*. *Appl. Environ. Microbiol.*, **2001**, *67*, 3046–3052.

Miethke, M., & Marahiel, M. A. Siderophore-based iron acquisition and pathogen control. *Microbiol. Mol. Biol. Rev.*, **2007**, *71*, 413–451.

Muangthong, A., Youpensuk, S., & Rerkasem, B. Isolation and characterization of endophytic nitrogen fixing bacteria in sugarcane. *Trop. Life Sci. Res.*, **2015**, *26*(1), 41–51.

Neilands, J. B. Siderophores. In: Eichhorn, L., & Marzilla, L. G., (eds.), *Advances in Inorganic Biochemistry* (pp. 137–166). Elsevier, **1983**.

Nga, N. T., Giau, N. T., Long, N. T., Lübeck, M., Shetty, N. P., De Neergaard, E., Thuy, T. T., Kim, P. V., & Jørgensen, H. J. Rhizobacterially induced protection of watermelon against *Didymella bryoniae*. *J. Appl. Microbiol.*, **2010**, *109*(2), 567–582.

Nongkhlaw, F. M., & Joshi, S. R. Epiphytic and endophytic bacteria that promote growth of ethnomedicinal plants in the subtropical forests of Meghalaya, India. *Rev. Biol. Trop.*, **2014**, *62*(4), 1295–1308.

O'Sullivan, D. J., & O'Gara, F. Traits of fluorescent *Pseudomonas* spp. involved in suppression of plant root pathogens. *Microbiol. Mol. Biol. Rev.*, **1992**, *56*, 662–676.

Okubo, T., Ikeda, S., Kaneko, T., Eda, S., Mitsui, H., Sato, S., Tabata, S., & Minamisawa, K. Nodulation-dependent communities of culturable bacterial endophytes from stems of field-grown soybeans. *Microbes Environ.*, **2009**, *24*(3), 253–258.

Partida-Martínez, L. P., & Heil, M. The microbe-free plant: Fact or artifact? *Front Plant Sci.*, **2011**, *2*, 100, doi: 10.3389/fpls.2011.00100.

Piromyou, P., Greetatorn, T., Teamtisong, K., Okubo, T., Shinoda, R., Nuntakij, A., Tittabutr, P., Boonkerd, N., Minamisawa, K., & Teaumroong, N. Preferential association of endophytic bradyrhizobia with different rice cultivars and its implications for rice endophyte evolution. *Appl. Environ. Microbiol.*, **2015a**, *81*(9), 3049–3061.

Piromyou, P., Songwattana, P., Greetatorn, T., Okubo, T., Kakizaki, K. C., Prakamhang, J., Tittabutr, P., Boonkerd, N., Teaumroong, N., & Minamisawa, K. The type III secretion system (T3SS) is a determinant for rice-endophyte colonization by non-photosynthetic *Bradyrhizobium. Microbes Environ.*, **2015b**, *30*(4), 291–300.

Pliego, C., Kamilova, F., & Lugtenberg, B. Plant growth-promoting bacteria: Fundamentals and exploitation. In: Maheshwari, D. K., (ed.), *Bacteria in Agrobiology: Crop Ecosystems* (pp. 295–343). Springer, Berlin, **2011**.

Posada, F., & Vega, F. E. Establishment of the fungal entomopathogen *Beauveria bassiana* (Ascomycota: Hypocreales) as an endophyte in cocoa seedlings (*Theobroma cacao*). *Mycologia*, **2005**, *97*, 1195–1200.

Pundir, R. K., Rana, S., Kaur, A., Kashyap, N., & Jain, P. Bioprospecting potential of endophytic bacteria isolated from indigenous plants of ambala (Haryana, India). *Int. J. Pharm. Sci. Res.*, **2014**, *5*(6), 2309–2019.

Rajkumar, M., Ae, N., Prasad, M. N., & Freitas, H. Potential of siderophore-producing bacteria for improving heavy metal phytoextraction. *Trends Biotechnol.*, **2010**, *28*(3), 142–149.

Rangjaroen, C., Rerkasem, B., Teaumroong, N., Sungthong, R., & Lumyong, S. Comparative study of endophytic and endophytic diazotrophic bacterial communities across rice landraces grown in the highlands of northern Thailand. *Arch. Microbiol.*, **2014**, *196*(1), 35–49.

Rashid, S., Charles, T. C., & Glick, B. R. Isolation and characterization of new plant growth-promoting bacterial endophytes. *Appl. Soil Ecol.*, **2012**, *61*, 217–224.

Reed, S. C., Cleveland, C. C., & Townsend, A. R. Functional ecology of free-living nitrogen fixation: A contemporary perspective. *Annu. Rev. Ecol. Evol. Syst.*, **2011**, *42*, 489–512.

Reinhold-Hurek, B., & Hurek, T. Living inside plants: Bacterial endophytes. *Curr. Opin. Plant Biol.*, **2011**, *14*(4), 435–443.

Rodríguez, H., & Fraga, R. Phosphate solubilizing bacteria and role in plant growth promotion. *Biotechnol. Adv.*, **1999**, *17*(4/5), 319–339.

Rozahon, M., Ismayil, N., Hamood, B., Erkin, R., Abdurahman, M., Mamtimin, H., Abdukerim, M., Lal, R., & Rahman, E. *Rhizobium populi* sp. nov., an endophytic bacterium isolated from *Populus euphratica. Int. J. Syst. Evol. Microbiol.*, **2014**, *64*, 3215–3221.

Ryan, R. P., Germaine, K., Franks, A., Ryan, D. J., & Dowling, D. N. Bacterial endophytes: Recent developments and applications. *FEMS Microbiol. Lett.*, **2008**, *278*, 1–9.

Santoyo, G., Moreno-Hagelsieb, G., Orozco-Mosqueda, M. C., & Glick, B. R. Plant growth-promoting bacterial endophytes. *Microbiol. Res.*, **2016**, *183*, 92–99.

Schouten, A., Van Der Berg, G., Edel-Hermann, V., Steinberg, C., Gautheron, N., Alabouvette, C., De Vos, C. H., Lemanceau, P., & Raaijmakers, J. M. Defense responses of *Fusarium oxysporum* to 2,4-diacetylphloroglucinol, a broad spectrum antibiotic produced by *Pseudomonas fluorescens. Mol. Plant Microbe Interact.*, **2004**, *17*, 1201–1211.

Sharma, S., Gaur, R. K., & Choudhary, D. K. Phenetic and functional characterization of endophytic root-nodule bacteria isolated from chickpea (*Cicer arietinum* L.) and mothbean (*Vigna aconitifolia* l.) of arid-and semi-arid regions of Rajasthan, India. *Pak. J. Biol. Sci.*, **2012**, *15*(18), 889–894.

Sharma, T., Kaul, S., & Dhar, M. K. Diversity of culturable bacterial endophytes of saffron in Kashmir, India. *SpringerPlus*, **2015**, *4*, 661, https://doi.org/10.1186/s40064–015–1435–3 (Accessed on 7 October 2019).

Shen, H., Li, Z., Han, D., Yang, F., Huang, Q., & Ran, L. Detection of indigenous endophytic bacteria in *Eucalyptus urophylla in vitro* conditions. *Front Agric. China*, **2010**, *4*, 37–41.

Sheng, H. M., Gao, H. S., Xue, L. G., Ding, S., Song, C. L., Feng, H. Y., & An, L. Z. Analysis of the composition and characteristics of culturable endophytic bacteria within subnival plants of the Tianshan Mountains, northwestern China. *Curr. Microbiol.*, **2011**, *62*(3), 923–932.

Sheng, X. F., Xia, J. J., Jiang, C. Y., He, L. Y., & Qian, M. Characterization of heavy metal-resistant endophytic bacteria from rape (*Brassica napus*) roots and their potential in promoting the growth and lead accumulation of rape. *Environ. Pollut.*, **2008**, *156*, 1164–1170.

Sheoran, N., Valiya, N. A., Munjal, V., Kundu, A., Subaharan, K., Venugopal, V., Rajamma, S., Eapen, S. J., & Kumar, A. Genetic analysis of plant endophytic *Pseudomonas putida* BP25 and chemo-profiling of its antimicrobial volatile organic compounds. *Microbiol. Res.*, **2015**, *173*, 66–78.

Shi, Y., Yang, H., Zhang, T., Sun, J., & Lou, K. Illumina-based analysis of endophytic bacterial diversity and space-time dynamics in sugar beet on the north slope of Tianshan mountain. *Appl. Microbiol. Biotechnol.*, **2014**, *98*(14), 6375–6385.

Shiferaw, B., Bantilan, M. C. S., & Serraj, R. Harnessing the potential of BNF for poor farmers: Technological policy and institutional constraints and research need. In: Serraj, R., (ed.), *Symbiotic Nitrogen Fixation; Prospects for Enhanced Application in Tropical Agriculture* (pp. 1–3). Oxford and IBH, New Delhi, **2004**.

Strobel, G. A., & Daisy, B. Bioprospecting for microbial endophytes and their natural products. *Microbiol. Mol. Biol. Rev.*, **2003**, *67*, 491–502.

Sun, L. N., Zhang, Y. F., He, L. Y., Chen, Z. J., Wang, Q. Y., Qian, M., & Sheng, X. F. Genetic diversity and characterization of heavy metal-resistant-endophytic bacteria from two copper-tolerant plant species on copper mine wasteland. *Bioresour. Technol.*, **2010**, *101*, 501–509.

Sun, L., Qiu, F., Zhang, X., Dai, X., Dong, X., & Song, W. Endophytic bacterial diversity in rice (*Oryza sativa* L.) roots estimated by 16S rDNA sequence analysis. *Microb. Ecol.*, **2008**, *55*(3), 415–424.

Tang, S. Y., Hara, S., Melling, L., Goh, K. J., & Hashidoko, Y. *Burkholderia vietnamiensis* isolated from root tissues of Nipa Palm (*Nypa fruticans*) in Sarawak, Malaysia, proved to be its major endophytic nitrogen-fixing bacterium. *Biosci. Biotechnol. Biochem.*, **2010**, *74*(9), 1972–1975.

Thomas, P., & Sekhar, A. C. Cultivation versus molecular analysis of banana (*Musa* sp.) shoot-tip tissue reveals enormous diversity of normally uncultivable endophytic bacteria. *Microb. Ecol.*, **2017**, *73*(4), 885–899.

Tiwari, S., Sarangi, B. K., & Thul, S. T. Identification of arsenic resistant endophytic bacteria from *Pteris vittata* roots and characterization for arsenic remediation application. *J. Environ. Manage.*, **2016**, *180*, 359–365.

Van Loon, L. C., Bakker, P. A. H. M., & Pieterse, C. M. J. Systemic resistance induced by rhizosphere bacteria. *Annu. Rev. Phytopathol.*, **2004**, *36*, 453–483.

Vendan, R. T., Yu, Y. J., Lee, S. H., & Rhee, Y. H. Diversity of endophytic bacteria in ginseng and their potential for plant growth promotion. *J. Microbiol.*, **2010**, *48*(5), 559–565.

Wang, K., Yan, P. S., Ding, Q. L., Wu, Q. X., Wang, Z. B., & Peng, J. Diversity of culturable root-associated/endophytic bacteria and their chitinolytic and aflatoxin inhibition activity of peanut plant in China. *World J. Microbiol. Biotechnol.*, **2013**, *29*(1), 1–10.

Wani, P. A., Khan, M. S., & Zaidi, A. Effect of metal tolerant plant growth promoting *Bradyrhizobium* sp. (*Vigna*) on growth, symbiosis, seed yield and metal uptake by greengram plants. *Chemosphere*, **2007a**, *70*, 36–45.

Wani, P. A., Khan, M. S., & Zaidi, A. Effect of metal-tolerant plant growth-promoting *Rhizobium* on the performance of pea grown in metal-amended soil. *Arch. Environ. Contam. Toxicol.*, **2007b**, *55*, 33–42.

Wilson, D. Endophyte: The evolution of a term, and clarification of its use and definition. *Oikos*, **1995**, *73*(2), 274–276.

Xiong, X. Q., Liao, H. D., Ma, J. S., Liu, X. M., Zhang, L. Y., Shi, X. W., Yang, X. L., Lu, X. N., & Zhu, Y. H. Isolation of a rice endophytic bacterium, *Pantoea* sp. Sd-1, with ligninolytic activity and characterization of its rice straw degradation ability. *Lett. Appl. Microbiol.*, **2014**, *58*(2), 123–129.

Yaish, M. W., Antony, I., & Glick, B. R. Isolation and characterization of endophytic plant growth-promoting bacteria from date palm tree (*Phoenix dactylifera* L.) and their potential role in salinity tolerance. *Antonie Van Leeuwenhoek*, **2015**, *107*(6), 1519–1532.

Yang, M. M., Mavrodi, D. V., Mavrodi, O. V., Bonsall, R. F., Parejko, J. A., Paulitz, T. C., Thomashow, L. S., Yang, H. T., Weller, D. M., & Guo, J. H. Biological control of take-all by fluorescent *Pseudomonas* spp. from Chinese wheat fields. *Phytopathology*, **2011**, *101*(12), 1481–1491.

Yasuda, M., Isawa, T., Shinozaki, S., Minamisawa, K., & Nakashita, H. Effects of colonization of a bacterial endophyte, *Azospirillum* sp. B510, on disease resistance in rice. *Biosci. Biotechnol. Biochem.*, **2009**, *73*(12), 2595–2599.

Yu, J., Zhou, X. F., Yang, S. J., Liu, W. H., & Hu, X. F. Design and application of specific 16S rDNA-targeted primers for assessing endophytic diversity in *Dendrobium officinale* using nested PCR-DGGE. *Appl. Microbiol. Biotechnol.*, **2013**, *97*(22), 9825–9836.

Zahran, H. H. Rhizobia from wild legumes: Diversity, taxonomy, ecology, nitrogen fixation and biotechnology. *J. Biotechnol.*, **2001**, *91*, 143–153.

Zhang, H. W., Song, Y. C., & Tan, R. X. Biology and chemistry of endophytes. *Nat. Prod. Rep.*, **2006**, *23*(5), 753–771.

Zhang, L., Gao, J. S., Kim, S. G., Zhang, C. W., Jiang, J. Q., Ma, X. T., Zhang, J., & Zhang, X. X. *Novosphingobium oryzae* sp. nov., a potential plant-promoting endophytic bacterium isolated from rice roots. *Int. J. Syst. Evol. Microbiol.*, **2016**, *66*(1), 302–307.

Zhang, L., Shi, X., Si, M., Li, C., Zhu, L., Zhao, L., Shen, X., & Wang, Y. *Rhizobium smilacinae* sp. nov., an endophytic bacterium isolated from the leaf of *Smilacina japonica*. *Antonie Van Leeuwenhoek*, **2014a**, *106*(4), 715–723.

Zhang, L., Wei, L., Zhu, L., Li, C., Wang, Y., & Shen, X. *Pseudoxanthomonas gei* sp. nov., a novel endophytic bacterium isolated from the stem of *Geum aleppicum*. *Antonie Van Leeuwenhoek*, **2014b**, *105*(4), 653–661.

Zhang, X. X., Gao, J. S., Cao, Y. H., Sheirdil, R. A., Wang, X. C., & Zhang, L. *Rhizobium oryzicola* sp. nov., potential plant-growth-promoting endophytic bacteria isolated from rice roots. *Int. J. Syst. Evol. Microbiol.*, **2015**, *65*(9), 2931–2936.

Zhang, X., Lin, L., Zhu, Z., Yang, X., Wang, Y., & An, Q. Colonization and modulation of host growth and metal uptake by endophytic bacteria of *Sedum alfredii*. *Int. J. Phytoremediat.*, **2013**, *15*(1), 51–64.

Zhang, Y. F., He, L. Y., Chen, Z. J., Zhang, W. H., Wang, Q. Y., Qian, M., & Sheng, X. F. Characterization of lead-resistant and ACC deaminase-producing endophytic bacteria and their potential in promoting lead accumulation of rape. *J. Hazard Mater.*, **2011**, *186*(2/3), 1720–1725.

Zhao, S., Zhou, N., Zhao, Z. Y., Zhang, K., Wu, G. H., & Tian, C. Y. Isolation of endophytic plant growth-promoting bacteria associated with the halophyte *Salicornia europaea* and evaluation of their promoting activity under salt stress. *Curr. Microbiol.*, **2016**, *73*(4), 574–581.

CHAPTER 4

ECCENTRICITY IN THE BEHAVIOR OF *PENICILLIUM* SPP. AS PHYTOPATHOGEN AND PHYTOAUGMENTOR

DHAVAL PATEL,[1] PRASAD ANDHARE,[2] SUDESHNA MENON,[1] SEBASTIAN VADAKAN,[1] and DWEIPAYAN GOSWAMI[1]

[1]*Department of Biochemistry and Biotechnology, St. Xavier's College (Autonomous), Ahmadabad, Gujarat–380009, India*

[2]*PD Patel Institute of Applied Sciences, Charotar University of Science and Technology, Changa, Gujarat–388421, India*

4.1 INTRODUCTION

A fungus is an associate of the assembly of eukaryotic organisms that includes unicellular microorganisms, for instance, yeasts and molds and multicellular fungi such as mushrooms and bracket fungi. These creatures are classified as a kingdom, 'Fungi.' Fungus is a Latin word of *'fungour'* which mean 'to flourish.' The branch of botany that deals with fungi is called 'mycology.' The scientist who is concerned with fungi is called a mycologist. Few fungi are most noteworthy organisms, in both the terms of their environmental as well as economic roles. Fungi are found in the most diverse habitats. Almost fungi can be found in every available habitat where organic matter (whether its' living or else dead) is existing. The fungus occurs in the soil, which abounds in dead decaying organic material. Many fungi nurture on our foodstuffs such as homemade bread, jams, pickles, vegetable, and even fruits. Some fungi are even present water that we drink and in the air that we respire.

Usually, the fungi are either dormant or they metabolize and nurture very slowly exploiting a series of carbon-based molecules generally organic (Bardgett, 2005). The existence of a living plant penetratingly changes the circumstance for fungi. Plants exude organic matter directly, and they feed arbuscular mycorrhizal fungi (AMF). The fungi distribute organic matter in the surrounding rhizospheric soil. In general, the attentiveness of the microbes is greatest adjacent to the surface of roots (rhizosphere). Dead organic materials are broken down by fungi, and the nutrient cycle of the ecosystem flourishes (Krantz, 2014). Moreover, the vascular plants need symbiotic fungi or other mycorrhizae, which would help them to provide essential nutrients, and in return, the fungi will be inhabited, which would result into the overall growth and development of the plant (Brundrett et al., 2009). Other fungi provide abundant drugs (for instance, penicillin), mushrooms, truffles, and morels that can be used as food, and the fermented bread, beverages are also brewed (champagne, wine, and beer).

Fungi are classified into cluster achlorophyllous (Gupta et al., 2017a), which means they do not have the chlorophyll pigments, which are generally present in plant cells, which are obligatory for carrying out photosynthesis. Fungi, therefore, are not capable of carrying out photosynthesis. Glycogen is the carbohydrate molecule in which fungi store energy, whereas in animals, the same molecule is used to store energy in muscle and liver cells, but in the plant, the energy storage molecule is starch instead of glycogen. Some fungi are heterotrophs, whose mode of the nutrient is through absorption. Saprophytic fungi are those fungi that release enzymes on to the organic material. These enzymes would breakdown dead organic material into chemicals which they can absorb easily and process them as their food source. The other type of fungi is parasitic that directly feed on living things for intaking their nutrients, e.g., trees or human (Athlete's Foot).

Fungi are also one of the sources to cause a number of the plant as well as animal diseases. Ringworm, an athlete's foot along with a few several more solemn diseases are caused by fungi in humans (Soltani et al., 2017). Genetically and chemically fungi resemble similar structures to animals, which is not observed in other organisms, and that makes it difficult to treat them. Rusts, smuts, and leaf rots, root rots, and stem rots are few plant diseases that are caused by fungi, and sometimes may they cause severe damage to crops. However, a few fungi, in specific the yeasts, they are imperative "model organisms" for studying difficulties in genetics and molecular biology (Bonini et al., 2017).

Fungi are simple to plant forms, and include mushrooms, molds, yeasts, and mildews (Patel et al., 2016). Unlike other plants, however, fungi do not have chlorophyll and are not capable of photosynthesis. Fungi have imperative culinary, medical, agricultural, and industrial uses as fungi are used to create dyes, medicines, and biodegradable building materials (Ncube, 2013). Fungi are tremendously advantageous organisms in biotechnology. Fungi construct inimitable complex molecules by means of recognized metabolic pathways. Different taxa produce sets of related molecules, with each different pathway resulting in slightly different final products. Metabolites formed along the metabolic pathway may also be biologically active (Sharma et al., 2016). In addition, the final compounds are often released into the environment. Manipulation of the genome, and environmental conditions during the formation of compounds, enable the optimization of product formation, which can be easily carried out through metabolic engineering (Bideaux et al., 2016).

In 1809 Johann Heinrich Friedrich described the genus *Penicillum* foe the first time in the scientific literature from his work *Observationes in ordines plantarum naturales*, writing "*Penicillium*." "Thallus e floccis caespitosis septatis simplicibus aut ramosis fertilibus erectis apice penicillatis," where penicillatis denoted to the "pencil-like." There are over 300 species revealed till the present day. Few species include *Penicillium albocoremium, Penicillium aurantiogriseum, Penicillium bilaiae, Penicillium camemberti, Penicillium candidum, Penicillium claviforme, Penicillium commune, Penicillium crustosum, Penicillium digitatum, Penicillium echinulatum*, and *Penicillium expansum*.

4.2 INTERACTION OF FUNGI WITH PLANTS AND ITS EFFECT ON AGRICULTURE

India is an agricultural country, and its economy mostly dependent on it. Fungi play both positive and negative roles in agriculture. The harmful activities are more than helpful activities. Some of the saprophytic fungi decompose the dead material of animals and plants in the soil. The enzymes secreted by these fungi convert the larger molecules like fats, carbohydrates, and nitrogen compounds of the dead animals and plants into simpler compounds such as carbon dioxide, hydrogen sulfide, ammonia, water, and some other nutrients that can be easily absorbed by green plants (Gupta et al., 2017b). These simpler compounds will get blend with the soil to form humus, and some few would go into the aerial atmosphere; the rest would be used as the raw material in the synthesis of food. By emancipating carbon dioxide, these fungi participate in sustaining the never-ending cycle of carbon in nature. The carbon dioxide is very significant for green plants in the preparation of food materials by photosynthesis. Some fungi are in symbiotic connotation with the roots of certain plants. Equitable development of the plant can be observed only when the specific fungal partner is present around the rhizosphere of the plant soil (Compant et al., 2016). This type of association of a fungus and plant is called mycorrhiza. Some nematodes are identified to cause severe losses to agricultural crops unswervingly, and some communicate through certain disease-causing viruses. A few fungi (e.g., *Dactylaria*) are capable of terminating the nematodes. These predatory fungi produce mycelial loops. Usually, these loops get tightened up when any nematodes pass through, and hence they are caught. Similar to PGPR (plant growth promoting rhizobacteria), some rhizosphere fungi are able to promote plant growth and its development upon root colonization and are known as 'plant growth promoting fungi' (PGPF) (Hossain et al., 2008). PGPF be a member of genera *Penicillium, Fusarium, Trichoderma*, and *Phoma*. Several species of PGPF have been shown to prompt systemic resistance (SR) counter to various pathogens in plants. PGPF that are soil inhabiting saprophytes, which are non-pathogenic have been testified to be beneficial to numerous crop plants not only by promoting their overall growth but also by playing a defensive role and defend them from diseases. Among these PGPF, roughly a few isolates of *Phoma* sp. and *Penicillium simplicissimum* were extremely effective in persuading SR in contradiction of cucumber anthracnose caused by *Colletotrichum orbiculare*. Besides this property, certain species of *Alternaria, Aspergillus, Cladosporium, Dematium, Gliocladium, Helminthosporium, Humicola*, and *Metarhizium* yield substances like humic substances in soil and hence may be significant in the conservation of soil organic matter (Table 4.1).

Availability of major/micro nutrient and water are facilitated by the extended root-zone of fungi which were formerly not accessible. Fungi facilitate mineralization, especially phosphate, along with the other nutrient that is easily utilized by the plants and overall reduces the use of chemical fertilizer by 35%. Uniform plant growth is observed owing to hydric status and soil aggregation that was formed as an optimistic impact of fungal hyphae. VAM fungi make the soil more porous that helps in deeper root penetration and increased soil microflora. It's because VAM fungi avert soil particles to get compact by releasing chemicals, which reduces soil erosion too. Transient stimulation of host-plant disease resistance is caused by VAM fungi, and the mechanisms have also been reported. They accumulate metal pollutants in the immediate surroundings of then plant roots by scavenging them in the hyphae. As the VAM fungi simplify augment uptake of water and other nutrients, they help the host plant to overcome adversative soil circumstances. As their aptitude to ease better root growth and uptake of nutrients, VAM fungi improve the persistence frequency of transplanted stocks and advance the hardening of tissue culture. VAM fungi intensify root biomass, which is prime necessities for other organisms such as *Rhizobium* spp. and *Trichoderma* spp. resulting in noteworthy improvement in their functionality and use efficiency.

TABLE 4.1 Fungal Species That Interacts With Plants

Fungal Genus	Species	Source	Function
Alternaria	*alternata*	Soil	Leaf spot disease
	eichhorniae	Water	Bioherbicide
	infectoria	Soil	Infect wheat
	raphani	Soil/Water	Unknown
Aspergillus	*acidus*	Soil	Food fermentation
	alliaceus	Soil/Water	Unknown
	clavatus	Soil/Animal manure	Produces patulin toxin
	flavus	Harvesting site	Produces mycotoxin
Cochliobolus	*lunatus*	Soil	Plant pathogen
Chaetomium	*cupreum*	Contaminated soil	Degradation of catechin
	globosum	Soil	Produces cellulase
	truncatulum	Soil/Water	Unknown
Mucor	*hiemalis*	Soil	Plant pathogen
	indicus	Soil/Food particle	Used in production of several valuable products
	racemosus	Soil	Plant pathogen
Penicillium	*bilaiae*	Soil	Agricultural inoculant
	digitatum	Soil/water	Plant pathogen
	echinulatum	Soil	Production of mycophenolic acid
	expansum	Soil	Plant pathogen
	glaucum	Soil	Used in making Gorgonzola cheese
	italicum	Soil	Plant pathogen
Trichoderma	*harzianum*	Soil	Biofertilizer
	longibrachiatum	Soil/Water	Produces xylanase
	ovalisporum	Soil/Water	Biocontrol
	pleurotum	Soil/Water	Unknown
	viride	Soil	PGPF

4.3   INTERACTION OF *PENICILLIUM* WITH PLANTS

Penicillium belongs to the kingdom 'Fungi,' phylum 'Ascomycota,' class 'Eurotiomycetes' order 'Eurotiales,' family 'Trichocomaceae,' genus '*Penicillium*.' *Penicillium* is repeatedly referred to as Deuteromycetes, or Fungi imperfecti. The name *Penicillium* derives from the word "brush," which denotes to the appearance of spores in *Penicillium*. There are over 300 species of *Penicillium*, and *Penicillium chrysogenum* (one of the species) is classified as a psychrotrophic microorganism, which has the best lipase enzyme activity. Moreover, it was also found that among all the other fungi studied in the artic tundra, *Penicillium chrysogenum* showed maximum production of lipase. *Penicillium chrysogenum* has the capability to produce alpha-amylase as it has high enzymatic activity. Secondary metabolites are also produced due to some component that is present in the genetical structure of the fungus. Species of *Penicillium* are omnipresent soil fungi favoring over cool and moderate climates, generally present wherever organic material is accessible. Saprophytic species of *Penicillium* and *Aspergillus* are among the best-recognized representatives of the Eurotiales; besides, they mainly feed on organic decomposable substances. *Penicillium* is filamentous fungi and has split conidiospores. Round conidia are present and are unicellular. Cell walls of *Penicillium* species are mainly composed of Glucans. *Penicillium* species tend to have minor hyphae due to which the protoplasmic movement challenging to perceive. The small hyphae also lead to reduced peripheral growth zones. *Penicillium* spores are capable of getting wet though they have a hydrophobic surface, and this is necessary for germination to occur. *Penicillium* is osmotolerant, making sense that although they nurture better with high water levels, and they are able to bear low water potential (Sharma et al., 2017).

Penicillium species are heterotrophic. The pathogenic species feed off of the natural product they wreck. *Penicillium* produces asexually and is incapable to have sporulation when submerged. However, they begin their reproduction straightforwardly when the hyphae arise into a gas phase. No species exhibit this exact mode of reproduction, and each fungus is classified on the bases of its reproduction. For example, in some species, conidia are borne on phialidies, which assembly out of the conidiophore. In others, the conidiophore bears metulae, where phialides are borne. In other fungi, the conidiophore may branch out before bearing metulae. Branching may or may not be symmetrical, liable on species. Sporulation is not encouraged by changes in oxygen, carbon dioxide, or water loss. Instead, it is associated with the modification in the physical environment at the hyphal surface (Jeevitha et al., 2012).

Penicillium fungi are versatile and opportunistic. They are post-harvest pathogens. *Penicillium* species are one of the well-known causes of fungal decomposition in fruits and vegetables. *P. italicum* and *P. digitatum* are the most communal invaders of citrus fruits, while *P. expansum* is identified to dose apples. *P. digitatum* works by emancipating ethylene to accelerate ripening. It then asylums fruit with green conidia initiating the fruit to shrink and dry out. *P. italicum* causes slimy rot and harvests blue-green conidia. These species prefer cooler temperatures, which clarifies why they are frequently found on foods left too long in the refrigerator. Many species yield mycotoxins; for example, *P. expansum* produces one termed patulin. Most of these species bear a resemblance to each other in color characteristics, decay style along with infection symptoms; they fall under a general group called blue mold. *P. expansum* is one of the most destructive species. These fungi live a long time and are quite hard-wearing, even under contrary conditions. Occasionally, *P. italicum* and *P. expansum* will adhere to each other to create synnemata. Synnemata also occurs in *Penicillum claviforme*. *Penicillium* growth typically occurs as a result of wound infections in the crop. The most common treatment is to use fungicide on the harvested crop. *Penicillium* species attack more than just fruit. For example, *Penicillium verrucosum* breeds on cereal products (Lund et al., 2003).

4.4   IDENTIFICATION OF *PENICILLIUM* STRAIN

A common classification system is used for living organisms to identify species and genus, and in this system, species is the core unit. Taxonomic rank is classified as family, order, class, phylum, and kingdom. So basically, the fungi are classified over 300 species and can be identified on the bases of their morphological characters (visualizing colony) and microscopic characters (under a microscope). The macroscopic observation includes colony diameter, conidium color and reverses mycelium color, degree of sporulation, degree of sulcation, degree of wooliness, and the appearances or nonappearance of exudates. Surface characters include texture velvety to powdery; green, blue-green, gray-green, white, yellow, or pinkish on the surface, and the reverse characters are usually white to yellowish, sometimes red or brown (El-Fadaly et al., 2015). Hyphae septate, hyaline, conidiophores simple or branched. Phialidies grouped in brush-like clusters (penicillin) at the ends of the conidiophores; conidia unicellular, round to ovoid, hyaline or pigmented, rough walled or smooth and in chains, are some of the typical microscopic characters of *Penicillium* spp. and molecular identification is done by isolating DNA of the fungi, purifying it, and going for 18S rRNA gene identification of them.

Mycologists used tools that were available to them at their time for identification, so it's quite obvious that the features used for classifying the fungi have altered over time. Hence, the features that were used once, which were significant at that time seem to be unreliable with the time. The primary classification of fungi was done on macroscopic structures of the growth of fungi. However, certainly by the late 1800s, it had turned out to be clearer that fungi could occasionally show dissimilarity in form, depending on peripheral conditions, and clearly, it is futile to base a classification on any features subject to alteration by peripheral conditions. Hence, to overcome such superior dilemma confidence was shown on microscopic features and was considered for identification, since these were not modified by peripheral factors. Of course, this feature added a little value in identification. Figure 4.1 shows a few morphological characters of *Penicillium* spp. and Figure 4.2 shows the microscopic structures of *Penicillium* spp.

4.5 MOLECULAR IDENTIFICATION OF *PENICILLIUM* SPP. USING 18S rRNA GENE SEQUENCE ANALYSIS

18S rRNA gene identification is generally carried out preliminary for strain identification. Precise primers are designed for diverse fungal groups. However, owing to the massive diversity in fungi, true group- and species-specific recognition can be problematic to achieve by using selective primer-based PCR amplification unaided. A two-step recognition approach, by which a cluster of target DNAs from the amalgamated samples is selectively amplified which is latter followed by additional precise examination through probe hybridization that to a specific target in the PCR amplicon, can provide improved specificity besides sensitivity for environmental sample screening (Fujita et al., 2016). Currently, the product obtains after PCR are blotted on nylon membranes and then are further screened with oligonucleotide probes, each intended to be specific to a genus or a species. This level of detection of compassion is accurate in the rapports of environmental samples. The sequences of the oligonucleotide probe are explicit for quite a few groups of fungi with diverse detection possibilities that have been well-known (Visagie et al., 2014).

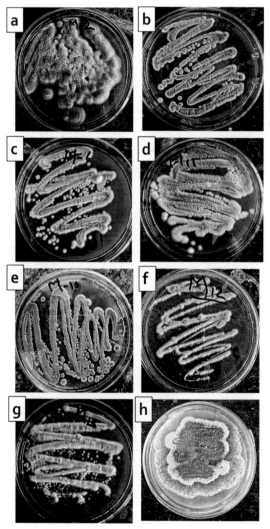

FIGURE 4.1 Morphological features of some (a-g) *Penicillium* spp. and (h) *Trichoderma viride*.

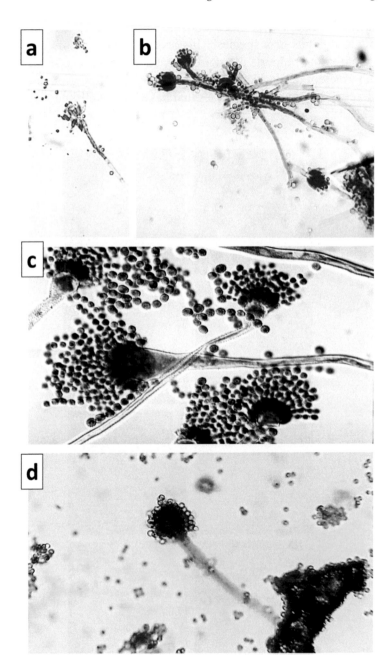

FIGURE 4.2   Microscopic features of some (a-c) *Penicillium* spp. and (d) *Aspergillus flavus.*

PCR amplification of 18S rRNA gene from the purified genomic DNA of any fungal isolate can be carried out by means of the following universal primers (forward primer 5'-GGAAGTAAAAGTCGTAACAAGG-3' and reverse primer 5'-TCCTCCGCTTATTGATATGC-3'). Thermal cycler (PCR machine) conditions involved an initial denaturation step at 95°C for 5 min, trailed by 30 cycles of 94°C for 1 min, 55°C for 1 min and 72°C for 1 min, and a final extension at 72°C for 1 min, followed by holding at 4°C. Amplified gene products are known as an amplicon, and they can be sequenced. Using the sequence data, the BLASTn search program (http://www.ncbi.nlm.nih.gov) can be used to look for nucleotide sequence homology for strain identification (Thakor et al., 2016).

DNA barcoding is an efficient way to identify the eukaryotic organism, by employing a standardized short DNA sequence and a curated locus database connected to convincingly recognized vouchers (Hebert et al., 2003; Blaxter et al., 2005; De Salle et al., 2005; Schoch et al., 2012). After 18S rRNA gene sequencing, this remains the secondary method to identify. For fungi, the internal transcribed spacer (ITS) region of rDNA are recognized as the authorized barcode. ITS is the furthermost extensively sequenced marker for fungi, in addition, the universal primers are also available (Schoch et al., 2012). In *Penicillium*, it works glowingly for categorizing the strains into a species complex or else in one of the 25 sections, besides it every so often delivers a species identification. The International Commission of *Penicillium* and *Aspergillus* (ICPA), in combination with the publication of an updated recognized species list presented below, categorical to include GenBank accession numbers to *Penicillium* as reference barcode sequences for individually species when accessible.

β-tubulin (*BenA*) as the best possibility for a tributary identification marker for *Penicillium* but has a constraint that *BenA* like paralogous genes are also located in *Aspergillus* and *Talaromyces* genes that can be misleadingly amplified (Peterson, 2008; Hubka and Kolarik, 2012; Peterson and Jurjević, 2013). Other than these, Calmodulin (*CaM*) or the second largest subunit of RNA polymerase II (*RPB2*) genes are also used as a tributary identification marker for *Penicillium* spp. Generally, a standard thermal cycle with an annealing temperature of 55°C is rummage-sale. For PCR amplification, amplification accomplishment is low for *CaM*, exclusively in sections *Canescentia* and *Ramosa*. In this circumstance, lowering the annealing temperature to 52°C gives good outcomes. The *RPB2* amplification is more intricate. A touch-up PCR (50–52–55°C) with primer pair

5Feur and 7CReur are recommended for the best amplification. Once amplification is challenging, the substitute touch-up PCR (48–50–52°C) outline can be used and/or another primer pair 5F and 7CR. After amplification, these amplicons are sequenced. The sequences so obtained are compared with the other sequence of the same genes in the databank repository using the BLAST tool on NCBI. Using this technique, one is able to pursuit all sequences in the database, but so far, it has been noted that there are several unidentified and misidentified sequences in the database. From the NCBI homepage (http://www.ncbi.nlm.nih.gov/refseq/), the ITS, the RefSeq data set, is handy, and it is now available to use for the query ITS sequences in contradiction of a verified ITS database (Schoch et al., 2012).

4.6   *PENICILLIUM* AS PLANT PATHOGEN

Major crop infections are caused by fungi secretarial for almost 70% of phyto diseases. Some fungal plant pathogen invades the plant tissues antagonistically, killing the host cells to obtain nutrients is known as necrotrophic (*necros* means death and *trophic* means feeding). Some of these fungi begin an intricate feeding with living host cells, and these fungal pathogens are known as biotrophic (McKeever and Chastagner, 2016). During the last two decades, it has been seen as a cumulative number of contagious infectious diseases in natural populations and accomplished landscapes. In both plants and animals, an extraordinary sum of fungal and fungal-like diseases has recently triggered some of the most spartan die-offs and exterminations ever perceived in wild species and are threatening food security. Human actions are escalating fungal disease spreading by altering natural environments and thus fashioning new opportunities for evolution. We argue that nascent fungal infections will cause collective attrition of biodiversity, with wider allegations for human, ecosystem health, and agriculture.

Penicillium causes a destructive fruit rot of citrus. Early symptoms include a soft water-soaked area on the peel, followed by the expansion of a circular colony of white mold, up to 4 cm diameter after 24–36 hours at 24°C. Green asexual spores (conidia) form at the epicenter of the colony, surrounded by a broad band of white mycelium. The lesion spreads more rapidly. The fruit rapidly spoils and collapses or in lower humidities shrinks and mummifies. Grey mold, blue mold, and green mold caused by *Botrytis cinerea, Penicillium italicum, Penicillium digitatum,* respectively, are common postharvest diseases of fruits and vegetables (Gatto et al., 2011).

4.6.1 PENICILLIUM ROT

Penicillium rot can be perceived in the field, but most often, it develops after harvest and can result in crop losses of up to 90%. It develops well at relatively high temperatures during storage, but it can continue to grow, although more slowly, even at temperatures close to freezing. More than a few species of *Penicillium* can be the source for causing blue mold. These fungi are common saprophytes on plant debris and senescent plant tissue (Naik et al., 2017). The incursion of pomegranate fruit can arise through wounds or bruises, but colonization habitually occurs on the surface of senescent fruit. The mycelium cultivates inside the fruit through the connective tissue and arils at progressive stages. High relative humidity and moderate temperatures of 70 to 77°F (21–25°C) favors the optimum growth. *Penicillium* rot or blue mold is one of the most communal and straightforwardly predictable post-harvest rots of apple but is not inevitably responsible for large losses. Its implication has amplified in current years since it produces a mycotoxin, *patulin,* which transpires in *Penicillium*-rotted fruit and afterward in fruit juice produced from discarded fruit (Oteiza et al., 2017). *Penicillium vermoesenii* is the contributory agent of a disease of palms, generally termed pink bud rot (Lopez-Llorca et al., 1994). The fungus has been observed in several species of palms (Hodel, 1985; Aragaki et al., 1991), normally causing necrosis and sunken lesions on new leaves. However, in moist conditions, the fungus sporulates heavily, forming pink pulverulent masses. If the infection reaches the central bud, it may even cause the death of the palm (Bliss, 1938; Hodel, 1985). *Penicillium oxalicum* grew on stubs left after fruit pick, leaf, and side-shoot trimming, which cause the stem softening and sneering (O'Neill et al., 1991). Brown ring tissue of the main stem is the general spot for the blue-green mold of *Penicillium* spp. and sporulation is perceived. Healthy fruits are generally resistance to infection for *P. oxalicum* but when the fruit is damaged, it becomes an aggressive pathogen and rot developing more speedily at 20°C and above than at 15°C (Hu et al., 2017).

The life cycle and epidemiology take in airborne or waterborne spores attacking fruit through wounds, bruises, or cracks anywhere they find a place on the surface of the fruit and is often the secondary trespasser of other rots. Though there are fallen fruits in the orchard under the trees, *Penicillium* rot is seldomly seen. Because of these, there are no predicting approaches that are technologically advanced or monitoring system applicable to it. The fungus is pervasive, and contamination will always occur if the fruit is spoiled or not handled appropriately. *P. implicatum* is an anamorph species that cause post-harvest rot on pomegranate fruit. All apple varieties are vulnerable, but it is most frequently perceived on Bramley in store. The fungus grounds a pale green to dark brown circular soft rot, which binges quickly over the fruit exterior and into the flesh, creating a sharp crossing point between the healthy and rotted tissue, such that the rot can be bundled out. Mature cuts are roofed in brilliant white furuncles, which swiftly turn blue. *P. expansum* subsists on desiccated fruit, or fruit jiffs stuck on loose bins or lying around storage or pack-house areas. Furthermost, wound infections in storage are consequences due to water-borne spores in post-harvest drench solutions (e.g., anti-scald agents) or in water cascades used to grade fruit (Bafort et al., 2017). *Penicillium digitatum* a phytopathogen causes *Penicillium* rot on papaya (Borrás and Aguilar, 1990; Bautista-Baños et al., 2003).

4.6.2   BLUE AND GREEN MOLD

Blue mold or bluish-green mold are generally belonging to *Penicillium,* the very same mold from which the penicillin antibiotic is prepared. Normally the blue colored mold grows on food, but it can be also found on the household belongings that have been damaged by water like wall-paper, insulation, and carpeting. Along with these household materials, blue mold can also be found on furnishings like couch cushions and mattresses that have formerly suffered water damage. Health problems can be caused by inhaling the spores from the blue colored mold, which include allergic reactions, inflammation of the lungs, and sinus infections. People with circumstances like asthma and emphysema are more prone to mold-related complaints, as are the very young and the very old, but anybody can be exaggerated (Tzortzakis et al., 2017).

You can get rid of mold from non-porous surfaces like glass, metal, bath-tubs, toilets, and tile floors with an antimicrobial cleanser (Foster 40–80). It's typically not likely to eliminate mold entirely from porous surfaces, though, such as wood, ceiling tiles, drywall, insulation, and carpeting. Those materials will prerequisite to be indifferent from the home-very cautiously, in order to prevent scattering of mold to other areas during the removal process-and substituted with noninfected materials. You should put on an N-95 respirator mask while washing up mold in order to defend yourself from breath in any microscopic mold spores, which can later cause health problems (Nolte et al., 2002).

Two species of *Penicillium* are mainly responsible for causing mold, i.e., *Penicillium digitatum,* which causes the green mold and *P. italicum* causes blue mold. In these diseases softening of damaged tissue occurs. After softening white fungal growth results, which progressively turns blue or green as spores develop. Postharvest fungicides can arrest spore development resulting in white only fungal growth (Eckert et al., 1985). Initially, the infections can occur through an open area and then results into the developing of damaged areas. The progress of mold proliferations is directly related with storage temperatures (up to an optimum of 27°C). Late season fruit more susceptible to have an infection, and even damaged rind is more susceptible (Nguyen et al., 2017).

Careful handling of fruits can reduce the damage to the rind. Good hygiene conditions and sorting reduces spore load and infection rates in the fruits. Regular sanitation helps in destroying the spores in recirculating water and packing line equipment. Postharvest fungicides applied within 24 hours of harvest can also reduce the chances. And even low-tempera-tures storage can slow down the growth of fungi (Harkema et al., 2017).

4.7   OVERCOMING *PENICILLIUM* INFECTIONS IN PLANTS

As we all know that safety concerns first in all fields. As one of them is pertaining towards fungi and the mycotoxins contamination in agriculture has been an issue of key apprehension. Today, with widely available reports and updated databases on fungal occurrence and mycotoxins contamination, we can overcome problems related with them. Farmers need to control the plant diseases in order to preserve the superiority and richness of food, fiber, and feed. Plant diseases can be prevented by different tactics that can even control that. Though there are decent agronomic and horticultural practices, farmers mostly rely heavily on pesticides and chemical fertilizer. Such involvements to agriculture have contributed expressively to the remarkable enhancements in crop productivity and value over the past decade (Miflin, 2000). There are lot of environmental pollution that have been caused by the disproportionate use and mismanagement of agrochemicals, which has led to significant changes in people's attitudes in the direction of the practice of pesticides in agriculture. At the moment, there are firm regulations on chemical pesticide usage, and there is political pressure to eliminate the most hazardous chemicals from the marketplace (Reddy et al., 2009). Moreover, the binge of plant diseases in natural ecosystems may exclude efficacious usage of chemicals, for the reason that the scale to which such practice might have to be functional. Subsequently, some pest management researchers have engrossed their efforts on evolving alternative inputs to artificial chemicals for controlling pests besides diseases. These substitutes are referred as biological controls. An assortment of biological controls is accessible for practice, nevertheless further progress and effective espousal are obligatory to realize greater thoughtful of the complex communications among plants, people, and the environment (Cadotte et al., 2017).

Biological control agents (BCA) are the organisms that overwhelm the pathogen or pest. Even the same term can be used for the natural product that might be extracted (or sometime even fermented) from various sources that can inhibit the pathogen or pest (Yousuf et al., 2016).

These preparations may be very simple combinations of natural components with specific activities or multifaceted assortments with multiple effects on the target pest or pathogen as well as on the host. Suppression of plant infection can be overcome in numerous ways, if growers' activities care made applicable, rotations, and planting of disease-resistant cultivars (naturally selected or genetically engineered) can be practiced. As biological control results from many diverse types of interactions among organisms, researchers have engrossed on characterizing how the mechanisms are operated. For instance, *Pseudomonas* is well-known to yield the antibiotic 2,4-diacetylphloroglucinol (DAPG) may also persuade host defenses (Iavicoli et al., 2003). Furthermore, DAPG-producers can rapidly colonize the roots, a trait that can additionally contribute their capability to overwhelm pathogen commotion in the rhizosphere of wheat through rivalry of organic nutrients (Weller et al., 2002).

Sunflower seeds (*Helianthus annuus* L.) are cultivated extensively for oil production, as the seeds are known to have rich polyunsaturated fatty-acids (65% linoleic acid) that contain low level of saturated fats and along with that sunflower seeds are also a decent source of dietary fiber, minerals, and vitamin E and it has also been testified to have harbor mold caused by *Penicillium chrysogenum* (Pozzi et al., 2005). Preharvest applications of thiophanate methyl (TM) controlled postharvest green mold consistently. Lessening of chemical pesticide usage including chemicals that control the soil-borne plant pathogens is extensively rummage-sale in agriculture. Biocontrol microbes can be applied to the seeds directly or to the soil former to planting possibly will take possession of the spermosphere along with or without rhizosphere of saplings and thus may thwart surrounding infection of soil-borne pathogens. Therefore, biocontrol agents may contribute in various ways of trophic and non-trophic interface mechanisms, which consist of production of antifungal compounds, restless parasitism of pathogens, the incentive of host plant defenses, or else competitive colonization of spermosphere and in addition rhizosphere substrates (Weller, 1998).

4.8   *PENICILLIUM* SPP. AS PHYTOAUGMENTOR

Rhizospheric fungi have the ability to stimulate plant growth are designated as 'PGPF (Hyakumachi, 1994). PGPF is nonpathogenic soil-inhabiting saprophytes, which have been reported for growth promotion in several crop plants and providing protection against diseases (Shivanna et al., 1996). Such PGPF belongs to various genera, including *Fusarium*, *Penicillium*, *Phoma*, and *Trichoderma*. Few species of PGPF have been stated to prompt the SR in contrast to numerous phytopathogens (Shoresh et al., 2005). *Hypocrea rufa* is a common inhabitant of the rhizosphere and decisively recognized as a bio-control agent of soil-borne plant pathogens (Harman et al., 2004). A praiseworthy amount of research has been focused on the mycoparasitic nature of *H. rufa* and its contribution in plant overall growth has been promoted. *H. rufa* controls the plant-pathogenic fungi by three different mechanisms, antibiosis, competition, and mycoparasitism (John et al., 2010). The multifaceted development of mycoparasitism necessitates the production of plenty of cell-wall digesting enzymes, for instance, chitinases, polysaccharide lyases, cellulases, and proteases in addition lipases which degrade the cell wall of fungi.

4.8.1   *PENICILLIUM BILAIAE*

Penicillium bilaiae is a species of inborn soil fungus that can be cast-off as a PGPF. R. Kucey was the first to identify that the organic acids are excreted by the microorganism, and then it assisted to solubilize soil-bound phosphate. The organism can provide feedstuff on plant waste foodstuffs and even can augment phosphate uptake by the root structure and establish symbiosis with several plant species. Inborn soil populations are often squat and can be amplified by using them as an agricultural inoculant. *Penicillium bilaiae* is a fungal microorganism applied to several crop species to enhance soil-bound phosphorous uptake (Kucey, 1983). It is also used to promote soil-bound phosphorous uptake in several crop species such as wheat, canola, and pulse crops. Phosphate-solubilizing soil and rhizosphere microorganisms have been distinguished by their relative abilities to dissolve calcium phosphate and apatite in pure culture and in connotation through plant roots (Gaur et al., 1973). *P. bilaiae* can also produce citric acid and it also helps to break down of $CaPO_4$. The use of efficient phosphate-solubilizing microorganisms (PSM) opens up a new horizon for better crop productivity besides sustaining soil health.

4.8.2   PENICILLIUM MENONORUM

Penicillium menonorum is a mono-verticillate, non-vesiculate species of the genus of *Penicillium* which was isolated from rhizosphere soil in Korea. The fungal isolate was found to exhibit plant growth-promoting activity through indole acetic acid (IAA) and siderophore production, as well as P solubilization. It produced 9.7 mg/L IAA and solubilized 408 mg of Ca_3PO_4/L. It also increased the dry biomass of cucumber roots by 57% and shoots by 52% in an experiment carried out. In the further experiment, it was also observed that the chlorophyll content was increased by 16%, starch by 45%, protein by 22%, and phosphate contents by 14%. This fungus also produces various useful enzymes like dehydrogenase and acid phosphatase. And it has increased its activities by 30% and 19% respectively. These results validate that the isolate has potential PGP attributes, and therefore, it can be well-thought-out as a new fungus to enhance soil fertility and promote plant growth.

4.8.3   PENICILLIUM SIMPLICISSIMUM

Penicillium simplicissimum is an anamorph species that belong to the genus of *Penicillium* that can be used for plant growth promotion. This species is present on food; besides, its primary habitation is in putrefying vegetation. *Penicillium simplicissimum* was evaluated for its ability to induce resistance against Cucumber mosaic virus (CMV) in *Arabidopsis thaliana* and tobacco plants. *P. simplicissimum* has the characteristic to degrade polyethylene as a sole carbon source.

4.8.4 PENICILLIUM CITRINUM

Penicillium citrinum is an anamorph, mesophilic fungus species that fit into genus of *Penicillium*, which produces tanzawaic acid and helps in inhibiting bacterial conjugation (Lopatkin et al., 2016). *Penicillium citrinum* is often found on moldy citrus fruits and occasionally it occurs in tropical spices and cereals. *Penicillium citrinum* is well-known for production of mycotoxin citrinin and cellulose digesting enzymes such as cellulose, endoglucanase, as well as xylulase. Gibberellins producing ability of this fungus is also been found. Numerous studies disclose that *Penicillum* spp. reinforced plant growth and development by its phosphate solubilization. Pandey et al. (2008) in their research displayed that eight species of *Penicillum* with the ability in making clear zone by *in vitro* airing. Many researches also testified the ability and effectiveness of *Penicillium* as bio-fertilizer and can be castoff in field (Hakim et al., 2017).

4.9 CONCLUSION

Despite strains of *Penicillium* being known as phytopathogen, several of its strains have shown promises to support sustainable agriculture acting as phytoaugmenter. Researchers have begun to develop a much more in complexity and detailed understanding of the *Penicillium* strains to facilitate their potentials as plant growth promotors. Despite of such lengthy research over the period of several decades, a lot more work, both basic and applied, remains to be done to unfold some hidden potentials of *Penicillium* strains, which may not be known yet. Focusing on the commercial market of fungi as bio-fertilizers, a lot of hard work is still to be done. Despite of several potential strains of *Trichoderma* been discovered by researchers, deeper efforts are not being employed to explore other strains of fungi as biofertilizers. The green revolution has intensified agriculture, and there is tremendous potential for biofertilizer and biocontrol agents in the agricultural industry. With only a few strains being commercialized, all belonging to the same genus '*Trichoderma*,' it is inevitable to introduce new strains that can be used in developing more efficient biofertilizers and biocontrol agents, and for this purpose strains of *Penicillium* have to be overlooked more like a friend than foe.

KEYWORDS

- 18S rRNA gene sequence analysis
- arbuscular mycorrhizal fungi
- biocontrol agent
- biofertilizer
- biopesticide
- blue mold
- calmodulin
- *Fusarium* spp.
- green mold
- leaf spot disease
- molecular identification
- mycophenolic acid
- *Penicillium* spp.
- *Penicillium bilaiae*
- *Penicillium citrinum*
- *Penicillium menonorum*
- *Penicillium* rot
- *Penicillium simplicissimum*
- *Phoma* spp.
- phytoaugmentor
- phytopathogen
- plant growth promoting fungi
- symbiotic fungi
- *Trichoderma* spp.
- β-tubulin

REFERENCES

Aragaki, M., Broschat, T. K., Chase, A. R., Ohr, H. D., Simone, G. W., & Uchida, J. *Diseases and Disorders of Ornamental Palms* (pp. 24–25). APS Press, St Paul, USA, **1991**.

Bafort, F., Parisi, O., Perraudin, J. P., & Jijakli, M. H. The lactoperoxidase system: A natural biochemical biocontrol agent for pre- and postharvest applications. *Journal of Phytopathology*, **2017**, *165*(1), 22–34.

Bardgett, R. *The Biology of Soil: A Community and Ecosystem Approach* (pp. 1–242). Oxford University Press, Oxford, UK, **2005**.

Bautista-Baños, S., Hernández-López, M., Bosquez-Molina, E., & Wilson, C. L. Effects of chitosan and plant extracts on growth of *Colletotrichum gloeosporioides*, anthracnose levels and quality of papaya fruit. *Crop Protection*, **2003**, *22*(9), 1087–1092.

Bideaux, C., Montheard, J., Cameleyre, X., Molina-Jouve, C., & Alfenore, S. Metabolic flux analysis model for optimizing xylose conversion into ethanol by the natural C5-fermenting yeast *Candida shehatae*. *Applied Microbiology and Biotechnology*, 2016, *100*(3), 1489–1499.

Blaxter, M., Mann, J., Chapman, T., Thomas, F., Whitton, C., Floyd, R., & Abebe, E. Defining operational taxonomic units using DNA barcode data. *Philosophical Transactions of the Royal Society B: Biological Sciences*, **2005**, *360*(1462), 1935–1943.

Bliss, D. E. The *Penicillium* disease of ornamental palms. *Fifth Western Shade Tree Conference* (pp. 20–27). University of California, Riverside, California, **1938**.

Bonini, N. M., & Berger, S. L. The sustained impact of model organisms-in genetics and epigenetics. *Genetics*, 2017, *205*(1), 1–4.

Borrás, A. D., & Aguilar, R. V. Biological control of *Penicillium digitatum* by *Trichoderma viride* on postharvest citrus fruits. *International Journal of Food Microbiology*, **1990**, *11*(2), 179–183.

Brundrett, M. C. Mycorrhizal associations and other means of nutrition of vascular plants: Understanding the global diversity of host plants by resolving conflicting information and developing reliable means of diagnosis. *Plant and Soil*, 2009, *320*(1/2), 37–77.

Cadotte, M. W., Barlow, J., Nuñez, M. A., Pettorelli, N., & Stephens, P. A. Solving environmental problems in the Anthropocene: The need to bring novel theoretical advances into the applied ecology fold. *Journal of Applied Ecology*, **2017**, *54*(1), 1–6.

Compant, S., Saikkonen, K., Mitter, B., Campisano, A., & Mercado-Blanco, J. Soil, plants and endophytes. *Plant and Soil*, **2016**, *405*(1/2), 1–11.

De Salle, R., Egan, M. G., & Siddall, M. The unholy trinity: taxonomy, species delimitation and DNA barcoding. *Philosophical Transactions of the Royal Society B: Biological Sciences*, **2005**, *360*(1462), 1905–1916.

Eckert, J. W., & Ogawa, J. M. The chemical control of postharvest diseases: Subtropical and tropical fruits. *Annual Review of Phytopathology*, **1985**, *23*(1), 421–454.

El-Fadaly, H. M., El-Kadi, S. M., Hamad, M. N., & Habib, A. A. Isolation and identification of Egyptian ras cheese (romy) contaminating fungi during ripening period. *Journal of Microbiology Research*, **2015**, *5*(1), 1–10.

Fujita, H., Kataoka, Y., Tobita, S., Kuwahara, M., & Sugimoto, N. Novel one-tube-one-step real-time methodology for rapid transcriptomic biomarker detection: Signal amplification by ternary initiation complexes. *Analytical Chemistry*, **2016**, *88*(14), 7137–7144.

Gatto, M. A., Ippolito, A., Linsalata, V., Cascarano, N. A., Nigro, F., Vanadia, S., Di Venere, D. Activity of extracts from wild edible herbs against postharvest fungal diseases of fruit and vegetables. *Postharvest Biol. Technol.*, **2011**, *61*(1), 72–82.

Gaur, A. C., Medan, M., & Ostwal, K. P. Solubilization of phosphatic compounds by native microflora of rock phosphates. *Indian J. Exp. Biol.*, **1973**, *11*, 427–429.

Gupta, A., Gupta, R., & Singh, R. L. Microbes and environment. In: Singh, R. L., (ed.), *Principles and Applications of Environmental Biotechnology for a Sustainable Future* (pp. 43–84). Springer, Singapore, **2017**.

Gupta, S., Rautela, P., Maharana, C., & Singh, K. P. Priming host defense against biotic stress by arbuscular mycorrhizal fungi. In: Singh, J. S., & Seneviratne, G., (eds.), *Agro-Environmental Sustainability*, *Managing Crop Health* (Vol. 1, pp. 255–270). Springer International Publishing, Basel, Switzerland, **2017**.

Hakim, S. S., Yuwati, T. W., & Nurulita, S. Isolation of peat swamp forest foliar endophyte fungi as biofertilizer. *Journal of Wetlands Environmental Management*, **2017**, *5*(1), 10–17.

Harkema, H., Paillart, M., Lukasse, L., Westra, E., & Hogeveen, E. *Transport and Storage of Cut Roses: Endless Possibilities?: Guide of Practice for Sea Freight of Cut Roses Developed within Green Change Project* (pp. 1–48). Wageningen Food and Biobased Research, Netherlands, **2017**.

Harman, G. E., Howell, C. R., Viterbo, A., Chet, I., & Lorito, M. *Trichoderma* species-opportunistic, avirulent plant symbionts. *Nature Reviews Microbiology*, **2004**, *2*(1), 43–56.

Hebert, P. D., Ratnasingham, S., & De Waard, J. R. Barcoding animal life: Cytochrome C oxidase subunit 1 divergences among closely related species. *Proceedings of the Royal Society of London B: Biological Sciences*, **2003**, *270*(S1), S96–S99.

Hodel, D. R. *Gliocladium* and *Fusarium* diseases of palms. *Principles*, **1985**, *29*, 85–88.

Hossain, M. M., Sultana, F., Kubota, M., & Hyakumachi, M. Differential inducible defense mechanisms against bacterial speck pathogen in *Arabidopsis thaliana* by plant-growth-promoting-fungus *Penicillium* sp. GP16–2 and its cell free filtrate. *Plant and Soil*, **2008**, *304*(1/2), 227–239.

Hu, Z., Xu, C., McDowell, N. G., Johnson, D. J., Wang, M., Luo, Y., Zhou, X., & Huang, Z. Linking microbial community composition to C loss rates during wood decomposition. *Soil Biology and Biochemistry*, **2017**, *104*, 108–116.

Hubka, V., & Kolarik M. β-tubulin paralogue tubC is frequently misidentified as the benA gene in *Aspergillus* section Nigri taxonomy: primer specificity testing and taxonomic consequences. *Persoonia-Molecular Phylogeny and Evolution of Fungi*, **2012**, *29*(1), 1–10.

Hyakumachi, M. Plant growth promoting fungi from turfgrass rhizosphere with potential for disease suppression. *Soil Microorganism*, **1994**, *44*, 53–68.

Iavicoli, A., Boutet, E., Buchala, A., & Métraux, J. P. Induced systemic resistance in *Arabidopsis thaliana* in response to root inoculation with *Pseudomonas fluorescens* CHA0. *Molecular Plant-Microbe Interactions*, **2003**, *16*(10), 851–858.

Jeevitha, M., & Sumathy, J. H. V. Biosorption of chromium by *Penicillium*. *International Journal of Environmental Sciences*, **2012**, *3*(1), 55–61.

John, R. P., Tyagi, R. D., Prevost, D., Brar, S. K., Pouleur, S., & Surampalli, R. Y. Mycoparasitic *Trichoderma viride* as a biocontrol agent against *Fusarium oxysporum* f. sp. *adzuki* and *Pythium arrhenomanes* and as a growth promoter of soybean. *Crop Protection*, **2010**, *29*, 1452–1459.

Krantz, L. The nutrient cycle. In: Wallander, H., (ed.), *Soil* (pp. 79–100). Springer International Publishing, Basel, Switzerland, 2014.

Kucey, R. M. N. Phosphate-solubilizing bacteria and fungi in various cultivated and virgin Alberta soils. *Canadian Journal of Soil Science*, **1983**, *63*, 671–678.

Lopatkin, A. J., Sysoeva, T. A., & You, L. Dissecting the effects of antibiotics on horizontal gene transfer: Analysis suggests a critical role of selection dynamics. *BioEssays*, **2016**, *38*(12), 1283–1292.

Lopez-Llorca, L. V., & Orts, S. Histopathology of infection of the palm *Washingtonia filifera* with the pink bud rot fungus *Penicillium vermoesenii*. *Mycological Research*, **1994**, *98*(10), 1195–1199.

Lund, F., & Frisvad, J. C. *Penicillium verrucosum* in wheat and barley indicates presence of ochratoxin A. *Journal of Applied Microbiology*, **2003**, *95*(5), 1117–1123.

Mc Keever, K. M., & Chastagner, G. A. A survey of *Phytophthora* species associated with *Abies* in U.S. Christmas tree farms. *Plant Disease*, **2016**, *100*(6), 1161–1169.

Miflin, B. Crop improvement in the 21st century. *Journal of Experimental Botany*, **2000**, *51* (342), 1–8.

Naik, M. K., Chennappa, G., Amaresh, Y. S., Sudha, S., Chowdappa, P., & Patil, S. Characterization of phyto-toxin producing *Alternaria* species isolated from sesame leaves and their toxicity. *Indian Journal of Experimental Biology*, **2017**, *55*, 36–43.

Ncube, T. *Development of a Fungal Cellulolytic Enzyme Combination for Use in Bioethanol Production Using Hyparrhenia* spp. *as a Source of Fermentable Sugars.* PhD Thesis, University of Limpopo (Turfloop Campus), Limpopo, South Africa, **2013**.

Nguyen, P. A., Strub, C., Fontana, A., & Schorr-Galindo, S. Crop molds and mycotoxins: Alternative management using biocontrol. *Biological Control*, **2017**, *104*, 10–27.

Nolte, K. B., Taylor, D. G., & Richmond, J. Y. Biosafety considerations for autopsy. *The American Journal of Forensic Medicine and Pathology*, **2002**, *23*(2), 107–122.

O'Neill, T. M., Bagabe, M., & Ann, D. M. Aspects of biology and control of a stem rot of cucumber caused by *Penicillium oxalicum*. *Plant Pathology*, **1991**, *40*(1), 78–84.

Oteiza, J. M., Khaneghah, A. M., Campagnollo, F. B., Granato, D., Mahmoudi, M. R., Santana, A. S., & Gianuzzi, L. Influence of production on the presence of patulin and ochratoxin A in fruit juices and wines of Argentina. *LWT-Food Science and Technology*, **2017**, *80*, 200–207.

Pandey, A., Namrata, D., Bhavesh, K., Rinu, K., & Pankaj, T. Phosphate solubilization by *Penicillium* spp. isolated from soil samples of Indian Himalayan region. *World Journal of Microbiology and Biotechnology*, **2008**, *24*(1), 97–102.

Patel, T. K., & Williamson, J. D. Mannitol in plants, fungi, and plant–fungal interactions. *Trends Plant Sci.*, **2016**, *21*(6), 486–497.

Peterson, S. W. Phylogenetic analysis of *Aspergillus* species using DNA sequences from four loci. *Mycologia*, **2008**, *100*(2), 205–226.

Peterson, S. W., & Jurjević, Ž. *Talaromyces columbinus* sp. nov., and genealogical concordance analysis in *Talaromyces* clade 2a. *PLoS One*, **2013**, *8*(10), e78084, doi: 10.1371/journal.pone.0078084.

Pozzi, C. R., Braghini, R., Arcaro, J. R., Zorzete, P., Israel, A. L., & Pozar, I. O. Mycoflora and occurrence of alternariol and alternariol monomethyl ether in Brazilian sunflower from sowing to harvest. *Journal of Agricultural and Food Chemistry*, **2005**, *53*(14), 5824–5828.

Reddy, K. R. N., Reddy, C. S., & Muralidharan, K. Potential of botanicals and biocontrol agents on growth and aflatoxin production by *Aspergillus flavus* infecting rice grains. *Food Control*, **2009**, *20*(2), 173–178.

Schoch, C. L., Seifert, K. A., Huhndorf, S., Robert, V., Spouge, J. L., Levesque, C. A., & Chen, W. Nuclear ribosomal internal transcribed spacer (ITS) region as a universal DNA barcode marker for fungi. *Proc. Natl. Acad. Sci. USA*, **2012**, *109*(16), 6241–6246.

Sharma, M., & Sharma, R. Drugs and drug intermediates from fungi: Striving for greener processes. *Critical Reviews in Microbiology*, **2016**, *42*(2), 322–338.

Sharma, V., Misra, S., & Srivastava, A. K. Developing a green and sustainable process for enhanced PHB production by *Azohydromonas australica*. *Biocatalysis and Agricultural Biotechnology*, **2017**, *10*, 122–129.

Shivanna, M. B., Meera, M. S., & Hyakumachi, M. Role of root colonization ability of plant growth promoting fungi in the suppression of take-all and common root rot of wheat. *Crop Protection*, **1996**, *15*, 497–504.

Shoresh, M., Yedida, I., & Chat, I. Involvement of jasmonic acid/ethylene signaling pathway in the systemic resistance induced in cucumber by *Trichoderma asperellum* T203. *Phytopathology*, **2005**, *95*, 76–84.

Soltani, E. K., Cerezuela, R., Charef, N., Mezaache-Aichour, S., Esteban, M. A., & Zerroug, M. M. Algerian propolis extracts: Chemical composition, bactericidal activity and *in vitro* effects on gilthead seabream innate immune responses. *Fish and Shellfish Immunology*, **2017**, *62*, 57–67.

Thakor, P., Goswami, D., Thakker, J., & Dhandhukia, P. Idiosyncrasy of local fungal isolate *Hypocrea rufa* strain P2: Plant growth promotion and mycoparasitism. *Journal of Microbiology, Biotechnology and Food Sciences*, **2016**, *5*(6), 593–598.

Tzortzakis, N., & Chrysargyris, A. Postharvest ozone application for the preservation of fruits and vegetables. *Food Reviews International*, **2017**, *33*(3), 270–315.

Visagie, C. M., Houbraken, J., Frisvad, J. C., Hong, S. B., Klaassen, C. H. W., Perrone, G., & Samson, R. A. Identification and nomenclature of the genus *Penicillium*. *Studies in Mycology*, **2014**, *78*, 343–371.

Weller, D. M. Biological control of soil borne plant pathogens in the rhizosphere with bacteria. *Annual Review of Phytopathology* **1998**, *26*(1), 379–407.

Weller, D. M., Raaijmakers, J. M., Gardener, B. B. M., & Thomashow, L. S. Microbial populations responsible for specific soil suppressiveness to plant pathogens. *Annual Review of Phytopathology*, **2002**, *40*(1), 309–348.

Yousuf, B., Gul, K., Wani, A. A., & Singh, P. Health benefits of anthocyanins and their encapsulation for potential use in food systems: A review. *Critical Reviews in Food Science and Nutrition*, **2016**, *56*(13), 2223–2230.

CHAPTER 5

ENVIRONMENTAL MANAGEMENT OF E-WASTE BY BIOLOGICAL PROCESS

APARNA GUNJAL,[1] MEGHMALA WAGHMODE,[2] NEHA PATIL,[3] and NEELU NAWANI[4]

[1]Department of Microbiology, Savitribai Phule Pune University, Pune–411007, Maharashtra, India

[2]Department of Microbiology, Annasaheb Magar Mahavidyalaya, Hadapsar, Pune–411028, Maharashtra, India

[3]Department of Microbiology, Waghire College, Saswad, Pune–412301, Maharashtra, India

[4]Dr D.Y. Patil Vidyapeeth's, Dr. D. Y. Patil Biotechnology and Bioinformatics Institute, Tathawade, Pune–411033, Maharashtra, India

5.1 INTRODUCTION

E-waste management can be defined as all the measures taken to protect humans and the environment from the toxic effects of constituents of electronic and other wastes. Active measures have to be taken to diminish its effects on health and the surroundings and capable of recovering valuable resources from them (Ogbuene et al., 2013). Electronic waste (e-waste) includes used electronic products that necessitate recycling or other proper forms of removal (Mohammed et al., 2013). Global production of e-waste has been reported to be 40 million tons per annum (Wath et al., 2010) and comprises 1–2% of the total solid waste. E-waste comprises remotes, compact discs, headphones, batteries, LCD, Plasma TVs, air conditioners, computer, and its accessories like monitors, printers, keyboards, central processing units; typewriters, mobile phones, chargers, refrigerators, and other household appliances. It contains more than 1000 different substances, which can be divided as 'hazardous' or 'nonhazardous.' Mostly, it consists of metals like copper (Cu), aluminum (Al) and expensive metals, viz., silver (Ag), gold (Au), platinum (Pt), and palladium (Pd). E-waste also found to contain miscellaneous substances (Bhat et al., 2012). E-waste contains approximately 20–50% heavy metals, which form a major part of the inorganic fraction. Every year, during the production of mobile phones and personal computers, 15% of cobalt, 13% of the palladium, and 3% of gold and silver are mined (Schluep et al., 2009). E-wastes are considered hazardous due to toxic materials in certain components. Discarded electronic equipments, if improperly disposed-off, can leach lead and many other harmful substances into the soil and groundwater. Leached heavy metals generated due to the wrong disposal of e-waste hamper the human health and ecosystem (Mohan and Bhamawat, 2008).

5.2   CLASSIFICATION OF E-WASTE

Domestic appliances (ovens, refrigerators, toasters, vacuum cleaners) are referred to as "white goods" which also contain printed circuit boards (PCBs), electronic components, connectors (containing copper, gold, and other conductive elements), other plastics, and silicon (in integrated circuits). Communication systems (PCs, phones, faxes)–computer monitors contain Cathode Ray Tubes. Lead, which is toxic metal, is present in computer monitors and TVs. Entertainment electronics (TV, CD players)–even the mode of entertainments such as TVs, heavy metal like lead is present. Light system (fluorescent tubes)–fluorescent tubes contain mercury and become hazardous wastes when they no longer work. During disposal of mercury-containing lamps or tubes, mercury can escape to the environment, which through the lungs can enter into the bloodstream. People who are involved in the handling of this waste are especially at risk. Mercury has been reported to affect the health of pregnant women and small children. E-tools (drilling machines)–if not properly handled are dangerous. Leisure equipment (electronic toys)–according to the report of Perez-Belis et al. (2013), 1014.25 kg of waste toys components and automatic issuing systems (ticket issuing machines) contain 31.83% electric and electronic fraction.

5.3   GLOBAL SCENARIO OF E-WASTE

In developed countries, 1% of solid waste comprises e-waste, whereas in developing countries, it ranges from 0.01% to 1% (Anonymous, 2006). Developed countries, namely the United States of America, the United Kingdom, Germany, Japan, and New Zealand, have modern processing techniques for recycling of e-waste. The Union Miniere Company in Belgium (Hageluken, 2006) and Boliden Mineral in Sweden (Theo, 1999) have recycling plants to process e-waste; while China (Tong et al., 2015), Taiwan (Lee et al., 2004) and South Korea (Lee et al., 2007) have implemented systems to recycle metals from e-waste; but in India no such measures are yet initiated. Each year, 20 to 50 million tons of electrical and electronic equipment wastes are discarded worldwide, while 12 million tons of electrical and electronic equipment wastes are discarded in Asian countries (Brigden, 2005). The share of the budding economies of China and India consumption of computers, in particular, is likely to increase, surpassing 178 and 80 million in the case of China and India, respectively (Anonymous, 2006).

E-waste generated in developed countries like the USA gets exported for recycling in developing countries. In 2003, approximately 23,000 metric tons of e-waste were exported from the UK to India, Africa, and China (Perkins et al., 2014). Significant quantities of e-waste like cell phone chargers, laptop computers, air-conditioners, printers, cameras, and other electronic refuse are in China. Developed countries like the USA, Europe, and Japan export e-waste to Hong Kong for dumping. Taiwan also needs adequate e-waste management facilities. Thailand generates a huge amount of e-waste from mobile phones and batteries, but relevant data on disposal of junk cell phones and batteries are not available.

In countries like Africa, there are about 62 million TVs, 200 million radios and about 7 million PCs. About 400,000 used PCs are imported through Lagos in Nigeria each month, and 1.2 to 1.5 million computers are made available in the market of South Africa each year. Universal Recycling Company, Desco Electronic Recyclers, and African Sky are three main recycling companies for e-waste in South Africa. Desco company is handling about 400 tons of PC boards and 2000 tons of e-waste, while Universal Recycling processes about 1800 tons e-waste in a year. Recycled e-waste is then exported to Asia and Europe by South Africa (Mohan and Bhamawat, 2008).

In the United Kingdom, the framework for legislation on the disposal of e-waste is the EU Waste Electrical and Electronic Equipment (WEEE) Directive. After the collection of WEEE, sorting depot separates its component parts and raw materials without incineration. Monitors are repaired or resale by an organization in Germany. From irreparable monitors, glass, plastics, PCBs, and other components are recycled (Mohan and Bhamawat, 2008).

5.4   INDIAN SCENARIO OF E-WASTE

In India, 146180 tons of electronic waste is generated per year. The e-waste problem in India seems to be compounding due to rapid changes taking place not only in computers and cell phones but also in domestic appliances. Environmental and health problems are the main issues due to uncontrolled burning, disassembly, and disposal, counting occupational safety and health hazards among those directly involved, due to unscientific methods for the processing of the waste. About 25,000 workers are working at a scrapyard in Delhi alone. Other e-waste scrap-yards exist in Meerut, Ferozabad, Bangalore, Chennai, and Mumbai. Most of the Municipal Corporation's destroy e-waste by burning in the open air along with garbage (Rao et al., 2014). The amount of e-waste generated in the top 9 states of India is shown in Table 5.1.

TABLE 5.1   E-Waste in Top 10 States in India

State	E-Waste (Tons)
Maharashtra	20270
Tamil Nadu	13486
Andhra Pradesh	12780
Uttar Pradesh	10381
West Bengal	10059
Karnataka	9118
Gujarat	8994
Madhya Pradesh	7800
Punjab	6958

Source: Jayapradha, 2015.

In India, each year, 10–20 thousand tons of e-waste is handled in Delhi alone, with 25% of this being the computers. Total WEEE waste generation in Maharashtra is 20270 tons, 646.48 tons, and 11017.06 tons from Navi Mumbai and Mumbai counterparts, 2584.21 tons, and 1032.37 tons from Pune and Pimpri-Chinchwad respectively. Mumbai and Pune fall under the top ten cities that are generating the highest quantities of e-waste, and Mumbai is the first in e-waste generation among all the cities of India (Borthakur and Sinha, 2013). Many developed nations, viz., USA, UK, and Germany, use India as a dumping ground for e-waste.

5.5 DISPOSAL METHODS OF E-WASTE

The disposal of e-waste is a major problem suffered in many regions across the globe and is of primary concern due to the toxicity and carcinogenicity of incompletely processed substances. The sources of e-waste are comparatively high-priced, and basically, long-lasting products applied for data processing, telecommunications, or entertainment in private households and businesses. In India, handling of computer scrap is done using diverse approaches in management alternatives such as product reuse, usual disposal in landfills, and incineration or open-air burning. The recycling of computer waste can be done using efficient and advanced processing technology, which are capital intensive, entails high-end effective skills and training of the processing personnel. Incineration of e-waste in open-air affects the environment.

5.5.1 PRODUCT REUSE

Refurbishing used computers and other electronic goods for reuse, apart from the customary trend of passing on the same to relatives and friends, is a general societal practice. Educational institutes or charitable institutions can accept old computers for their use. Such deemed damaging practice adopted for product reuse contributes significantly to the rapidly increasing burden of computer waste.

5.5.2 CONVENTIONAL DISPOSAL IN LANDFILLS

Dumping of the waste is at landfill sites where it may remain for an uncertain period. According to the report of the Environmental Protection

Agency (EPA), more than 3.2 million tons of e-waste was dumped in US landfills in 1997 (Anonymous, 2004). The tremendously low biodegradable properties of plastic components in computers get further compounded in dry conditions, which complements landfills, and in strictly regulated landfill sites, the degradation rate is found to be slower. The highly toxic constituents found in computers contribute to metal leaching, which causes large-scale soil and groundwater pollution. In landfills, the multi-layered configuration of computer waste becomes a congregation of plastic and steel casings, circuit boards, glass tubes, wires, and other assorted parts and materials. Electronic discards contribute about 70% of heavy metals (including mercury and cadmium) found in landfills (Tachi et al., 2001).

5.5.3   INCINERATION OR OPEN-AIR BURNING

After manual separation of components, motherboards are applied to open-pit burning for the extraction of the thin layer of copper foils laminated on the circuit board. After charring, the material is distilled through a simple froth floating process. Ash is removed, and the copper with some carbon impurity are processed in the recycling stage. Worthless, defective integrated circuit chips and condensers are burned in small enclosures with chimneys for the extraction of metals (Anonymous, 2003).

5.6   E-WASTE TREATMENT BY THERMAL PLASMA TECHNIQUE

There is a report on e-waste (cell phone) treatment by thermal plasma technique where the particularly high temperature in an oxygen-starved environment was used for the complete decomposition of input plastic waste into syngas (fuel gas). The generation of hydrogen gas of high concentration is from the hydrogen of the plastic part of the cell phone waste (Ruji and Chang, 2013). Plasma treatment of cell phone waste in a reduced environment produces gaseous components, viz. hydrogen (H_2), carbon monoxide (CO), and hydrocarbons. It reduces the quantity of the e-waste from which recovery of precious metals is possible. E-waste treatment by thermal plasma technique decreases the requirement of the landfill and is one of the safe disposal techniques for cell phone waste. The principal advantage of plasma technology includes the low exhaust gas flow rates, installations with smaller footprints and low investment costs, and faster start-up and shut-down times.

5.7   MANAGEMENT OF E-WASTE

Non-renewable nature of the metallic elements necessitates their recovery from the e-waste stream. To reuse e-waste in an eco-friendly, efficient, and proper way is a need of the hour by recycling of valuable metals from e-waste stream. The e-waste management is mostly done by bioleaching. This method employs leaching of metals from e-waste using microorganisms. In microbiological leaching, the natural tendency of microorganisms is used for the transformation of the metals present in the waste in a solid form to a dissolved form. Among the chief groups of bacteria, the most frequently used bacteria for e-waste management are *Acidithiobacillus ferrooxidans*, *Acidithiobacillus thiooxidans*, *Leptospirillum ferrooxidans* and heterotrophs, for example, *Sulfolobus* sp. Fungi such as *Penicillium* sp. and *Aspergillus niger* are used in the leaching of metals from industrial wastes.

There is a report on the bioleaching of metals (Cu, Ni, and Zn) by *Acidiphilium acidophilum* using computer PCBs. *A. acidophilum* has been reported to leach Cu and Zn from shredded PCBs (Hudec et al., 2009). Bioleaching is the method where precious metals are recovered using microorganisms. PCBs are the main component of most e-waste, and some metals in PCBs are hazardous. If not disposed-off properly, these metals can leach into the soil and water and leak into watersheds. According to the literature, metal extraction from e-waste and PCBs has been studied using mesophilic chemolithotrophic bacteria (*Acidithiobacillus thioxidans* and *Acidithiobacillus ferroxidan*) (Choi et al., 2004), cyanogenic bacteria (*Chromobacterium violaceum* and *Pseudomonas fluorescens*) (Ting et al., 2008) and acidophilic moderately thermophilic bacteria (*Sulfobacillus thermosulfidooxidans* and *Thermoplasma acidophilum*) (Ilyas et al., 2010). Study has been done on metal bioleaching from e-waste by *Chromobacterium violaceum* and *Pseudomonas* sp. (Pradhan and Kumar, 2012).

5.8   PHYTOREMEDIATION

Phytoremediation involves the use of vegetation or plant for removal of toxicity from soil and water. After disposal of e-waste in landfills, heavy metals and persistent organic pollutants are found in the leachate. E-waste generated leachate can be decreased by using aquatic weed water hycianth (*Eichhornia crassipes*) (Omondi et al., 2015). Microorganisms colonizing

with plants have a major contribution in removal of polychlorinated biphenyls from e-waste leachate in landfills (Chen et al., 2010). Plant species
widely used for the phytoremediation include *Arabidopsis* sp., *Alyssum
bertolonii*, *Amaranthus retroflexus*, *Chenopodium album*, *Brassica juncea*
and *Helianthus annuus* (Rajiv et al., 2009).

5.9   VERMIREMEDIATION

Landfilling of e-waste leads to a decrease in the nutrient levels, which is
the primary requisite of the microorganisms for their activity. Microbial
activity can be obtained through the activity of earthworms. Earthworms
are responsible for the soil aeration and casting of nutrients. Earthworms
accumulate chemicals and cause their biotransformation to less harmful
products (Sinha et al., 2010). Some of the earthworm species used for
vermiremediation are *Aporrectodea tuberculata*, *Eisenia fetida*, *Lumbricus
terrestris*, *Dendrobaena rubida*, and *Eiseniella tetraedra*.

5.10   RECOVERY OF METALS FROM E-WASTE

The metal fractions recovered from e-waste are further processed using
hydrometallurgical, pyrometallurgical, electrometallurgical, biometallurgical processes, and their combinations. The major methods for processing
e-waste are hydrometallurgical and pyrometallurgical processes, which
are followed by electrometallurgical/electrochemical processes (e.g.,
electrorefining or electrowinning) for the separation and recovery of
selected metal. At present, there are limited laboratory studies for e-waste
processing through biometallurgical routes like bioleaching of metals
from e-waste. The pretreatment of e-waste is not always mandatory for
pyrometallurgical routes. For example, mobile phones and MP3 players
can be treated directly using smelting processes. Preprocessing is required
to segregate metal fractions from other fractions for hydrometallurgical
routes (Khaliq et al., 2014).

5.10.1   PYROMETALLURGICAL PROCESS

The pyrometallurgical processing has been a conventional technology for the extraction of precious metals from e-waste. The smelter

processes and recycles 100,000 tonnes of electronics per annum, representing 14% of total throughput. In the reactor, materials are kept in a molten metal bath (1250°C), and churned by a mixture of 39% oxygen. Combustion of plastics and flammable materials in the feeding reduces energy cost. Iron, lead, and zinc are converted into oxides in the oxidation zone where it becomes fixed in a silica-based slag. Before disposal, the slag is cooled and milled to recover metals. The precious metal like copper is separated and transferred to the converter. Liquid blister copper is upgraded in the converters and then refined in anode furnaces, which will cast into anodes with the purity of 99.1%. The residual 0.9% contains valuable metals like gold, silver, platinum, and palladium, as well as recoverable metals, such as tellurium, selenium, and nickel (Cui and Zhang, 2008).

5.10.2 HYDROMETALLURGICAL PROCESS

The hydrometallurgical process involves a series of acid or caustic leaching of solid stuff. The solutions are then subjected for isolation and purification procedures like impurities precipitation, adsorption, solvent extraction, and ion-exchange to concentrate the metals. Concentrated metals are then treated by chemical reduction, electrorefining process, or crystallization for the recovery of metals. The patents available on hydrometallurgical processing of ores are shown in Table 5.2. The hydrometallurgical recovery of valuable metals from e-waste is shown in Table 5.3.

TABLE 5.2   Patents on Biohydrometallurgical Processing of Ores

Ores	Metals Extracted	Microbes	References
Iron	Fe	*Pseudomonas* sp.	Hoffmann et al., 1989
Gold	Au	*Chromobacterium violoceum, Chlorella vulgaris*	Kleid et al., 1990
Sulfidic ore	Au, Ag, Cu	*Thiobacillus* sp., *Leptospirillum ferrooxidans*	Hill, 1992
Sulfidic ore	Au, Ag, Pt	Bacteria reducing sulfate and hydrogen	Hunter et al., 1996

TABLE 5.3   Hydrometallurgical Process for Extraction of Metals

Agent for Leaching	Process Conditions	Metals Extracted
H_2SO_4, chloride, thiourea, and cyanide leaching	Leaching and metal extraction using processes viz., cementation, precipitation, ion exchange, and carbon adsorption	Au, Ag, Pd, and Cu
HCl, $MgCl_2$, H_2SO_4, and H_2O_2	Dissolution of electronic wastes in solvents and leaching and metal extraction	Al, Sn, Pb, and Zn (stage 1); Cu and Ni (stage 2) and Au, Ag, Pd, and Pt (last stage)
HCl, H_2SO_4, and $NaClO_3$	Burning e-waste at high temperature, i.e., 400–500°C and then leaching	Ag, Au, and Pd
Aqua regia and H_2SO_4	Mechanical treatment and dissolution of e-waste in solvents	Cu

Source: Khaliq et al., 2014.

5.10.2.1   CYANIDE LEACHING

Mining industries use cyanide as leaching lixiviant for gold extraction because of its high efficiency and low cost (Syed, 2012). The mechanism involved in the dissolution of gold in cyanide solution is an electrochemical process. The overall reactions are (Dorin and Woods, 1991):

$$4Au + 8CN \rightarrow 4Au\,(CN)_2^- + 4e \qquad (1)$$

$$O_2 + 2H_2O + 4e \rightarrow 4OH^- \qquad (2)$$

5.10.2.2   HALIDE LEACHING

All halogens have been tested for gold extraction, except fluorine and astatine (Hilson and Monhemius, 2006). With chloride, bromide, and iodide, gold forms Au(I) and Au(III) complexes. Aqua regia is used as a traditional medium for dissolving gold. Aqua regia is a mixture of hydrochloric acid and concentrated nitric acid in 1:3 proportions. The reactions involved in the aqua regia leaching are as follows (Sheng and Etsell, 2007):

$$2HNO_3 + 6HCl \rightarrow 2NO + 4H_2O + 3Cl_2 \qquad (3)$$

$$2Au + 11HCl + 3HNO_3 \rightarrow 2HAuCl_4 + 3NOCl + 6H_2O \qquad (4)$$

5.10.2.3   THIOUREA LEACHING

Thiourea can dissolve gold under acidic conditions and lead to the formation of a cationic complex. A rapid reaction occurs, which extracts gold up to 99% (Hilson and Monhemius, 2006). The following reactions occur:

$$Au + 2CS(NH_2)_2 \rightarrow Au(CS(NH_2)_2)_2^+ + e \qquad (5)$$

5.10.2.4   THIOSULFATE LEACHING

An electrochemical reaction for the dissolution of gold in an ammonical thiosulfate solution is catalyzed in the presence of cupric ions. On the anodic surface of gold, ammonia or thiosulfate ions react with Au^+ ions and form either $Au(NH_3)_2^+$ or $Au(S_2O_3)_2^{3-}$. $Cu(NH_3)_2^+$ converts to $Cu(S_2O_3)_3^{5-}$ ions, and the same is for $Au(NH_3)_2^+$. The $Cu(S_2O_3)_3^{5-}$ and $Cu(NH_3)_2^+$ species in solution form oxidized product, i.e., $Cu(NH_3)_4^{2+}$. It is reported that at higher temperatures and low pH values, thiosulfate stability is low.

5.11   BIOMETALLURGY PROCESS

Microbes utilize metal species for structural and catalytic functions. Microbes can bind with the metal ions present in the external environment at the cell surface or to transport them into the cell for various intracellular functions. Bioleaching and biosorption are the main areas of biometallurgy to recover the metals.

5.12   BIOLEACHING PROCESS

The acidophilic microbes which take part in the dissolution of metals from the e-waste grow in an inorganic medium having low pH values and can tolerate high metal ion concentrations. The first mechanism is redoxolysis (direct and indirect). The second mechanism of metal solubilization is by the formation of organic or inorganic acids. For example, production of citric acid or gluconic acid by *A. niger* and *P. simplicissimum*, and H_2SO_4 by *T. ferrooxidans* and *T. thiooxidans*. The acid supplies the protons which contribute to the solubilization process. The third mechanism of metal extraction involves complex formation between metabolites produced by the microorganisms and the metal ions, which can increase their mobility. Bioaccumulation is another important mechanism in fungal bioleaching (Burgstaller and Schinner, 1993). Oxidation of Fe (II) to Fe (III) and S to H_2SO_4 are the main functions of the acidophilic microorganisms. These acidophilic microorganisms are *Acidithiobacillus*, *Sulfolobus*, *Acidianus*, and *Leptospirillium*, which oxidizes Fe (II) to Fe (III) and S to H_2SO_4 (Okibe et al., 2003). Bioleaching has been successfully applied for the extraction of valuable metals and copper from ores (Ehrlich, 1997). Bioleaching is bacteria assisted reaction for the extraction of metals from metallic sulfides (Morin et al., 2006). In the case of Cu:$Fe_2(SO_4)_3$ created by *Acidithiobacillus ferrooxidans* oxidizes elemental copper contained in the waste to the copper in the form of ion, according to the reactions:

$$Cu + Fe_2(SO_4)_3 \rightarrow Cu^{2+} + 2Fe^{2+} + 3SO_4^{2-} \tag{6}$$

$$2FeSO_4 + H_2SO_4 + 0.5\,O_2 + bacteria \rightarrow Fe_2(SO_4)_3 + H_2O \tag{7}$$

The e-waste containing base metals subjected to bioleaching is represented in Table 5.4. Bacterial metal sulfide oxidation involves both direct and indirect mechanisms. In direct mechanism, oxidation of minerals and solubilization of metals is achieved by microorganisms, whereas, in the indirect mechanism, ferric ion (Fe^{3+}) is the oxidizing agent for minerals and microorganisms regenerate Fe^{3+} from Fe^{2+}. The reactions are,

$$MS + 2Fe^{3+} \rightarrow M^{2+} + 2Fe^{2+} + S^0 \tag{8}$$

$$2Fe + 0.5O_2 + 2H^+ \rightarrow 2Fe^{3+} + H_2O \tag{9}$$

TABLE 5.4 E-Waste Subjected for Bioleaching

Metals Extracted	Microbes
Cu, Sn, Ni, Pb, Zn, Al	*A. ferrooxidans, A. thiooxidans, A. niger, P. simplicissimum*
Cu, Zn, Al, Ni	*Sulfobacilllus thermosulfidooxidans + Thermoplasma acidophilum*
Cu, Pb, Zn	*A. ferrooxidans, A. thiooxidans* and mixture
Cu	*A. ferrooxidans, A. ferrooxidans + A. thiooxidans*
Au	*Chromobacterium violaceum*

Source: Willner et al., 2015.

Thiosulfate and polysulfide pathways are operated in the leaching of metal sulfides. Proton attack and oxidation processes are involved in the dissolution of metal sulfides.

5.13 BIOSORPTION

Biosorption is the physico-chemical interaction between the charged surface groups of microorganisms and ions in solution. Physico-chemical mechanisms like ion-exchange, coordination, complexation, and chelation between metal ions and ligands. This interaction depends on the specific properties of the biomass (alive, or dead, or as a derived product). Other metal-removal mechanisms include metal precipitation, sequestration by metal-binding proteins or siderophores, transport, and internal compartmentalization (White et al., 1995). Biosorption of metals from solutions includes chemical and physical sorption mechanisms. Chemical sorption mechanisms involve the complexation, chelation, microprecipitation, and microbial reduction, and physical sorption mechanisms involve electrostatic forces and ion exchange (Kuyucak and Volesky, 1988).

E-waste is considered as a secondary ore for the recovery of gold, and cyanidation is widely used for gold recovery. In current years, the gold cyanidation process is carried out by using cyanogenic bacteria such as *Chromobacterium violaceum*, *Pseudomonas fluorescens*, *Escherichia coli*, and *Pseudomonas aeruginosa*. All these species are involved in the gold dissolution with their metabolic processes. The mechanism of gold recovery involves assemblage of the chemical knowledge (interaction of metals and cyanide) with microbiological principles (biological cyanide formation) for solubilization of metals from waste PCBs and the formation of water-soluble cyanide complexes.

5.14 CONCLUSION

The extraction of precious metals like copper, gold, silver, and palladium is the incentive for e-waste recycling. Cyanogenic bacteria, in combination with the hydrometallurgy process, will be useful in the extraction of metals from electronic waste and thus will help in the management of e-waste. Pollution of soil due to the dumping of e-waste in landfills can be prevented through the synergistic action of plants, earthworms, and their associated microorganisms. Compared to biosorption and bioaccumulation, bioleaching will be most effective in the reuse and recycling of e-waste.

KEYWORDS

- bioaccumulation
- bioleaching
- biometallurgy
- biosorption
- chelation
- cyanogenic
- E-waste
- extraction
- hydrometallurgical
- incineration
- landfill
- phytoremediation
- plasma technique
- pollution
- pyrometallurgical
- scrap
- solubilization
- toxicity
- vermiremediation

REFERENCES

Anonymous. *Poison PCS and Toxic TVs*. Silicon Valley Toxics Coalition, San Francisco, **2004**, http://svtc.org/wp-content/uploads/ppc-ttv1.pdf (Accessed on 7 October 2019).

Anonymous. *Quarterly Newsletter of Envis Centre*. Pondicherry Pollution Control Committee, Pondicherry, **2006**.

Anonymous. *Scrapping the Hi-Tech Myth: Computer Waste in India* (pp. 6–57). Toxics Link, New Delhi, India, **2003**.

Anonymous. *The E-Waste Problem*. Greenpeace International, Amsterdam, Netherlands, **2006**, www.greenpeace.org/archive-international/en/campaigns/detox/electronics/the-e-waste-problem/ (Accessed on 7 October 2019).

Bhat, V., Rao, P., & Patil, Y. Development of an integrated model to recover precious metals from electronic scrap–a novel strategy for e-waste management. *Procedia–Social and Behavioral Science*, **2012**, *37*, 397–406.

Borthakur, A., & Sinha, A. Generation of electronic waste in India: Current scenario, dilemmas and stakeholders. *Afr. J. Environ. Sci. Technol.*, **2013**, *7*(9), 899–910.

Brigden, K., Labunska, I., Santillo, D., & Allsopp, M. *Recycling of Electronic Wastes in China and India: Workplace and Environmental Contamination*. Greenpeace International,

Amsterdam, Netherlands, **2005**. www.greenpeace.org/india/Global/india/report/**2005**/8/recycling-of-electronic-wastes.pdf (Accessed on 7 October 2019).

Burgstaller, W., & Schinner, F. Leaching of metals with fungi. *J. Biotechnol.*, **1993**, *27*, 91–116.

Chen, Y., Tang, X., Cheema, A., Liu, W., & Shen, C. Cyclodextrin enhanced phytoremediation of aged PCBs–contaminated soil from e-waste recycling area. *J. Environ. Monit.*, **2010**, *12*(7), 1482–1489.

Choi, M., Cho, K., Kim, D. S., & Kim, D. J. Microbial recovery of copper from printed circuit boards of waste computer by *Acidithiobacillus*. *J. Environ. Sci. Health. Part, A., Environ. Sci. Eng. Toxic Hazard Subst. Control*, **2004**, *39*(11/12), 2973–2982.

Cui, J., & Zhang, L. Metallurgical recovery of metals from electronic waste: A review. *J. Hazard Mater.*, **2008**, *158*(2/3), 228–256.

Dorin, R., & Woods, R. Determination of leaching rates of precious metals by electrochemical techniques. *J. Appl. Electrochem.*, **1991**, *21*(5), 419–424.

Ehrlich, H. Microbes and metals. *J. Appl. Microbiol. Biotechnol.*, **1997**, *48*(6), 687–692.

Hageluken, C. Recycling of electronic scrap at Umicore's integrated metals smelter and refinery. *Erzmetall*, **2006**, *59*(3), 152–161.

Hill, D. L., & Brierley, J. A. Biooxdiation process for recovery of metal values from sulfur-containing ore minerals. *Eur. Pat. Appl.*, **1992**, 0 522 978 A1.

Hilson, G., & Monhemius, A. Alternatives to cyanide in the gold mining industry: What prospects for the future. *J. Cleaner Prod.*, **2006**, *14*(12/13), 1158–1167.

Hoffmann, M., Arnold, R., & Stephanopoulos, G. Microbial reduction of iron ore. *US Patent*, 4880740, **1989**, http://electronicrecyclers.com/ewaste-defined.aspx (Accessed on 7 October 2019).

Hudec, R., Sodhi, M., & Goglia-Arora, D. Biorecovery of metals from electronic waste. *7th Latin American and Caribbean Conference for Engineering and Technology*, San Cristobal, Venezuela, **2009**.

Hunter, R., Stewart, F., Darsow, T., & Fogelsong, M. Method and apparatus for extracting precious metals from their ores and the product thereof. *Int. Pat. Appl.*, WO 96/00308, **1996**.

Ilyas, S., Ruan, C., Bhatti, N., Ghauri, M., & Anwar, M. Column bioleaching of metals from electronic scrap. *Hydrometallurgy, 101*(3/4), 135–140.

Jayapradha, A. Scenario of e-waste in India and application of new recycling approaches for e-waste management. *J. Chem. Pharmaceut. Res.*, **2015**, *7*(3), 232–238.

Khaliq, A., Rhamdhani, M., Brooks, G., & Masood, S. Metal extraction processes for electronic waste and existing industrial routes: A review and Australian perspective. *Resources*, **2014**, *3*(1), 152–179.

Kleid, D., Kohr, W., & Thibodeau, F. Processes to recover and reconcentrate gold from its ores. *Eur. Pat. Appl.*, 0 432 935 A1, **1990**.

Kuyucak, N., & Volesky, B. Biosorbents for recovery of metals from industrial solutions. *J. Biotechnol. Lett.*, **1988**, *10*(2), 137–142.

Lee, C., Song, T., & Yoo, M. Present status of the recycling of waste electrical and electronic equipment in Korea. *Resour. Conserv. Recycl.*, **2007**, *50*(4), 380–397.

Lee, H., Chang, T., Fan, S., & Chang, C. An overview of recycling and treatment of scrap computers. *J. Hazard Mater.*, **2004**, *114*(1–3), 93–100.

Mohammed, A., Oyeleke, B., Ndamitso, M., Achife, E., & Badeggi, M. The impact of electronic waste disposal and possible microbial and plant control. *The Inter J. Eng. Sci.*, **2013**, *2*(10), 35–42.

Mohan, D., & Bhamawat, P. E-waste management–global scenario: A review. *J. Environ. Res. Develop.*, **2008**, *2*(4), 817–823.

Morin, D., Lips, A., Pinches, T., Huisman, J., Frias, C., Norberg, A., & Forssberg, E. BioMinE–integrated project for the development of biotechnology for metal-bearing materials in Europe. *Hydrometallurgy*, **2006**, *83*(1–4), 69–76.

Ogbuene, B., Igwebuike, H., & Agusiegbe, M. The impact of open solid waste dump sites on soil quality: A case study of Ugwuajiln Enugu. *British J. Adv. Acad. Res.*, **2013**, *2*, 43–53.

Okibe, N., Gericke, M., Hallberg, K., & Johnson, D. Enumeration and characterization of acidophilic microorganisms isolated from a pilot plant stirred-tank bioleaching operation. *Appl. Environ. Microbiol.*, **2003**, *69*(4), 1036–1043.

Omondi, E., Ndiba, P., & Njuru, P. Phytoremediation of polychlorobiphenyls (PCB's) in landfill e-Waste leachate with Water Hyacinth (*E. crassipes*). *Intern. J. Scient. Technol. Res.*, **2015**, *4*(5), 147–156.

Perez-Belis, V., Bovea, M., & Gómez, A. Waste electric and electronic toys: Management practices and characterization. *Resour. Conserv. Recycl.*, **2013**, *77*, 1–12.

Perkins, D., Brune, M., Nxele, T., & Sly, P. E-waste: A global hazard. *Annals of Global Health*, **2014**, *80*(4), 286–295.

Pradhan, J., & Kumar, S. Metals bioleaching from electronic waste by *Chromobacterium violaceum* and *Pseudomonads* sp. *Waste Manage Res.*, **2012**, *30*(11), 1151–1159.

Rajiv, K., Dalsukh, V., Shanu, S., Shweta, S., & Sunil, H. *Bioremediation of Contaminated Sites: A Low-Cost Nature's Biotechnology for Environmental Clean Up by Versatile Microbes, Plants and Earthworms* (pp. 1–73). Nova Science Publishers, USA, **2009**.

Rao, N., Shaik, F., & Kamalakar, D. E-waste supervision–need of the hour. *J. Chem. Biol. Phy. Sci.*, **2014**, *4*, 3809–3818.

Ruji, B., & Chang, S. E-waste (cellphone) treatment by thermal plasma technique. *International Symposium on Feedstock Recycling of Polymeric Materials*. New Delhi, India, **2013**.

Schluep, M., Hageluken, C., Kuehr, R., & Wang, F. *Sustainable Innovation and Technology Transfer: Industrial Sector Studies; Recycling–from E-Waste to Resources*. United Nations Environment Programme, Nairobi, Kenya, **2009**.

Sheng, P., & Etsell, T. Recovery of gold from computer circuit board scrap using aqua regia. *Waste Manage Res.*, **2007**, *25*(4), 380–383.

Sinha, K., Heart, S., & Valani, D. Earthworms–the environmental engineers: Review of vermiculture technologies for environmental management and resource development. *Inter J. Global Environ Issues*, **2010**, *10*, 265–292.

Suzuki, I. Microbial leaching of metals from sulfide minerals. *Biotechnol. Adv.*, **2001**, *19*(2), 119–132.

Syed, S. Recovery of gold from secondary sources–a review. *Hydrometallurgy*, **2012**, *115 & 116*, 30–51.

Tachi, K., Cate, G., Steve, C., & Bill, S. *Computers, E-Waste, and Product Stewardship: Is California Ready for the Challenge?–A Menu of Policy Options for Computer Extended Product Responsibility*. Global Futures Foundation, San Francisco, USA, **2001**, http://infohouse.p2ric.org/ref/41/40164.htm (Accessed on 7 October 2019).

Theo, L. Integrated recycling of non-ferrous metals at Boliden Ltd. Ronnskar smelter. *Proceedings of IEEE International Symposium on Electronics and the Environment*, **1999**, pp. 42–47.

Ting, P., Tan, C., & Pham, A. Cyanide-generating bacteria for gold recovery from electronic scrap material. *J. Biotechnol.*, **2008**, *136*, 647–677.

Tong, X., Li, J., Tao, D., & Cai, Y. Re-making spaces of conservation: deconstructing discourses of e-waste recycling in China. *Area*, **2015**, *47*(1), 31–39.

Wath, S. B., Vaidya, A. N., Dutt, S., & Chakrabarti, T. A roadmap for development of sustainable e-waste management system in India. *Science of the Total Environment*, **2010**, *409*(1), 19–32.

White, C., Wilkinson, S., & Gadd, G. The role of microorganisms in biosorption of toxic metals and radionuclide. *Int. Biodeterioration. Biodegrad.*, **1995**, *35*(1–3), 17–40.

Willner, J., Kadukova, J., Fornalczyk, A., & Saternus, M. Biohydrometallurgical methods for metals recovery from waste materials. *Metalurgija*, **2015**, *54*(1), 255–258.

CHAPTER 6

MICROBIAL DEGRADATION OF WASTES FOR ENVIRONMENTAL PROTECTION

MANOBENDRO SARKER[1,2] and MD. MAKSUDUR RAHMAN[2]

[1]*Department of Food Engineering and Technology, State University of Bangladesh, Dhanmondi, Dhaka–1205, Bangladesh*

[2]*Biomass Energy Engineering Research Center, School of Agriculture and Biology, Shanghai Jiao Tong University, 800 Dongchuan Road, Shanghai–200240, PR China*

6.1 INTRODUCTION

At present, meeting the growing demand of food for the rapidly increasing population is a major challenge around the world. According to Alexandratos (2011), crop production would need to increase by 60% in 2050 to meet the growing demand of food. However, higher production of agricultural commodities is generating a huge amount of agricultural residues, which is one of the major causes of environmental pollution when discarded as waste. In addition, postharvest loss of agricultural commodities is almost 49 to 80% in tropical countries (Gustavo et al., 2003) and one-third of world food production (approximately 1.3 billion metric tons) is being wasted every year that causing environmental pollution (Gustavsson et al., 2011). On the other hand, a large volume of industrial effluents with organic, highly toxic compounds and synthetic chemicals is being disposed directly and indirectly into the environment, which has become one of the major global concerns. Environmental scientists have studied on several processes to remove or detoxify organic and hazardous pollutants, like lead, copper, and cadmium, while bioconversion has aroused as an effective and environmental-friendly technique (Singh and Tripathi, 2007; Wasi et al., 2008). In bioconversion, chemical reactions are catalyzed biologically to reduce pollution in terms of yielding biofuels and other high-value products (Srirangan et al., 2012). At present, bioethanol and biogas are the most fermented products, while intrinsic and extrinsic factors exert a significant effect on the microbial growth and the yield of renewable energy from the biodegradable wastes of different sources.

In general, biodegradable waste is known as biomass, which refers to the organic material and can be converted into high-value products by microorganisms or their enzymes (Table 6.1). Long et al. (2013) stated that biomass is one of the leading renewable energy sources that contribute 9–13% to the total global energy demand. In Europe, the production of biomass energy is almost double in the last decade, and 6% of the total amount of primary energy (Gabrielle et al., 2014). Although biogas is not used as gaseous vehicle fuel, it can be easily used to produce heat and electricity (Oleskowicz-Popiel et al., 2012).

TABLE 6.1   Organic Wastes from Different Sources and Possible Use

Type of Process/ Product	Organic Waste	Possible Use	References
Agricultural waste	Straw, stem, leaves, husk, hull, stalk, chaff, bagasse, nutshell, bran	Biogas, ethanol, enzyme	Barakat et al., 2014
Fruits and vegetable processing	Skin, peel, seed, pomace	Biofuel, enzyme	Reddy et al., 2011; Zhengyun et al., 2013
Fish processing	Fish, eye, viscera, scale, bone, gonad, gut	Enzyme and bioactive compound production	Kim and Mendis, 2006; Zampolli et al., 2006
Poultry processing and slaughterhouse	Livestock manner, rumen, feathers, skin, blood, horn, shell, deboning residue, dead body	Biofuel, enzyme	Abraham et al., 2014; Yazid et al., 2016
Milk processing industry wastes	Fat, casein, lactose	Biofuel, enzyme	Porwal et al., 2015
Oil processing industry	Shells, husks, fibers, sludge, press-cake	Enzyme	Jørgensen et al., 2010
Sugar mill	Molasses	Enzymes, oligosaccharides	Ghazi et al., 2006
Paper/wood industry	Pulp, sawdust	Enzymes, biogas	Rathna et al., 2014

According to Weiland (2010), anaerobic digestion is the most efficient and sustainable process for biomass energy production, refers to the
biological degradation of organic compounds by mesophilic or thermophilic anaerobic or facultative bacteria and results in biogas. The anaerobic
digestion favors several advantages, like reducing odors and pathogens
as well as decreasing greenhouse gas and other undesirable air emissions. Therefore, various biodegradable materials can be treated by using
anaerobic digestion (Nallathambi, 1997; Ward et al., 2014). Zheng et al.
(2014) reported that the agricultural residue is lignocellulosic and easily
biodegraded, which primarily consists of three major structural elements,
namely cellulose, hemicelluloses, and lignin. Meanwhile, biogas has
obtained a very good response due to its economical and eco-friendly
usage of agricultural residues.

Several researches have also motivated to enhance the biogas production through co-digestion, which refers to a combination of different wastes
as substrates in the fermentation process. Carucci et al. (2005) noted that
synergistic effect in the co-digestion process enhances the biogas production. Interestingly, co-digestions of slaughterhouse wastes with municipal
solid wastes (Cuetos et al., 2008), and fruit and vegetables wastes with
slaughterhouse wastes (Alvarez and Liden, 2008) exert significant effect
in biogas production. In addition, effluent of anaerobic digestion can be
further used as substrate for ethanol production, and solid byproducts can
be used as fertilizer because of incompletely degraded organic and inorganic matter content (Alburquerque et al., 2012).

In the context of diminishing fossil fuel reserves, bioethanol has
emerged as an attractive and promising renewable energy. On the other
hand, abundant amount of lignocellulosic biomass, like rice straw, wheat
straw, corn stover, and sugarcane bagasse has motivated researchers to
the production of bioethanol (Gruno et al., 2004). Usually, bioethanol
production from lignocellulosic biomass takes place in three steps. The
first step is delignification in which cellulose and hemicellulose are liber-
ated from lignin, the second step is depolymerization of cellulose in which
hemicellulose converts into free sugar (hexose and pentose), and the third
step includes the fermentation of free sugar to produce ethanol. Several
microorganisms have been found to use in bioethanol production from
different wastes, like *Saccharomyces cerevisiae* for sugarcane bagasse
(Irfan et al., 2014), for sorghum stalk juice (Shen et al., 2011), for algae,
fish waste and pineapple (Hossain et al., 2008), *Aspergillus ellipticus*

and *A. fumigatus* for banana pseudostem (Ingale et al., 2014). Therefore, *Saccharomyces cerevisiae* is used for bioethanol production on an industrial scale from sugar and starch-based materials (Hahn-Hagerdal et al., 2007; Tian et al., 2009).

6.2   ORGANIC WASTES

Organic waste refers to biodegradable materials like agricultural wastes, industrial wastes, household wastes, fruits, and vegetable wastes, human, and animal wastes, while biodegradability largely depends on the composition and microbial strain associated with degradation. In this sense, organic wastes show potentiality in the production of various valuable products through solid-state fermentation (SSF) (Pandey et al., 2000a, b, c).

6.3   BIOCONVERSION AND BIOREMEDIATION OF WASTES FROM DIFFERENT SOURCES

6.3.1   AGRICULTURAL WASTES

Agricultural sector produces huge amount of lignocellulosic residues (stalk, stem, root, leaves, stover, groundnut shell, groundnut straw, and bagasse) per year in the world, which consist of cellulose (35–50%), hemicellulose (25–30%) and lignin (25–30%). The main constituent of cellulose is glucose, while hemicellulose is a polymer of five different sugars, including L-arabinose, D-galactose, D-glucose, D-mannose, and D-xylose and organic acids. Lignin is complex, amorphous, and branched polymer of different phenylpropane (Mussatto et al., 2012). Half of the lignocellulosic residue is used as a roofing material, fuel, animal feed, and packing material, whereas the other half is disposed of as waste and of burning. Both burning and disposal of agricultural wastes are causing severe pollution. However, lignocellulosic wastes have a great potential as biomass feedstock and animal feedstuff, but the presence of lignin reduces the digestibility efficiency (Pandey et al., 2007; Graminha et al., 2008).

Lignin is the most abundant organic polymer, which is being degraded by ligninolytic enzymes. Ligninolytic enzymes are a group of extracellular enzymes including peroxidases, laccases, and oxidases (Ruiz-Duenas and Martinez, 2009). It is also capable of hydrolyzing xenobiotic substances such as lignin, hydrocarbons, phenols, carbon tetrachloride aromatics, perchloroethylene, azo dyes, pesticides, humic substances which are discarded into the environment as waste from the industrial area. Disposal of xenobiotic substances is another reason for environmental pollution and threatening to several important organisms (Danzo, 1997). However, ligninolytic enzymes can be used in various sectors, like agricultural, chemical, food, paper, textile, cosmetic, and fuel industry for bioremediation purposes (Rodriguez and Toca, 2006). Production cost of ligninolytic enzymes depends on raw material used in the process. In this perspective, low cost organic wastes are suggested to use as raw material for producing ligninolytic enzymes. Apple pomace (Vendruscolo et al., 2008; Gassara et al., 2010), brewery by-product (Bartolome et al., 2003; Gassara et al., 2010), municipal, and industrial sludge (Gassara et al., 2010) have been studied as raw material for the production of ligninolytic enzymes. The only study on the production of ligninolytic enzymes using fish processing waste was carried out by Gassara et al. (2010). They used *Phanerochaete chrysosporium* BKM-F-1767 strain in producing lignin peroxidase, manganese peroxidase and laccase from fish waste. A slight inferior result was revealed compared to apple pomace. In addition, they recommended using other microbial strains for this purpose.

It has been reported that hundreds of fungi are capable of degrading lignocellulosic residues (Kirk, 1983). Soft rot fungi, such as *Ascomycetes* and fungi imperfecti can decompose cellulose efficiently, but degrade lignin slowly and incompletely. On the other hand, white rot fungi is found to degrade lignin efficiently by producing ligninolytic enzymes, such as lignin peroxidase, manganese peroxidase and laccase (Isroi et al., 2011). In addition, *Trichoderma*, and *Phanerochaete* have been reported as the most efficient lignocellulolytic fungi (Ander and Eriksson, 2006). Most of the bacteria, specially *Cytophaga* and *Sporocytophaga* can degrade cellulose under proper conditions, but limited to lignin degradation. Mesophilic *Bacillus* both of aerobic and anaerobic, including *B. subtilis*, *B. polymyxa*, *B. licheniformis*, *B. pumilus*, *B. brevis*, *B. firmus*, *B. circulans*, *B. megaterium* and *B. cereus* are reported as efficient cellulose and hemicellulose degraders (Ball et al., 1989). In the meantime, several pre-treatments have been suggested for cellulose and lignin degradation as well as yielding fermentable sugar (Chang et al., 2012). In this sense, SSF can be potentially applied for increasing digestibility and producing renewable energy (Lio and Wang, 2012).

6.3.1.1   BIOETHANOL PRODUCTION

Saccharification is an important step in bioethanol production where complex carbohydrates are converted into simple sugars. Therefore, enzymatic hydrolysis shows several advantages, like less toxicity, lower energy consumption, and negligible corrosion than chemical hydrolysis (Sun and Cheng, 2002; Taherzadeh and Karimi, 2007). Then hydrolyzed biomass is subjected to microbial fermentation for ethanol production.

In several studies, simultaneous saccharification and fermentation have been reported better than separate saccharification and fermentation (Bjerre et al., 1996; Balat et al., 2008). At the instance of ethanol production from lignocellulosic waste by hydrolysis or fermentation separately is not cost-effective and associated with growth retardation of yeast due to higher concentration of reducing sugar (Kim et al., 2008). They also suggested using the simultaneous saccharification and fermentation process in which reducing sugar formed by hydrolysis will be converted into ethanol during fermentation. Taherzadeh and Karimi (2007) stated that longer time duration in separate saccharification and fermentation process leads to contamination. In a study, Mutreja et al. (2011) produced ethanol from different agricultural wastes using recombinant cellulase and *Saccharomyces cerevisiae* in simultaneous saccharification and fermentation process while maximum ethanol yield has been found from *Syzygium cumini* (Jamun) at 30°C.

6.3.1.2   BIOGAS PRODUCTION

Biogas is an important renewable energy produced by anaerobic digestion in which organic matter is degraded into reduced methane. In addition, nitrogen, ammonia, hydrogen, and hydrogen sulfide are generated with methane (Angelidaki et al., 2003). Anaerobic digestion for biogas production is divided into four phases, namely hydrolysis, acidogenesis, acetogenesis, and methanogenesis. At each phase, different facultative or anaerobic bacteria use carbon as their energy source (Gerardi, 2003; Schnurer et al., 2009). During hydrolysis, organic matter is degraded into monomers by extracellular hydrolytic enzymes include cellulase, hemicellulase, amylase, lipase, and proteases (Parawira et al., 2008). During acidogenesis, monomers produced in hydrolysis are further degraded into short-chain organic acids, alcohols, hydrogen, ammonia, and carbon dioxide (Schink, 1997;

Schnurer and Jarvis, 2009). In the acetogenic phase, fatty acid with more than two carbon atoms, and alcohols with more than one carbon atom are further degraded into acetic acid, hydrogen, and carbon dioxide by aceto-genic microorganisms (Schink, 1997). Methanogenesis is the last stage of biogas production where methanogenic microorganisms use acetate, hydrogen, alcohol, and carbon dioxide for producing methane under strictly anaerobic conditions (Liu and Whitman, 2008).

Biogas production through anaerobic digestion is easier than bioeth-anol production due to transformation of inhibiting compounds into methane during bioethanol production (Barakat et al., 2014; Benjamin et al., 2014). However, monodigestion of lignocellulosic wastes often results in low methane yield (Sawatdeenarunat et al., 2015). Co-substance, such as animal manure has been suggested to use with lignocellulosic biomass to increase the buffering capacity by supplying macro and micronutri-ents (Mata-Alvarez et al., 2014). Several researchers have investigated co-digestion, like wheat straw with cattle and chicken manure (Wang et al., 2012), rice straw with kitchen waste and pig manure (Ye et al., 2013) and oat straw with cattle manure (Lehtomäki et al., 2007). It showed a better yield of biogas from lignocellulosic waste.

6.3.2 FRUITS AND VEGETABLES PROCESSING WASTES

In 2013, the total production of fresh fruits and vegetables was 1790 million tons, where 950 million tons from vegetables and 790 million tons from fruits (FAO, 2014). Citrus, mango, apple, grape, pineapple, olive, tomato, and potato are mostly processed fruits and vegetables. In the meantime, processing industry is growing faster, and fruits and vegetables are processed into a variety of products, like jam, jelly, squash, pickle, sauce, and to make available them throughout the year. Thus, a processing plant generates a large fraction of solid and liquid wastes especially at the early stage of processing. Fruit and vegetable processing plants produce about 10–60% solid wastes that include peel, pits, seeds, pomace, trimmings, and spoiled food. In contrast, liquid wastes include pulp, caustic peeling water, wash water, and liquid chemicals (Chakraverty et al., 2003). Generally, open field dumping of organic wastes is practiced, and this is another reason for environmental pollution. However, wastes from the agro-based industry are rich in sugar, and bioremediation or bioconversion is found to be a potential way in managing wastes and reducing environmental pollution (Chakraverty et al., 2003).

Anaerobic digestion can be operated under mesophilic or thermophilic conditions, while thermophilic digestion is found to be more efficient. On the one hand, anaerobic digestion of fruits and vegetable wastes was studied by Das and Mondal (2013) for the production of biogas. On the other hand, predigested waste has been studied as a source of mixed bacteria, while dried fruits and vegetable wastes can be used as substrates in anaerobic digestion. They reported that the biogas from organic wastes through anaerobic digestion is mainly methane. Researcher's effort has also now focused on co-digestion to improve the biogas production by controlling carbon to nitrogen ratio (Chellapandi, 2004; Gelegenis et al., 2007). Anaerobic co-digestion of orange peel waste and jatropha deoiled cake was studied by Elaiyaraju and Partha (2012) for biogas production. Inhibit acid, like volatile fatty acid is observed when fruit and vegetable waste is digested alone (Jiang et al., 2012), but the amount decreases when it is co-digested with swine feces and urine (Feng et al., 2008). Dahunsi et al. (2015) produced biogas from watermelon peels, pineapple peels, and food wastes through anaerobic co-digestion.

Banana is a widely consumed tropical fruit. Nearly 200 tons of banana peel is produced daily in Thailand, and a small fraction of it is used as feed while the maximum amount is being rotted (Pangnakorn, 2006). In a study, Sang et al. (2006) noted that nearly 6 million tons of banana stems and leaves and 1.8 million tons of banana peels are discarded annually as waste in China. Interestingly, Zhengyun et al. (2013) conducted a research on biogas production from banana waste (banana stalk and peel), used as a substrate in mesophilic anaerobic fermentation. In this study, the higher biogas production from banana peel was also noted compared to banana stalk. In another research, the production of hydrogen and methane gas through anaerobic digestion was studied by Nathoa et al. (2014). They reported that banana peel is an attractive substrate for producing hydrogen and methane gas through two-phase anaerobic digestion.

Mango is the second most consumed tropical fruit (Joseph and Abolaji, 1997), grown in more than 90 countries with a global production of 26 million tons in 2004 (FAOSTAT, 2004). The edible portion accounts to 33–85%, while peel and kernel are almost 7–24% and 9–40%, respectively (Wu et al., 1993). However, mango peel is not being used commercially, and a large volume of waste is produced in the processing industry, thus contributing to environmental pollution (Berardini et al., 2005). Researchers are looking for economic raw material for lactic acid production, and mango peel has been suggested by several researchers. Recently,

Jawad et al. (2013) conducted a research work on the lactic acid production from mango peel using bio-fermentation. Reddy et al. (2011) used dried mango peel as a substrate for *Saccharomyces cerevisiae* CFTRI101 in producing ethanol. Direct fermentation of dried mango peel substrate exerted slow fermentation while the addition of yeast extract, peptone, and wheat bran increased the fermentation rate.

In addition, biodegradable organic matter with higher moisture content in vegetable waste facilitates biological treatment through anaerobic digestion (Bouallagui et al., 2003). Limitations of anaerobic digestion include rapid acidification due to lower pH and production of free fatty acids (Bouallagui et al., 2003, 2005). In spite of the fact that aerobic digestion is not preferred for vegetable waste (Landine et al., 1983). Biomethanation of vegetable waste was studied by Kameswari et al. (2007). On the other hand, Velmurugan and Ramanujam (2011) carried out biomethanation of vegetable waste under mesophilic conditions by using a fed-batch laboratory-scale reactor. Sagagi et al. (2009) studied on the biogas production from some selected fruits and vegetable waste and revealed that higher biogas production from cow dung (control) followed by fruit waste and lastly, vegetable waste. Jiang et al. (2012) also stated that cattle slurry could be suggested to use as co-substrate with vegetable waste in anaerobic co-digestion.

6.3.3 FISH PROCESSING WASTES

The consumption of fish resources is more than the previous time, and the amount is almost 105.6 million tons, which is 75% of total production around the world while the remaining 34.8 million tons are being wasted (FAO, 2007). Fish processing industries produce a large amount of solid waste (fish head, eye, viscera, scales, bones, liver, gonads, guts, and some muscle tissues), and liquid waste includes blood and effluent (Awarenet, 2004). In most countries, incineration or dumping of these wastes is common, which causes environmental pollution (Bozzano and Sarda, 2002).

Advantages in biotechnology exploit the production of value-added products from fish industry wastes. Therefore, fish oil with higher polyunsaturated fatty acids is now extracted and integrated into beverages and other food items (Chen et al., 2006; Kim and Mendis, 2006; Zampolli et al., 2006; Rubio-Rodriguez et al., 2010). Several studies also have suggested that fish processing waste may have potential use as a rich source of protein, lipid, and minerals (Toppe et al., 2007; Kacem et al., 2011). Moreover, fish scale and cartilage could be used in producing gelatin for using in food, cosmetic, and medical industry (Blanco et al., 2006; Karim and Bhat, 2009). Gao et al. (2006) stated that a fish viscus is one of the low-cost substrates for the production of lactic acid by lactic acid bacteria. Interestingly, fish wastes are also good substrates for microbial growth in producing a number of important metabolites (Coello et al., 2000; Vazquez et al., 2006, 2008). Meanwhile, various strains of bacteria, yeast, and mold have been studied to produce protease and lipase enzymes, having importance in food, textile, chemical, and pharmaceutical industries.

Fishmeal is produced conventionally from fish waste, but the conventional method has not drawn the attention due to excess production cost and indigestibility of conventional fishmeal. However, biological fermentation can be proposed for yielding fishmeal from fish waste. *Aspergillus awamori* has been suggested to use in the low-cost fermentation process in producing high-quality fish feed from fish waste (Yamamoto et al., 2005).

Afonso and Borquez (2002) could not find any toxic or carcinogenic material in fish processing wastewater unlike other types of effluents, including industrial and municipal effluent. Considering this outcome, fermented effluent from fish processing area has been suggested to use in agriculture. Several studies have focused on the production of liquid from fishery effluent through the fermentation process (Kim and Lee, 2009; Kim et al., 2010; Dao and Kim, 2011). Kim et al. (2010) used five strains namely, *Brevibacillus agri, Bacillus cereus, Bacillus licheniformis*, and *Brevibacillus parabrevis* in the fermentation of fish waste for producing liquid fertilizer. This liquid fertilizer has the comparable fertilizing ability and proposed to use in hydroponic culture.

Protease is one of the most important enzymes that can be produced from fish waste. Previous studies have stated that *Penicillium* sp., *Serratia marcescens, Streptomyces* sp., *Rhizopus oryzae, Pseudomonas, Bacillus* sp. and *Vibrio* are widely used in protease enzyme production (De Azeredo et al., 2004; Joo and Chang, 2005; Vazquez et al., 2006). Several researchers have stated that wastes from fish processing industry are inexpensive feedstocks having potential for producing protease (Triki-Ellouz et al., 2003; Vazquez et al., 2006; Wang and Yeh, 2006; Haddar et al., 2010). Defatted tuna fish waste was reported to favor higher protease activity by *Bacillus cereus* (Esakkiraj et al., 2009). They also revealed that fat-free nature of fish waste is a good substrate for microbial species for the production of proteases enzyme. A similar investigation was also observed by *Rhizopus oryzae* for lipase production (Ghorbel et al., 2005).

Production of lipase enzyme by various microorganisms including fungi, yeast, and bacteria has been received much attention in modern enzymology (Sharma et al., 2001). Various studies have reported that fish processing waste is economic substrate for producing lipase enzyme. In addition, pre-treatments like physical and chemical treatment are found effective before using fish processing waste as culture media. Effluent from fish waste boiling was found potential to produce lipase enzyme by *Saccharomyces xylosus* (Ben et al., 2008; Esakkiraj et al., 2010a). Compositions of waste like carbon source, nitrogen source and free fatty acids have great effect on microbial lipase enzyme production. According to Esakkiraj et al. (2010b), higher lipase production from cod liver oil was found by *Staphylococcus epidermidis* CMST-Pi 1. It has been scientifically proved that well-balanced nutrients in growth media ensure higher microbial lipase production (Ghorbel et al., 2005; Ben et al., 2008).

Chitinolytic enzyme is another important enzyme can be produced from fish processing waste by bacteria, fungi, yeasts, and viruses. Chitinolytic enzyme has drawn an interest in the food and pharmaceutical industry due to various functional activities, including antimicrobial, antifungal, antitumor, and immunoenhancer (Tsai et al., 2000; Wen et al., 2002; Shen et al., 2009). This enzyme also has potential in the production of single-cell protein (Dahiya et al., 2006). Shrimp, shellfish, and crab shell powder are effective substrates for *Pseudomonas* sp., *Bacillus* sp., *Serratia* sp. and *Monascus* sp. in the production of chitinolytic enzyme (Wang et al., 2002).

6.3.4 POULTRY PROCESSING AND SLAUGHTERHOUSE WASTES

The increasing consumption of egg, chicken, broiler, and turkey meat due to better nutritional quality than other sources has made the poultry industry an important and growing sector around the world. Processed poultry products are convenient to prepare, and consequently, the meat-processing industry is growing rapidly (USDA, 2014). Therefore, a large volume of waste is produced daily from the processing industry and slaughterhouse, which causes environmental pollution unless, managed them properly. The major wastes from meat processing industry and slaughterhouse include livestock manner, rumen, feathers, skin, blood, horn, shell, deboning residue, soft meat and dead body. However, a small amount of bone, feather, and eggshell is processed as a source of calcium, phosphorus, and amino acids to make animal feed (Salminen and Rintala, 2002; Lasekan et al., 2013) and rest of waste materials are discarded in the environment.

Several authors have utilized animal fleshing as a substrate in SSF for the production of protease. A high yield of protease production from animal hair was also reported with activated sludge or anaerobically digested sludge (Abraham et al., 2014; Yazid et al., 2016). Feather is a major waste in the poultry industry since it accounts for 8% of total weight and consists of 90% keratin protein (Onifade et al., 1998). Keratin is insoluble structural protein and difficult to degrade because of strong hydrogen bonds as well as hydrophobic interactions among protein chains, and because of this, feather meal becomes low-grade feed. Considering environmental pollution, several countries banned disposal of feather through burning. On the other hand, some microbial enzymes have been extensively investigated to hydrolyze the insoluble keratin into digestible proteins (Kornillowicz-Kowalska and Bohacz, 2011; Gupta et al., 2013) for producing feedstuffs, fertilizers, and films (Onifade et al., 1998; Gupta and Ramnani, 2006; Jayathilakan et al., 2012).

Therefore, keratinase producing microorganisms having the ability to decompose feather and have been listed as follows: *Absidia* sp., *Alternaria radicina*, *Aspergillus* sp., *Doratomyces microsporus*, *Onygena* sp., *Stachybotrys atra*, *Trichurus spiralis* and *Rhizomucor* sp. (Friedrich et al., 1999), *Streptomyces albs*, *S. fradiae*, *S. pactum*, *S. thermoviolaceus* (Noval and Nickerson, 1959), *S. thermonitrificans* (Mohamedin, 1999), *Flavobacterium pennavorans* (Yamamura et al., 2002), *Bacillus* sp., *Stenotrophomonas* sp., *Bacillus licheniformis* and *B. pumilus* (Nitisinprasert et al., 1999) and *Vibrio* sp. (Sangali and Brandelli, 2000). Several research works have suggested to use feather hydrolysates for the production of biofuel, fuel pellets for heating (Ichida et al., 2001; Dudynski et al., 2012), and biohydrogen (Bálint et al., 2005).

In general, litter and livestock manner are used as fertilizer and fish feed. However, litter and manure can be converted into biogas, mainly methane by anaerobic digestion (Salminen and Rintala, 2002). Anaerobic bacteria are found to degrade the litter and chicken manner at a high rate. In a previous study, it was found that anaerobically digested solid poultry waste is rich in nitrogen but potentially phytotoxic (Salminen et al., 2001). They also observed a positive effect of aerobic post-treatment in reducing the phytotoxicity and other inhibitory compounds including organic acids and ammonia produced in anaerobic digestion of solid poultry waste. Budiyono et al. (2011) conducted a research on anaerobic digestion of cattle manner and liquid rumens for biogas production. Dried rumen content was also studied for producing fed through fermentation with *Trichoderma harzianum* (Nova, 2000). According to the Environmental Protection Agency (EPA), wastewater from the slaughterhouse and meat processing industry is more harmful to the environment. Likely, slaughter-house wastewater is found to be feasible in anaerobic digestion (Pozo Del et al., 2000). Seif and Moursy (2001) treated slaughterhouse wastewater by using a laboratory-scale bioreactor. They also suggested that aerobic digestion should be practiced after anaerobic digestion to improve the effluent quality before discharging into the environment.

6.3.5 DAIRY INDUSTRY WASTES

The dairy industry is one of the important food industries and regarded as a wet industry because of using large volumes of freshwater for different diverse purposes. Dairy plants in varying sizes also discharge a large amount of wastewater than any other food industries. Studies elucidated that production characteristics, and technical cycle of processing line have great impact on the quantity and quality of waste (Table 6.2) discharged from dairy processing area (Wildbrett, 2002; Munavalli and Saler, 2009). Almost 2.5–3.0 liters of effluent per liter of processed milk are produced from receiving station to processing area in a large dairy plant, which contains fat, casein, lactose, inorganic salts, detergents, and various sanitizers used for cleaning and washing purposes. All of these pollutants can be characterized by Biological Oxygen Demand (BOD) and Chemical Oxygen Demand (COD) (Vidal et al., 2000; Singh et al., 2014). In earlier, physical methods include sedimentation, aeration, flotation, degasification, chlorination, ozonation, neutralization, coagulation, and sorption ion exchange are found to treat industrial wastewater. Partial treatment, higher cost, secondary pollutants generation, higher quantity solids and use of chemical agents are possible limitations of physico-chemical methods (Rodrigues et al., 2007).

TABLE 6.2 Compositions of Effluent of a Typical Dairy Industry

Specification	Value
pH	7.2
Alkalinity	600 mg/l as $CaCO_3$
Total dissolved solids	1060 mg/l
Suspended solids	760 mg/l
BOD	1240 mg/l
COD	84 mg/l
Total nitrogen	84 mg/l
Phosphorous	11.7 mg/l
Oil and Grease	290 mg/l
Chloride	105 mg/l

Source: Rao and Datta, 1978.

Bioremediation is a process in which living microorganisms or their enzymes are used for the treatment of wastewater. Environment-friendly microorganisms are commonly used in bioremediation, which has proved to be effective in lowering industrial waste (Ojo, 2006). Degradation of organic waste into nonhazardous end products by using naturally occurring microorganisms is more accepted than any other physical methods. The bacteria including *Pseudomonas aeruginosa, P. fluorescens, Bacillus cereus, B. subtilis, Enterobacter, Streptococcus faecalis, Escherichia coli* and the yeasts belonging to *Saccharomyces, Candida,* and *Cryptococcus* genus are common in dairy wastewater (Madigan et al., 2000).

However, biological treatment is also a cost-effective method for the degradation of dairy wastewater. Several studies have suggested that activated sludge process, aerated lagoons, trickling filters, sequencing batch reactor (SBR), upflow anaerobic sludge blanket reactor and anaerobic filters can be used to manage dairy and other industrial effluents effectively (Stover and Campana, 2003; Demirel et al., 2005). Porwal et al. (2015) conducted a research by utilizing microbial isolates of activated sludge for the biodegradation of dairy effluent. They used the single and mixed culture of yeast isolates (DSI1) and two bacterial isolates (DSI3) obtained from activated dairy sludge to treat the effluent. The addition of any single microbial culture to the activated sludge process was found to increase the overall efficiency of the treatment system. Furthermore, the mixed culture has proved to be more effective and beneficial compared to a single culture. A laboratory-scale reactor comprising aeration tank and final clarifier was studied continuously for three months to treat dairy waste using activated sludge. Activated sludge at retention time of 5 days was found effective with BOD percentage removal efficiency of 95% for dairy effluent (Lateef et al., 2013).

6.3.6   CEREAL PROCESSING EFFLUENT

Parboiling process of rice generates almost two liters of effluent per kilo of raw rice. In addition, parboiled rice effluent contains high amount of organic matter as well as nitrogen and phosphorus (Queiroz and Koetz, 1997). Recently, several yeast stains have been used for treating industrial effluent specially for removing phosphorous (Choi and Park, 2003). On the contrary, use of *Candida utilis* is not encouraged to reduce nitrogen content in parboiled rice effluent (Rodrigues and Koetz, 1996). Bastos et al. (2014) treated parboiling effluent by using cyanobacterium, *Aphanothece microscopica* Nägeli. It was revealed that this bacterium is associated with the production of single-cell protein (SCP) from wastewater. This bacterium usually uses photosynthesis and removes a high amount of nitrogen and organic matter from parboiling rice effluent. Furthermore, Schneid et al. (2004) suggested using *Saccharomyces boulardii* and *Pichia pastoris* for treating parboiling rice effluent with the addition of carbon source. De los Santos et al. (2012) evaluated the effectiveness of *Pichia pastoris* X-33 as bioremediator to reduce organic matter in parboiled rice supplemented with glycerol.

Corn is another important cereal for producing starch, gluten, dextrin, glucose, fructose, and corn syrup. In addition, corn-based ingredients are very important for food processing, pharmaceuticals, and biochemical industries. According to CRAR (2009), in the period of 2008–2009, the total production of corn was 789 million tons throughout the world. Two distinct processes, namely wet-milling and refinery are commonly used while both processes generate huge amount of waste. Ersahin et al. (2007) stated that refinery process produces more effluent than wet milling while wet milling produces effluent with higher COD. On the other hand, the biodegradability of corn processing effluent is comparatively high due to rich in protein and starch content (Howgrave-Graham et al., 1994; Ereme-ktar et al., 2002). A three-stage advanced wastewater treatment plant including an EGSB reactor, intermittently aerated activated sludge system and chemical post-treatment unit, was used for treating corn processing effluent (Ersahin et al., 2007). In another research, Duran de Bazua et al. (2007) evaluated that both of aerobic and anaerobic processes in a two-stage biological treatment system for the treatment of wastewater from a corn processing industry manufacturing tortillas.

6.3.7   OIL PROCESSING INDUSTRY EFFLUENT

Waste from the edible oil refining industry is a major concern in several countries around the world. Soybean, groundnut, rapeseed, sunflower, safflower, cottonseed, coconut, mustard, and rice bran are the main sources of edible oil. Oil refining industry produces spent earth as solid and waste-water as liquid waste. In an edible oil refining industry, huge amount of wastewater generated from processing sections including degumming, de-acidification, deodorization, neutralization, equipment cleaning, and floor cleaning (Thompson et al., 1994). Oil refining effluent causes serious environmental problem specially life treating to aquatic animals. Hence, the treatment of oil refining industry waste is essential because of high amount of organic content. Rupani et al. (2010) reported that the treatment of oil refining industry depends on the amount and type of organic matter present in waste stream.

Wastewater of edible oil refining industry can be treated by chemical and biological means. Chemical treatment process combines chemical purchasing cost and production of chemical sludge, whereas biological treatment is convenient and cost-effective. Mkhize et al. (2000) stated that activated sludge and SBR could be used in the biological treatment of oil industry effluent. Interestingly, Aslan et al. (2009) revealed that 95% of BOD in soybean oil refining effluent could be removed by using an activated sludge.

6.3.8   SUGAR MILL EFFLUENT

Sugar industry plays a vital role in the economic development of a country but effluent generated from sugar mill contains high amount of organic matter. According to Baskaran et al. (2009), 110–115 m^3 water is required in a sugar mill with a capacity of producing 35,000 kg sugar daily while 87% of water comes out as wastewater. It has been reported that sugar mill effluent causes environmental pollution when discharged into water, consequently, it results in high rate of fish and aquatic animal mortality and various diseases of people when they use for agricultural and domestic purposes (Baruah et al., 1993).

Out of several treatments, bioremediation offers a cheap and environmental friendly alternative for biological degradation of industrial effluents. In a study, Saranraj and Stella (2012) used *Bacillu subtilis*, *Serratia marcescens* and *Enterobacter asburiae* for treating sugar mill effluent. Bioremediating of sugar mill effluent showed a drastic reduction in organic matter content in terms of COD, TSS, and TDS, heavy metal content, and other physical properties. In addition, Buvaneswari et al. (2013) identified five native bacterial isolates, namely *Staphylococcus aureus*, *Bacillus cereus*, *Klebsiella pneumoniae*, *Enterobacter aeruginosa,* and *Escherichia coli* that are capable of bioremediating of sugar mill effluent.

6.3.9 BAKER'S YEAST PRODUCTION EFFLUENT

Baker's yeast, *Saccharomyces cerevisiae* is commonly used in the production of bakery products. A baker's yeast industry is important where molasses is used as substrate for the production of *Saccharomyces cerevisiae* (Catalkaya and Sengul, 2006). The production process includes cultivation, fermentation, separation, rinsing, and pressurized filtration. Fermentation process produces a large volume of wastewater with high amount of organic substances, nitrogen, trimethylglycine, sulfate, and phosphorous content (Blonskaja et al., 2006; Liang et al., 2009).

In an earlier study, Gulmez et al. (1998) studied the feasibility of anaerobic treatment for baker's yeast industry wastewater. Feasibility was found comparable with the pharmaceutical industry. Kalyuzhnyi et al. (2005) treated the baker's yeast industry effluent through anaerobic treatment by using an UASB reactor followed by aerobic-anoxic biofilter and coagulation process. Moreover, Krapivina et al. (2007) used anaerobic SBR for treating sulfate-rich baker's yeast industry effluent.

6.3.10 TANNERY INDUSTRY EFFLUENT

The leather industry is economically important for a country because of increasing demand for leather products in which 65% leather is used in footwear. Nowadays, leather industry uses tanning process due to easier procedure, low cost, light color with greater stability of leather, which generates large volume of wastewater (Jenitta et al., 2013). In the tanning process, huge amount of chromium sulfate Cr(III) and pentachlorophenol (PCP) are used for the conversion of hides and skin into the leather (Ackerley et al., 2004). At least 300 kg of chemicals has been reported to use per ton of hides during the tanning process (Verheijen et al., 1996). It is found that the amount of Cr(III) varies from 500 to 7000 ppm and PCP varies from 10 to 90 ppm in tannery effluent which persist longer period in the environment having negative effect on the flora and fauna. Additionally, Cr(III) converts into soluble Cr(VI) during longer persistence in the environment which is toxic and carcinogenic (Ackerley et al., 2004).

Physicochemical methods, such as precipitation with hydroxide, carbonates, and sulfides, adsorption on the activated carbon, use of resins, and membrane separation technique can be applied to remove metals from tannery wastewater, but all of them are expensive (Park et al., 2000). On the contrary, several researchers report that biotransformation and biosorption are emerging technologies in which microorganisms are used to transform or to adsorb metal from effluent (Kovacevic et al., 2000; Eddy, 2003). Aerobic digestion of tannery effluent has been reported to reduce the COD and BOD by 60–80% and 95%, respectively (Ganesh et al., 2006). Again, anaerobic digestion produces large amount of biogas by converting the organic pollutants of tannery effluent into a small amount of sludge. Furthermore, chromate reductases are a group of enzymes, which catalyze the reduction of toxic and carcinogenic Cr(VI) to the less soluble and less toxic Cr(III) (Park et al., 2000). Bioremediation can be applied to degrade pollutants in tannery effluent (Mohapatra, 2006; Movahedin et al., 2006). Hence, both intracellular and extracellular enzymes produced by microorganisms are efficient catalysts for bioconversion of organic materials in tannery effluent (Nelson and Cox, 2004). Biosorption of chromium from tannery effluent by fungal strain (*Aspergillus niger* FIST1) and bacterial strain (*Acinetobacter* sp. IST3) were studied by Thakur and Srivastava (2011). SEM-EDX analysis revealed that both of fungi and bacterium contained chromium within the cell. In another study, Tolfa da Fontoura et al. (2015) treated tannery wastewater with *Scenedesmus* sp. The cultivation of microalgae (*Scenedesmus* sp.) in wastewater of tannery industry was found to be a prominent treatment process. This microalga is also capable of removing of N-NH3, TKN, phosphorus, BOD, and COD up to 80% from effluent.

6.4   PRODUCTION OF VALUABLE PRODUCTS FROM DIFFERENT WASTES

6.4.1   SINGLE CELL PROTEIN (SCP)

SCP refers to the dried cells of microorganisms including algae, fungi, yeast, and bacteria which easily grow on waste substrates, rich in carbon (Ware, 1977). The increasing demand of protein is one of the major challenges, and meeting the protein demand will be a major concern in poor and developing countries in near future. SCP can be extracted by cultivating microbial biomass on different biodegradable wastes, which can be used as protein supplement and substitute of traditional protein source in human foods or animal feeds. Despite the name SCP, protein is not only one component of microbial cell, which also contains carbohydrates, lipids, vitamins, and minerals (Litchfield, 1983). Jamel et al. (2008) stated that SCP contains a significant amount of phosphorous and potassium. However, nutritional quality of SCP varies with microorganisms used, method of harvesting, drying, and processing to be used as food (Bhalla et al., 2007).

6.4.1.1   MICROORGANISMS USED IN SCP PRODUCTION

Fungi, algae, yeast, and bacteria are found to use in the production of SCP. Filamentous fungi are ease to harvest but exerts low amount of protein and retard growth rate. Algae protein is comparable, and in many cases protein quality is higher compared to traditional plant sources. Several studies revealed that two major species, the unicellular green alga, *Chlorella*, and filamentous blue-green alga, *Spirulina* are the most cultivated for SCP (Raja et al., 2008). Algae poses several disadvantages including concentrate heavy metals, higher cellulose content, and higher production cost with technical difficulties. In addition, the celluloid cell wall of algae remains indigestible in humans and other non-ruminants. Extraction of SCP from algae is still complex process (Rasoul-Amini et al., 2009). However, yeast is more suitable for the production of SCP compared to bacteria, mold, and algae (Nigam, 2000). Not only higher nitrogen content in yeast, but also larger size, higher lysine, lower nucleic acid and ability to grow in acidic environment are major advantages to use yeast for the production of SCP (Nasseri et al., 2011). Different species of algae, fungi, yeasts, and bacteria used in the production of SCP in commercial-scale are given in Table 6.3.

TABLE 6.3   Microorganism and Substrates Used in the Production of Single Cell Protein

Microorganism	Scientific Name	Substrate
Bacteria	*Aeromonas hydrophila*	Lactose
	Aeromonas delvacuate	n-Alkanes
	Acinetobacter calcoaceticus	Ethanol
	Bacillus megaterium	Non-protein nitrogenous compounds
	Bacillus subtilis	Cellulose, hemicellulose
	Cellulomonas sp.	
	Flavobacterium sp.	
	Thermomonospora fusca	
	Lactobacillus sp.	Glucose, amylose, maltose
	Methylomonas methylotrophus	Methanol
	Methylomonas clara	Uric acid and non-protein
	Pseudomonas fluorescens	Nitrogenous compounds
	Rhodopseudomonas capsulata	Glucose
Fungi	*Aspergillus fumigatus*	Maltose, glucose
	Aspergillus niger	Cellulose, hemicellulose
	Aspergillus oryzae	
	Cephalosporium eichhorniae	
	Chaetomium cellulolyticum	
	Penecillum cyclopium	Glucose, lactose, galactose
	Rhizopus chinensis	Glucose, maltose
	Scytalidium acidophilum	Cellulose, pentose
	Trichoderma viridae	
	Trichoderma alva	
Yeast	*Amoco torula*	Ethanol
	Candida tropicalis	Maltose, glucose
	Candida itilis	Glucose
	Candida novellas	n-Alkanes
	Candida intermedia	Lactose
	Saccharomyces cerevisiae	Lactose, pentose, maltose
Algae	*Chlorella pyrenoidosa*	Carbon dioxide through photosynthesis
	Chlorella sorokiniana	
	Chondrus crispus	
	Scenedesmus sp.	
	Spirulina sp.	
	Porphyrium sp.	

Source: Bhalla et al., 2007.

6.4.1.2   SCP PRODUCTION FROM WASTES

Agricultural and industrial waste as a source of hydrocarbon, nitrogenous compounds, polysaccharides, and slaughterhouse waste (horn, feather, nail, and hair) as fibrous protein have been studied for SCP production (Azzam, 1992; Zubi, 2005). Agricultural waste has great potentiality as substrate for bioconversion into protein. Cellulose is attractive substrate compared to other constituents of agricultural waste. In nature, cellulose is found in matrix form with hemicellulose and lignin and because of this, pretreatment (chemical or enzymatic) is recommended for lignocellulosic substrate where cellulose converts into fermentable sugar (Callihan and Clemmer, 1979). Hongzhang Chen et al. (1999) extracted hemicellulose hydrolysate from wheat straw and used as substrate for the production of SCP (*Trichosporon cutaneum* 851). In addition, lignocellulosic agricultural waste, like sugarcane bagasse is an ideal substrate for the production of SCP (Chandel et al., 2012). Samadi et al. (2016) used SSF in a tray bioreactor to produce SCP from sugarcane bagasse using yeast (*Saccharomyces cerevisiae*). Results also revealed that carbonate-bicarbonate is the most effective buffer for protein extraction with fermentation time (72 hr), relative humidity (85%) and bioreactor temperature 35°C. The highest protein yield (13.41%) was attained at optimum conditions containing all essential amino acids with some non-essential ones.

The bioconversion of fruit wastes into SCP has the potential to meet the worldwide demand of food protein. The production of SCP from fruit waste may be an effective way to manage organic waste and lower the environmental pollution as well (Barton, 1999). Sufficient amount of carbohydrate and other nutrients make the fruit waste as ideal substrate for growth of microorganisms (Adoki, 2008; Yabaya and Ado, 2008). Furthermore, submerged fermentation with *Saccharomyces cerevisiae* was found effective for bioconversion of fruit wastes into SCP, whereas a higher amount of protein (53.4%) from cucumber peel followed by orange peel (30.5%) per 100 g of the substrate was reported by Mondal et al. (2012). Dhanasekaran et al. (2011) used two strains of yeast namely, *Saccharomyces cerevisiae* and *Candida tropicalis* for the bioconversion of pineapple waste into SCP. Results revealed that the highest protein content was attained on 3rd day of fermentation for both of strains. Pineapple waste has been suggested to use as effective substrate for yeasts in the production of protein. In another study, banana was found to be the best substrate for *Saccharomyces cerevisiae* and followed by rind of pomegranate, apple waste, mango waste and sweet orange peel. Mondal (2006) stated that SCP production depends on both compositions of the substrate and cultural factors.

Extraction of natural pigment, capsanthin from dry red capsicum produces a large amount of capsicum powder (Uquiche et al., 2004). This amount is almost 98.6% of the inputted capsicum, which is being discarded as waste in China because of its hot and pungent flavor. However, this waste contains capsaicin and can be used as substrate by yeast in the production of SCP. Zhao et al. (2010) used four yeast strains for bioconversion of capsicum waste into SCP. Inhibition of cell growth of yeast by capsaicin in the medium (CPM) was not found in this study. However, *Candida utilis* 1769 was found to be potential for the highest SCP formation.

Molasses is commonly used as animal feed or in alcohol production or discarded as waste. Chemical composition, absence of toxic material, and availability with low cost make the molasses as suitable substrate for the production of SCP (Bekatorou et al., 2006). Though molasses is rich of carbon, but it requires ammonia and phosphorus salt supplementation (Ugalde and Castrillo, 2002).

Several studies identified that *Cryptococcus albidus, Lipomyces lipofera, L. starkeyi, Rhodosporidium toruloides, Rhodotorula glutinis, Trichosporon pullulan*, and *Yarrowia lipolytica* (formerly known *Candida lipolytica*) the most specific oily yeasts for bioconversion of waste produced in oil industry (Li et al., 2008; Ageitos et al., 2011). Wastewaters from olive oil and animal fat treatment industry have been also studied to produce SCP (Papanikolaou et al., 2001). Begea et al. (2011) used sunflower oil as a carbon source for *Candida* yeast in the production of biological protein. They suggested that the oil industry waste or oily wastewater of the food industry can be used to produce SCP and reduce environmental pollution as well.

A large amount of slaughterhouse wastes are discarded as waste in Turkey. A study shows that almost 25 tons of waste are discarded per year from a medium-size slaughterhouse. These wastes can be converted into valuable products, like protein or amino acid through bioconversion (Atalo and Gashe, 1993). Physical and chemical methods can be applied to produce crude horn hydrolysate (CHH), which is a suitable substrate for the production of SCP. Kurbanoulu (2001) conducted an experiment to produce SCP from CHH by using *Candida utilis* NRRL Y-900. Extracted protein from CHH showed good profile of essential amino acids recommended by FAO.

6.4.2 MUSHROOM

Mushroom is a fruiting body of saprophytic, mycorrhizal, and parasites fungi that belongs to Basidiomycetes or Ascomycetes order. In general, mushroom grows on moist agricultural waste, animal waste and soil of higher organic matter. After identifying the nutritional value of mushroom, it has been started to cultivate commercially. Mushroom cultivation in commercial scale is one of the prominent biotechnological processes for valorizing wastes produced from different sources. On the other hand, higher amount of protein, minerals, phytochemicals with medicinal and pharmacological properties are remarkable reasons of cultivating mushroom (Diego et al., 2011). Recent research works stated that some medicinal attributes of mushroom include antiviral, antibacterial, antiparasitic, antitumor, anti-inflammatory, antihypertension, antidiabetic, antiatherosclerosis, hepatoprotective, and immune-modulating effects (Wasser and Weis, 1999; Wasser, 2002; Daba and Ezeronye, 2003; Paterson, 2006).

Furthermore, ecological advantages of the cultivation of edible mushrooms include efficient utilization of agricultural waste like chicken and horse manure, cereal straw and leaves, sugarcane bagasse. In addition, solid and liquid waste from food, paper, chemical, and pharmaceutical industries can be used to cultivate edible mushrooms. It was reported that more than 2000 edible mushroom species are found around the world. Recently, Renganathan et al. (2008) reported that three mushrooms, namely *Agaricus bisporus* (white button mushroom), *Pleurotus* spp. (oyster mushroom), and *Volvariella volvacea* (Paddy straw mushroom) are commercially cultivated around the world. They also noted that *Agaricus bisporus* is not cultivated in tropical and sub-tropical regions due to suitable climatic condition. Pleurotus species including *P. ostreatus*, *P. sajor-caju*, *P. pulmonarius*, *P. eryngii*, *P. cornucopiae*, *P. tuber-regium*, *P. citrinopileatus,* and *P. flabellatu* are common to cultivate, which contain all essential amino acids, minerals (Ca, P, K, Fe, Na) and vitamins C, thiamine, riboflavin, niacin, folic acid and antioxidants (Çaglarirmak, 2007; Regula and Siwulski, 2007). This species is also known as oyster mushroom, and found to be capable of utilizing complex organic matters with their extensive enzymatic system (Yalinkilic et al., 1994). In addition, oyster mushroom has probiotic properties and relatively high nutritive value than others; this is because it has been recommended to add in the dietary chart for people of all ages (Bernas et al., 2006).

According to a survey, almost 200 billion tons of organic waste is produced annually around the world (Zhang, 2008). The maximum amount of produced waste is inedible to humans and animals, causing environmental pollution (Laufenberg et al., 2003). Considerable emphasis has given on the valorization of waste, particularly produced from the agro-food industry (Lal, 2005). However, lignocellulosic waste including cereals straw, rice husks, corn husks and cobs, cotton stalks, maize, and sorghum stover, vine prunings, sugarcane bagasse, coconut, and banana residues, coffee husk and pulp, seed hulls, peanut shells, sunflower seed hulls, paper waste, wood sawdust can be biologically processed through SSF to valuable and useful products (Fan et al., 2000a; Pandey et al., 2000a, b; Webb et al., 2004).

Mushroom cultivation is the most efficient utilization and value-addition through biotransformation of lignocellulosic residues among various applications of SSF (Chiu et al., 2000; Zervakis and Philippoussis, 2000; Chang, 2001, 2006). In several studies, mushroom production through SSF has been noted as biodegradation and bioremediation of hazardous residue (Perez et al., 2007) and biological detoxification of lignocellulosic residues (Fan et al., 2000b; Pandey et al., 2000c; Soccol and Vandenberghe, 2003). Moreover, *Lentinula edodes* and *Pleurotus* sp. are found to be potential in the bioconversion of lignocellulosic residues including wheat straw, cotton wastes, coffee pulp, peanut shells, oilseed hull, wood chips, sawdust, and vine prunings (Ragunathan et al., 1996; Campbell and Racjan, 1999; Philippoussis et al., 2000, 2001a, b; Poppe, 2000; Stamets, 2000). It has been reported that chemical compositions of biomass have a great influence on the yield and quality of mushroom (Kues and Liu, 2000; Philippoussis et al., 2001c, 2003; Baldrian and Valaskova, 2008). Several studies showed that apart from physicochemical composition of substrate, strain, and length of incubation play an important role in the production of *L. edodes* (Sabota, 1996; Chen et al., 2000; Philippoussis et al., 2002, 2003).

Randive (2012) studied on the cultivation of oyster mushroom on paddy and wheat wastes. Nutritional analysis of cultivated mushroom revealed some differences within mushrooms of the same species cultivated on different substrates. According to Philippoussis et al. (2000), cultivation of *P. ostreatus*, *P. eryngii* and *P. pulmonarius* demonstrated a higher colonization rate on wheat straw and cotton waste substrates. They also reported that ratio of cellulose to lignin has a positive correlation with mycelium growth along with yield of *P. ostreatus*, and *P. pulmonarius*. Interestingly, wheat straw supplemented with cottonseed and soybean cake has proved to enhance the productivity of *P. ostreatus* (Upadhyay et al., 2002; Shah et al., 2004). It is a fact that Hasan et al. (2010) prepared substrates by mixing rice straw, poultry litter, banana leaf midribs, horse dung and lime at different ratio for the cultivation of oyster mushroom (*Pleurotus ostreatus*) and noted the highest mycelium running from banana leaf midribs + 10% horse dung + 1% lime.

Cassava is the most consumed food in African countries, and almost 700 million people depend on cassava. As a result, a large quantity of cassava waste is generated after processing, and the maximum amount is tuber peels. Cassava waste is usually used as a goat feed in Africa (Sonnenberg et al., 2014). However, several researchers have focused on the mushroom production from cassava waste. Onuoha et al. (2009) utilized sawdust, oil palm fiber, dry cassava peel, and mixture of these agro-wastes for mushroom cultivation (*Pleurotus pulmonarius*). Adebayo et al. (2009) prepared substrates by mixing cotton waste and cassava peel for the production of *Pleurotus pulmonarius*. Results revealed that the yield of *Pleurotus pulmonarius* from cassava peel was comparatively lower than other substrates due to insufficient nitrogen content (Adebayo et al., 2009; Onuoha et al., 2009). Furthermore, Sonnenberg et al. (2014) used cassava peels and stems with different ratios of rice or wheat bran for cultivation of *Pleurotus ostreatus* and *P. pulmonarius*. They reported that bran supplementation increased the production of oyster mushroom.

Pineapple is one of the most consumed fruits in the world, which can be processed into juice, squash, jam, jelly, and slice can. As a consequent, a large volume of waste is produced from processing industry (Bresolin et al., 2013). Pineapple waste can be used in human nutrition through bioconversion (Martin et al., 2012). According to Fortkamp and Knob (2014), pineapple peel is nutritionally superior to edible portion because of higher dietary fiber and protein content, but it is commonly used as animal feed and in the soil amendment. In a recent work, Souza et al. (2016) cultivated three types of mushroom, namely *Pleurotus albidus*, *Lentinus citrinus*, and *Pleurotus florida* from pineapple waste. In addition, they reported all essential amino acids with no toxic element in cultivated mushrooms. Nwachukwu and Adedokun (2014) conducted a research with king tuber mushroom (*Pleurotus tuber-regium*) cultivation on several wastes including sawdust, mixture of sawdust and paper waste, and mixture of fluted pumpkin stem and paper waste. They reported that the maximum production of king tuber mushroom on mixture of fluted pumpkin stem and paper waste compared with other mixtures. Obodai et al. (2003) used banana leaves, cocoyam peelings and oil-palm pericarp for the production of mushroom. As a pretreatment, dried banana leaves and cocoyam peelings were shocked in water overnight, whereas the oil-palm pericarp was kept in water for 20–30 minute. Two strains of *Volvariella volvacea* were evaluated, and the highest production of both strains was noted for banana leaves.

Lakshmi and Sornaraj (2014) conducted an experiment to eliminate seafood waste through mushroom cultivation (*Pleurotus flabellatus*) in laboratory condition. They mixed cooked fish waste with agro-industrial wastes, such as sugarcane bagasse, woodchips, and pith at the ratio of (1:1) and allowed to decompose for 15 days. Interestingly, Chang, and Miles (1993) stated that the utilization of liquid inoculum could reduce the production cost and make the mushroom cultivation process easier. Furthermore, Friel, and McLoughlin (2000) stated that instead of solid inoculum, liquid inoculum could be directly used as substrate to cultivate mushroom. Silveira et al. (2008) cultivated *Pleurotus ostreatus* in banana straw using liquid inoculum and compared with the results of using solid medium. In addition, they mixed 5% rice bran with dried banana straw powder for using as substrate in the production of *Pleurotus ostreatus*.

6.5   CONCLUSION

The increasing amount of waste from different sources ranging from agricultural sector to processing area will be environmental and health issue if necessary actions are not be implied immediately. In this sense, biodegradation is the most effective way of waste management than other methods while production of bioenergy through biodegradation is promising. The increasing demand of energy will be a prime concern in the near future, and hence, the organic waste has aroused as an important resource for yielding bioenergy. If we can ensure the proper utilization of organic waste in producing energy, it will be the most significant dealings with waste management. Furthermore, the production of renewable energy through anaerobic digestion by using organic wastes has significantly remarked around the world due to its availability, economical, and renewable advantages. Several scholars have motivated their attention to bioenergy production in terms of biogas and bioethanol by anaerobic digestion as well as co-digestion. Researchers have also developed a number of microbial strains for bioconversion of different wastes. The future work regarding environmental protection through waste management will focus on the development of new microbial strain for specific organic waste.

KEYWORDS

- **agricultural residue**
- **anaerobic digestion**
- **bioconversion**
- **biodegradation**
- **bioethanol**
- **biogas**
- **bioremediation**
- **dairy industry waste**
- **effluent**
- **environmental pollution**
- **fish processing waste**
- **fruit and vegetable waste**
- **lignocellulosic biomass**
- **microbial degradation**
- **mushroom production**
- **renewable energy**
- **slaughterhouse waste**

REFERENCES

Abraham, J., Gea, T., & Sanche, A. Substitution of chemical dehairing by proteases from solid-state fermentation of hair wastes. *J. Clean Prod.*, **2014**, *74*, 191–198.

Ackerley, D. F., Gonzalez, C. F., Park, C. H., Blake, R., Keyhan, M., & Matin, A. Chromate reducing properties of soluble flavoproteins from *Pseudomonas putida* and *Escherichia coli*. *Applied and Environmental Microbiology*, **2004**, *70*, 873–882.

Adebayo, G. J., Omolara, B. N., & Toyin, A. E. Evaluation of yield of oyster mushroom (*Pleurotus pulmonarius*) grown on cotton waste and cassava peel. *Afr. J. Biotechnol.*, **2009**, *8*(2), 215–218.

Adoki, A. Factors affecting yeast growth and protein yield production from orange, plantain and banana waste processing residues using *Candida* sp. *Afr. J. Biotechnol.*, **2008**, *7*(3), 290–295.

Afonso, M. D., & Borquez, R. Review of the treatment of seafood processing wastewaters and recovery of proteins therein by membrane separation processes: Prospects of the ultrafiltration of wastewaters from the fish meal industry. *Desalination*, **2002**, *142*, 29–45.

Ageitos, J. M., Vallejo, J. A., Veiga-Crespo, P., & Villa, T. G. Oily Yeasts as oleaginous cell factories. *Appl. Microbiol. Biotechnol.*, **2011**, *90*, 1219–1227.

Ahring, B. K., Mladenovska, Z., & Iranpour, R. State of the art and future perspectives of thermophilic anaerobic digestion. In: *Proceedings of 9th World Congress on Anaerobic Digestion* (pp. 455–460). Antwerp, Belgium, **2001**.

Alburquerque, J. A., De La Fuente, C., Ferrer-Costa, A., Carrasco, L., Cegarra, J., Abad, M., & Bernal, M. P. Assessment of the fertilizer potential of digestates from farm and agro-industrial residues. *Biomass Bioenergy*, **2012**, *40*, 181–189.

Alexandratos, N. *World Food and Agriculture to, 2030/50 Revisited* (pp. 24–26). FAO, Rome, **2011**.

Alvarez, R., & Liden, G. Semi-continuous co-digestion of solid slaughterhouse waste, manure, and fruit and vegetable waste. *Renew Energ.*, **2008**, *33*, 726–734.

Ander, P., & Eriksson, K. E. Selective degradation of wood components by white-rot fungi. *Physiologia Plantarum*, **2006**, *41*, 239–248.

Angelidaki, I., Ellegaard, L., & Ahring, B. K. Applications of the anaerobic digestion process. In: Scheper, T., (ed.), *Biomethanation, II* (pp. 1–33). Springer-Verlag, Berlin, Heidelberg, **2003**.

Aslan, S., Alyuz, B., Bozkurt, Z., & Bakaoğlu, M. Characterization and biological treatability of edible oil wastewaters. *Polish J. Environ. Stud.*, **2009**, *18*(4), 533–538.

Atalo, K., & Gashe, B. A. Protease production by a thermophilic *Bacillus* species P-001A which degrades various kinds of fibrous proteins. *Biotechnology Letters*, **1993**, *15*, 1151–1156.

AWARENET. *Handbook for the Prevention and Minimization of Waste and Valorization of By-Products in European Agro-Food Industries* (pp. 1–7). Agro-Food Waste Minimization and Reduction Network Grow Programme, European Commission, **2004**.

Azzam, A. M. Production of metabolites, industrial enzymes, amino acids. *J. Environ. Sci. Eng.*, **1992**, *56*, 67–99.

Balat, M., Balat, H., & Öz, C. Progress in bioethanol processing. *Progress in Energy and Combustion Science*, **2008**, *34*(5), 551–573.

Baldrian, P., & Valaskova, V. Degradation of cellulose by Basidiomycetous fungi. *FEMS Microbiol. Rev.*, **2008**, *32*, 501–521.

Bálint, B., Bagi, Z., Toth, A., Rakhely, G., Perei, K., & Kovacs, K. L. Utilization of keratin-containing biowaste to produce biohydrogen. *Applied Microbiology and Biotechnology*, **2005**, *69*, 404–410.

Ball, A. S., Betts, W. B., & McCarthy, A. J. Degradation of lignin-related compounds by actinomycetes. *Applied and Environmental Microbiology*, **1989**, *55*(6), 1642–1644.

Barakat, A., Chuetor, S., Monlau, F., Solhy, A., & Rouau, X. Eco-friendly dry chemo-mechanical pretreatments of lignocellulosic biomass: Impact on energy and yield of the enzymatic hydrolysis. *Applied Energy*, **2014**, *113*, 97–105.

Bartolome, B., Gomez-Cordoves, C., Sancho, A. I., Diez, N., Ferreira, P., Soliveri, J., & Copa-Patino, J. Growth and release of hydroxycinnamic acids from brewer's spent grain by *Streptomyces avermitilis* CECT**3339**. *Enzym. Microb. Technol.*, **2003**, *32*, 140–144.

Barton, A. F. M. Industrial and agricultural recycling processing. In: *Resources Recovery and Recycling*. John Wiley and Sons Ltd., New York, **1999**.

Baruah, A. K., Sharma, R. N., & Borah, G. C. Impact of sugar mill and distillery effluent on water quality of the River Galabil, Assam. *Indian Journal of Environmental Health*, **1993**, *35*, 288–293.

Baskaran, L., Ganesh, K. S., Chidambaram, A. L. A., & Sundaramorthy, P. Amelioration of sugar mill effluent polluted soil and its effect of green gram. *Botany Research International*, **2009**, *2*(2), 131–135.

Bastos, R. G., Bonini, M. A., Zepka, L. Q., Jacob-Lopes, E., & Isabel, M. Treatment of rice parboiling wastewater by cyanobacterium *Aphanothece microscopica* Nageli with potential for biomass products. *Desalination and Water Treatment*, **2014**, *56*(3), 608–614.

Begea, M., Sirbu, A., Kourkoutas, Y., & Dima, R. Single-cell protein production of *Candida* strains in culture media based on vegetal oils. *Romanian Biotechnological Letters*, **2011**, *17*(6), 7776–7786.

Bekatorou, A., Psarianos, C., & Koutinas, A. A. Production of food grade yeasts. *Food Technol. Biotechnol.*, **2006**, *44*, 407–415.

Ben, R. F., Frikha, F., Kammoun, W., Belbahri, L., Gargouri, Y., & Miled, N. Culture of *Staphylococcus xylosus* in fish processing by-product-based media for lipase production. *Lett. Appl. Microbiol.*, **2008**, *47*, 549–554.

Benjamin, Y., Cheng, H., & Görgens, J. F. Optimization of dilute sulfuric acid pretreatment to maximize combined sugar yield from sugarcane bagasse for ethanol production. *Appl. Biochem. Biotechnol.*, **2014**, *172*, 610–630.

Berardini, N., Knodler, M., Schieber, A., & Carle, R. Utilization of mango peels as a source of pectin and polyphenolics. *Innovative Food Sci. Emerg. Technol.*, **2005**, *6*, 442–452.

Bernas, E., Jaworska, G., & Lisiewska, Z. Edible mushrooms as a source of valuable nutritive constituents. *Acta. Sci. Pol. Technol. Aliment*, **2006**, *5*(1), 5–20.

Bhalla, T. C., Sharma, N. N., & Sharma, M. Production of metabolites, industrial enzymes, amino acids, organic acids, antibiotics, vitamins and single cell proteins. *J. Environ.*, **2007**, *6*, 34–78.

Bjerre, A. B., Olesen, A. B., Fernqvist, T., Plöger, A., & Schmidt, A. S. Pretreatment of wheat straw using combined wet oxidation and alkaline hydrolysis resulting in convertible cellulose and hemicellulose. *Biotechnology and Bioengineering*, **1996**, *49*(5), 568–577.

Blanco, M., Sotelo, C. G., Chapela, M. J., & Perez-Martin, R. I. Towards sustainable and efficient use of fishery resources: Present and future trends. *Trends Food Sci. Technol.*, **2006**, *18*, 29–36.

Blonskaja, V., Kamenev, I., & Zub, S. Possibilities of using ozone for the treatment of wastewater from the yeast industry. *Proceedings of the Estonian Academy of Sciences*, **2006**, *55*(1), 29–39.

Bouallagui, H., Ben Cheikh, R., Marouani, L., & Hamdi, M. Mesopholic biogas production from fruit and vegetable waste in a tubular digester. *Bioresource Technology*, **2003**, *86*, 85–89.

Bouallagui, H., Touhami, Y., Ben, C. R., & Hamdi, M. Bioreactor performance in anaerobic digestion of fruit and vegetable wastes. *Process Biochemistry*, **2005**, *40*(3/4), 989–999.

Bozzano, A., & Sarda, F. Fishery discard consumption rate and scavenging activity in the northwestern Mediterranean Sea. *ICES J. Mar. Sci.*, **2002**, *59*, 15–28.

Bresolin, I. R., Bresolin, I. T., Silveira, E., Tambourgi, E. B., & Mazzola, P. G. Isolation and purification of bromelain from waste peel of pineapple for therapeutic application. *Brazilian Archives of Biology and Technology*, **2013**, *56*(6), 971–979.

Budiyono, B., Widiasa, I. N., Johari, S., & Sunarso, S. Study on slaughterhouse wastes potency and characteristic for biogas production. *International Journal of Waste Resources*, **2011**, *1*(2), 4–7.

Buvaneswari, S., Muthukumaran, M., Damodarkumar, S., & Murugesan, S. Isolation and identification of predominant bacteria to evaluate the bioremediation in sugar mill effluent. *Int. J. Curr. Sci.*, **2013**, *5*(1), 123–132.

Çaglarirmak, N. The nutrients of exotic mushrooms (*Lentinula edodes* and *Pleurotus* species) and an estimated approach to the volatile compounds. *Food Chem.*, **2007**, *105*(3), 1188–1184.

Callihan, C. D., & Clemmer, J. E. Biomass from cellulosic materials. In: Rose, A. H., (ed.), *Microbial Biomass-Economic Microbiology* (pp. 271–273). Academic Press, London, UK, **1979**.

Campbell, A. C., & Racjan, M. The commercial exploitation of the white-rot fungus *Lentinula edodes* (Shiitake). *Int. Biodeteriorat. Biodegrad.*, **1999**, *43*, 101–107.

Carucci, G., Carrasco, I., Trifoni, K., Majone, M., & Beccari, M. Anaerobic digestion of food industry wastes: effect of digestion on methane yield. *J. Environ. Eng. – ASCE*, **2005**, *131*, 1037–1045.

Catalkaya, E. C., & Sengul, F. Application of Box-Wilson experimental design method for the photodegradation of baker's yeast industry with UV/H_2O_2 and UV/H_2O_2/Fe (II) process. *Journal of Hazardous Materials*, **2006**, *28*, 201–207.

Chakraverty, A., Mujumdar, A. S., & Ramaswamy, H. S. *Handbook of Postharvest Technology: Cereals, Fruits, Vegetables, Tea, and Spices* (pp. 819–838). CRC Press, Florida, **2003**.

Chandel, A. K., Da Silva, S. S., Carvalho, W., & Singh, O. V. Sugarcane bagasse and leaves: Foreseeable biomass of biofuel and bioproducts. *Journal of Chemical Technology and Biotechnology*, **2012**, *87*(1), 11–20.

Chang, J., Cheng, W., Yin, Q. Q., Zuo, R. Y., Song, A. D., Zheng, Q. H., Wang, P., Wang, X., & Liu, J. X. Effect of steam explosion and microbial fermentation on cellulose and lignin degradation of corn stover. *Bioresource Technology*, **2012**, *104*, 587–592.

Chang, S. T. A 40-year journey through bioconversion of lignocellulosic wastes to mushrooms and dietary supplements. *Int. J. Med. Mush.*, **2001**, *3*(4), 299–310.

Chang, S. T. The world mushroom industry: trends and technological development. *Int. J. Med. Mush.*, **2006**, *8*(4), 297–314.

Chang, S. T., & Miles, P. G. Mushrooms: Trends in production and technological development. *Genetic Engineering and Biotechnology Monitor*, **1993**, *41*(42), 73–81.

Chellapandi, P. Enzymes and microbiological pretreatments of oil industry wastes for biogas production in batch digesters. In: Pathade, G. R., & Goel, P. K., (eds.), *Biotechnology and Environmental Management* (pp. 39–74). ABD Publishers, Jaipur, **2004**.

Chen, A. W., Arrold, N., & Stamets, P. Shiitake cultivation systems. In: Van Griensven, J. L. J. D., (ed.), *Proceedings of the XVth International Congress on the Science and Cultivation of Edible Fungi*. Balkema, Rotterdam, **2000**.

Chen, C. C., Chaung, H. C., Chung, M. Y., & Huang, L. T. Menhaden fish oil improves spatial memory in rat pups following recurrent pentylenetetrazole-induced seizures. *Epilepsy Behav.*, **2006**, *8*, 516–521.

Chiu, S. W., Law, S. C., & Ching, M. L. Themes for mushroom exploitation in the 21st century: Sustainability, waste management, and conservation. *J. Gen. Appl. Microbiol.*, **2000**, 46, 269–282.

Choi, M. H., & Park, Y. H. Production of yeast biomass using waste Chinese cabbage. *Biomass and Bioenergy*, **2003**, *25*(2), 221–226.

Coello, N., Brito, L., & Nonus, M. Biosynthesis of L-lysine by *Corynebacterium glutamicum* grown on fish silage. *Bioresour. Technol.*, **2000**, *73*, 221–225.

CRAR. *Annual Report*. Corn Refiners Association, Washington, D.C., USA, 2009.

Cuetos, M. J., Gomez, X., Otero, M., & Moran, A. Anaerobic digestion of solid slaughterhouse waste (SHW) at laboratory scale: Influence of co-digestion with the organic fraction of municipal solid waste (OFMSW). *Biochem. Eng. J.*, **2008**, *40*, 99–106.

Daba, A. S., & Ezeronye, O. U. Anti-cancer effect of polysaccharides isolated from higher basidiomycetes mushrooms. *African Journal of Biotechnology*, **2003**, *2*(12), 672–678.

Dahiya, N., Tewari, R., & Hoondal, G. S. Biotechnological aspects of chitinolytic enzymes: A review. *Appl. Microbiol. Biotechnol.*, **2006**, *71*, 773–782. Dahunsi, S. O., Owolabi, J. B., & Oranusi, S. Biogas generation from watermelon peels, pineapple peels and food wastes. *International Conference on African Development Issues*, Covenant University, Ota, Nigeria, **2015**.

Danzo, B. J. Environmental xenobiotics may disrupt normal endocrine function by interfering with the binding of physiological ligands to steroid receptors and binding proteins. *Environ Health Perspect*, **1997**, *105*, 294–301.

Dao, V. T., & Kim, J. K. Scaled-up bioconversion of fish waste to liquid fertilizer using a 5 L ribbon-type reactor. *J. Environ. Manage*, **2011**, *92*, 2441–2446.

Das, A., & Mondal, C. Catalytic effect of tungsten on anaerobic digestion process for biogas production from fruit and vegetable wastes. *International Journal of Scientific Engineering and Technology*, **2013**, *2*(4), 216–221.

De Azeredo, L. A. I., Freire, D. M. G., Soares, R. M. A., Leite, S. G. F., & Coelho, R. R. R. Production and partial characterization of thermophilic proteases from *Streptomyces* sp. isolated from Brazilian cerrado soil. *Enzym. Microb. Technol.*, **2004**, *34*, 354–358.

De Los Santos, D. G., Turnes, C. G., & Conceicao, F. R. Bioremediation of parboiled rice effluent supplemented with biodiesel-derived glycerol using *Pichia pastoris* X-33. *The Scientific World Journal*, **2012**, 492925, doi:10.1100/2012/492925.

Demirel, B., Yenigun, O., & Onay, T. T. Anaerobic treatment of dairy wastewaters: A review. *Proc. Biochem.*, **2005**, *40*, 2583–2595.

Dhanasekaran, D., Lawanya, S., Saha, S., Thajuddin, N., & Panneerselvam, A. Production of single cell protein from pineapple waste using yeast. *Innovative Romanian Food Biotechnology*, **2011**, *8*, 26–32.

Diego, C. Z., Emilio, P. G. J., Marla, T. A. M., & Arturo, P. G. A reliable quality index for mushroom cultivation. *J. Agric. Sci.*, **2011**, *3*(4), 50–56.

Dudynski, M., Kwiatkowski, K., & Bajer, K. From feathers to syngas – technologies and devices. *Waste Management*, **2012**, *32*, 685–691.

Duran de Bazua, C., Sanchez-Tovar, S. A., Hernandez-Morales, M. R., & Bernal-Gonzalez, M. Use of anaerobic-aerobic treatment systems for maize processing installations: Applied microbiology in action. In: Mendez-Vilas, A., (ed.), *Communicating Current Research and Educational Topics and Trends in Applied Microbiology* (pp. 3–12). Formatex Research Center, Badajoz, Spain, **2007**.

Eddy, M. *Wastewater Engineering, Treatment Disposal and Reuse*, McGraw Hill, New York, **2003**.

Elaiyaraju, P., & Partha, N. Biogas production from co-digestion of orange peel waste and jatropha de-oiled cake in an anaerobic batch reactor. *African Journal of Biotechnology*, **2012**, *11*(14), 3339–3345.

Eremektar, G., Karahan-Gul, O., Babuna, F. G., Ovez, S., Uner, H., & Orhon, D. Biological treatability of a corn wet mill. *Water Science and Technology*, **2002**, *45*(12), 339–346.

Ersahin, M. E., Insel, G., Dereli, R. K., Ozturk, I., & Kinaci, C. Model based evaluation for the anaerobic treatment of corn processing wastewaters. *Clean-Soil Air Water*, **2007**, *35*(6), 576–581.

Esakkiraj, P., Austin, J. D. G., Palavesam, A., & Immanuel, G. Media preparation using tuna-processing wastes for improved lipase production by shrimp gut isolate *Staphylococcus epidermidis* CMST Pi 2. *Appl. Biochem. Biotechnol.*, **2010a**, *160*, 1254–1265.

Esakkiraj, P., Immanuel, G., Sowmya, S. M., Iyapparaj, P., & Palavesam, A. Evaluation of protease-producing ability of fish gut isolate *Bacillus cereus* for aqua feed. *Food Bioprocess Technol.*, **2009**, *2*, 383–390.

Esakkiraj, P., Rajkumarbharathi, M., Palavesam, A., & Immanuel, G. Lipase production by *Staphylococcus epidermidis* CMST-Pi 1 isolated from the gut of shrimp *Penaeus indicus*. *Ann. Microbiol.*, **2010b**, *60*, 37–42.

Fan, L., Pandey, A., & Mohan, R. Use of various coffee industry residues for the cultivation of *Pleurotus ostreatus* in solid state fermentation. *Acta Biotechnol.*, **2000a**, *20*(1), 41–52.

Fan, L., Pandey, A., & Soccol, C. R. Solid state cultivation–an efficient method to use toxic agro-industrial residues. *J. Basic Microbiol.*, **2000b**, *40*(3), 187–197.

FAO. The State of World Fisheries and Aquaculture. Food and Agriculture Organization of the United Nations, Rome, **2007**.

FAO. Towards the future we want: End hunger and make the transition to sustainable agricultural and food systems. Food and Agriculture Organization of the United Nations, Rome, **2014**.

FAOSTAT. Statistics Division, Food and Agriculture Organization of the United Nations, Rome, Italy, **2004**.

Feng, C., Shimada, S., Zhang, Z., & Maekawa, T. A pilot plant two-phase anaerobic digestion system for bioenergy recovery from swine wastes and garbage. *Waste Management*, **2008**, *28*, 1827–1834.

Fortkamp, D., & Knob, A. High xylanase production by *Trichoderma viride* using pineapple peel as substrate and its application in pulp biobleaching. *African Journal of Biotechnology*, **2014**, *13*(22), 2248–2259.

Friedrich, J., Gradisar, H. D., & Mandin, C. J. P. Screening fungi for synthesis of keratinolytic enzymes. *Letters in Applied Microbiology*, **1999**, *28*, 127–130.

Friel, M. T., & Mcloughlin, A. J. Production of a liquid inoculum/spawn of *Agaricus bisporus*. *Biotechnology Letters*, **2000**, *22*(5), 351–354.

Gabrielle, B., Bamiere, L., Caldes, N., De Cara, S., Decocq, G., Ferchaud, F., Loyce, C., Pelzer, E., Perez, Y., & Wohlfahrt, J. Paving the way for sustainable bioenergy in Europe: technological options and research avenues for large-scale biomass feedstock supply. *Renew Sustain Energy Rev.*, **2014**, *33*, 11–25.

Ganesh, R., Balaji, G., & Ramanujam, R. A. Biodegradation of tannery wastewater using sequencing batch reactor – respirometric assessment. *Bioresource Technol.*, **2006**, *97*, 1815–1821.

Gao, M. T., Hirata, M., Toorisaka, E., & Hano, T. Acid-hydrolysis of fish wastes for lactic acid fermentation. *Bioresour. Technol.*, **2006**, *97*, 2414–2420.

Gassara, F., Brar, S. K., Tyagi, R. D., Verma, M., & Surampalli, R. Y. Screening of agro-industrial wastes to produce ligninolytic enzymes by *Phanerochaete chrysosporium*. *Biochem. Eng. J.*, **2010**, *49*, 388–394.

Gelegenis, J., Georgakakis, D., Angelidaki, I., & Mavris, V. Optimization of biogas production by co-digestion whey with diluted poultry manure. *Renew Energy*, **2007**, *32*, 2147–2160.

Gerardi, M. H. Anaerobic digestion stages. In: *The Microbiology of Anaerobic Digesters* (pp. 51–57). John Wiley and Sons Inc., Hoboken, 2003.

Ghazi, I., Fernandez-Arrojo, L., Gomez, D. S. A., Alcalde, M., Plou, F. J., & Ballesteros, A. Beet sugar syrup and molasses as low-cost feedstock for the enzymatic production of fructo-oligosaccharides. *J. Agric. Food Chem.*, **2006**, *54*, 2964–2968.

Ghorbel, S., Soussi, N., Ellouz, Y. T., Duffosse, L., Guerard, F., & Nazri, M. Preparation and testing of Sardinella protein hydrolysate as nitrogen source for extracellular lipase production by *Rhizopus oryzae*. *World J. Microbiol. Biotechnol.*, **2005**, *21*, 33–38.

Graminha, E. B. N., Gonçalves, A. Z. L., Pirota, R. D. P. B., Balsalobre, M. A. A., Da Silva, R., & Gomes, E. Enzyme production by solid-state fermentation: Application to animal nutrition. *Animal Feed Science and Technology*, **2008**, *144*(1/2), 1–22.

Gruno, M., Vaeljamaee, P., Pettersson, G., & Johansson, G. Inhibition of the *Trichoderma reesei* cellulases by cellobiose is strongly dependent on the nature of the substrate. *Biotechnol. Bioeng.*, **2004**, *86*, 503–511.

Gulmez, B., Ozturk, I., Alp, K., & Arikan, O. A. Common anaerobic treatability of pharmaceutical and yeast industry wastewater. *Water Science and Technology*, **1998**, *38*(4/5), 37–44.

Gupta, R., & Ramnani, P. Microbial keratinases and their prospective applications: An overview. *Applied Microbiology and Biotechnology*, **2006**, *70*, 21–33.

Gupta, R., Sharma, R., & Beg, Q. K. Revisiting microbial keratinases: Next generation of proteases for sustainable biotechnology. *Critical Reviews in Biotechnology*, **2013**, *33*, 216–228.

Gustavo, B. C. V., Juan, F. M. J., Stella, M., Maria, S. T., Aurelio, L. M., & Jorge, W. C. *Handling and Preservation of Fruits and Vegetables by Combined Methods for Rural Areas*. Food and Agriculture Organization of the United Nations, Rome, Italy, **2003**.

Gustavsson, J., Cederberg, C., Sonesson, U., Van Otterdijk, R., & Meybeck, A. *Global Food Losses and Food Waste*. Food and Agriculture Organization of the United Nations, Rome, Italy, **2011**.

Haddar, A., Fakhfakh-Zouari, N., Hmidet, N., Frikha, F., Nasri, M., & Sellami, K. A. Low-cost fermentation medium for alkaline protease production by *Bacillus mojavensis* A21 using hulled grain of wheat and Sardinella peptone. *J. Biosci. Bioeng.*, **2010**, *110*, 288–294.

Hahn-Hagerdal, B., Karhumaa, K., Fonseca, C., Spencer-Martins, I., & Gorwa-Grauslund, M. F. Towards industrial pentose-fermenting yeast strains. *Appl. Microbiol. Biotechnol.*, **2007**, *74*, 937–953.

Hasan, M. N., Rahman, M. S., Nigar, S., Bhuiyan, M. Z. A., & Ara, N. Performance of oyster mushroom (*Pleurotus ostreatus*) on different pretreated substrates. *Int. J. Sustain. Crop Prod.*, **2010**, *5*(4), 16–24.

Hongzhang, C., Jian, L., & Zuohu, L. Production of single cell protein by fermentation of extract from hemicellulose autohydrolyzate. *Engineering Chemistry and Metallurgy*, **1999**, *20*(4), 428–431.

Hossain, A. B. M. S., Saleh, A. A., Aishah, S., Boyce, A. N., Chowdhury, P. P., & Naqiuddin, M. Bioethanol production from agricultural waste biomass as a renewable bioenergy resource in biomaterials. *4th Kuala Lumpur International Conference on Biomedical Engineering*. Kuala Lumpur, Malaysia, **2008**.

Howgrave-Graham, A. R., Isherwood, H. I., & Wallis, F. M. Evaluation of two up flow anaerobic digesters purifying industrial wastewaters high in organic matter. *Water Science and Technology*, **1994**, *29*(9), 225–229.

Ichida, J. M., Krizova, L., Le Fevre, C. A., Keener, H. M., Elwell, D. L., & Burtt, E. H. Bacterial inoculum enhances keratin degradation and biofilm formation in poultry compost. *Journal of Microbiological Methods*, **2001**, *47*, 199–208.

Ingale, S., Joshi, S. J., & Gupte, A. Production of bioethanol using agriculture waste: Banana pseudo stem. *Braz. J. Microbiol.*, **2014**, *45*(3), 885–892.

Irfan, M., Nadeem, M., & Syed, Q. Ethanol production from agriculture wastes using *Saccharomyces cerevisiae*. *Braz. J. Microbiol.*, **2014**, *45*(2), 457–465.

Isroi, M. R., Syamsiah, S., Niklasson, C., Cahyanto, M. N., Ludquist, K., & Taherzadeh, M. J. Biological pretreatment of lignocellulose with with-rot fungi and its application: A review. *BioResources*, **2011**, *6*, 5224–5259.

Jamel, P., Alam, M. Z., & Umi, N. Media optimization for bio proteins production from cheaper carbon source. *J. Engi. Sci. Technol.*, **2008**, *3*(2), 124–130.

Jawad, A. H., Alkarkhi, A. F. M., Jason, O. C., Easa, A. M., & Norulaini, N. A. N. Production of lactic acid from mango peel waste, factorial experiment. *Journal of King Saud University – Science*, **2013**, *25*, 39–45.

Jayathilakan, K., Sultana, K., Radhakrishna, K., & Bawa, A. S. Utilization of byproducts and waste materials from meat, poultry and fish processing industries: A review. *Journal of Food Science and Technology*, **2012**, *49*, 278–293.

Jenitta, X. J., Gnanasalomi, V. D. V., & Gnanadoss, J. J. Treatment of leather effluents and waste using fungi. *International Journal of Computing Algorithm*, **2013**, *2*, 294–298.

Jiang, Y., Heaven, S., & Banks, C. J. Strategies for stable anaerobic digestion of vegetable waste. *Renewable Energy*, **2012**, *44*, 206–214.

Joo, H. S., & Chang, C. S. Production of protease from a new alkalophilic *Bacillus* sp. I-312 grown on soybean meal: Optimization and some properties. *Proc. Biochem.*, **2005**, *40*, 1263–1270.

Jørgensen, H., Sanadi, A. R., Felby, C., Lange, N. E. K., Fischer, M., & Ernst, S. Production of ethanol and feed by high dry matter hydrolysis and fermentation of Palm kernel press cake. *Appl. Biochem. Biotechnol.*, **2010**, *161*, 318–332.

Joseph, J. K., & Abolaji, J. Effects of replacing maize with graded levels of cooked Nigerian mango-seed kernels (*Mangifera indica*) on the performance, carcass yield and meat quality of broiler chickens. *Bioresour. Technol.*, **1997**, *61*, 99–102.

Kacem, M., Sellami, M., Kammoun, W., Frikha, F., Miled, N., & Ben, R. F. Seasonal variations of chemical composition and fatty acid profiles of viscera of three marine species from the Tunisian coast. *J. Aquat. Food Prod. Technol.*, **2011**, *20*, 233–246.

Kalyuzhnyi, S., Gladchenko, M., Starostina, E., Shcherbakov, S., & Versprille, A. Combined biological and physico-chemical treatment of baker's yeast wastewater. *Water Science and Technology*, **2005**, *52*(1/2), 175–181.

Kameswari, S. B. K., Velmurugan, B., Thirumaran, K., & Ramanujam, R. A. Biomethanation of vegetable market waste–untapped carbon trading opportunities. *Proceedings of the International Conference on Sustainable Solid Waste Management* (pp. 415–420). Chennai, India, **2007**.

Karim, A. A., & Bhat, R. Fish gelatin: properties, challenges, and prospects as an alternative to mammalian gelatins. *Food Hydrocolloids*, **2009**, *23*, 563–576.

Kim, J. K., & Lee, G. Aerobically biodegraded fish-meal wastewater as a fertilizer. *Environ. Res. J.*, **2009**, *3*, 219–236.

Kim, J. K., Dao, V. T., Kong, I. S., & Lee, H. H. Identification and characterization of microorganisms from earthworm viscera for the conversion of fish wastes into liquid fertilizer. *Bioresour. Technol.*, **2010**, *101*, 5131–5136.

Kim, S. K., & Mendis, E. Bioactive compounds from marine processing by-products – a review. *Food Res. Int.*, **2006**, *39*, 383–393.

Kirk, T. K. Degradation and conversion of lignocelluloses. In: Smith, J. E., Berry, D. R., Kristiansen, B., (eds.), *The Filamentous Fungi, Fungal Technology* (pp. 266–295). Edward Arnold, London, **1983**.

Kornillowicz-Kowalska, T., & Bohacz, J. Biodegradation of keratin waste: Theory and practical aspects. *Waste Management*, **2011**, *31*, 1689–1701.

Kovacevic, F. Z., Sipos, L., & Briski, F. Biosorption of chromium, copper, nickel and zinc ions onto fungal pellets of *Aspergillus niger* 405 from aqueous solutions. *Food Technology and Biotechnology*, **2000**, *38*, 211–216.

Krapivina, M., Kurissoo, T., Blonskaja, V., Zub, S., & Vilu, R. Treatment of sulfate containing yeast wastewater in an anaerobic sequence batch reactor. *Proceedings of the Estonian Academy of Sciences*, **2007**, *56*(1), 38–52.

Kues, U., & Liu, Y. Fruiting body production in basidiomycetes. *Appl. Microbiol. Biotechnol.*, **2000**, *54*, 141–152.

Kurbanoúlu, E. B. Production of single cell protein from ram horn hydrolysate. *Turk. J. Biol.*, **2001**, *25*, 371–377.

Lakshmi, S. S., & Sornaraj, R. Utilization of seafood processing wastes for cultivation of the edible mushroom *Pleurotus flabellatus*. *African Journal of Biotechnology*, **2014**, *13*(17), 1779–1785.

Lal, R. World crop residues production and implications of its use as a biofuel. *Environ. Int.*, **2005**, *31*(4), 575–584.

Landine, R. C., Brown, G. J., Cocci, A. A., & Virara, H. Anaerobic treatment of high strength, high solids potato waste. *Agric. Wastes*, **1983**, *17*, 111–123.

Lasekan, A., Bakar, A., & Hashim, D. Potential of chicken by-products as sources of useful biological resources. *Waste Management*, **2013**, *33*, 552–565.

Lateef, A., Chaudhry, M. N., & Ilyas, S. Biological treatment of dairy wastewater using activated Sludge. *Science Asia*, **2013**, *39*, 179–185.

Laufenberg, G., Kunz, B., & Nystroem, M. Transformation of vegetable waste into value added products: (A) the upgrading concept; (B) practical implementations. *Bioresour. Technol.*, **2003**, *87*, 167–198.

Lehtomäki, A., Huttunen, S., & Rintala, J. A. Laboratory investigations on co-digestion of energy crops and crop residues with cow manure for methane production: effect of crop to manure ratio. *Resources, Conservation and Recycling*, **2007**, *51*(3), 591–609.

Li, Q., Du, W., & Liu, D. Perspectives of microbial oils for biodiesel production. *Appl. Microbiol. Biotechnol.*, **2008**, *80*, 749–756.

Liang, Z., Wang, Y., Zhou, Y., & Liu, H. Coagulation removal of melanoidins from biologically treated molasses wastewater using ferric chloride. *Chemical Engineering Journal*, **2009**, *152*, 88–94.

Lio, J., & Wang, T. Solid-state fermentation of soybean and corn processing coproducts for potential feed improvement. *Journal of Agricultural and Food Chemistry*, **2012**, *60*(31), 7702–7709.

Litchfield, J. H. Single-cell proteins. *Science*, **1983**, *219*, 740–746.

Liu, Y., & Whitman, W. B. Metabolic, phylogenetic, and ecological diversity of the methanogenic archaea. *Annals of the New York Academy of Sciences*, **2008**, *1125*(1), 171–189.

Long, H., Li, X., Wang, H., & Jia, J. Biomass resources and their bioenergy potential estimation: A review. *Renew. Sustain. Energy Rev.*, **2013**, *26*, 344–352.

Madigan, T. M., Martinko, J. M., & Parker, J. *Brock's Biology of Microorganisms* (pp. 1–29). Prentice-Hall, Englewood Cliffs, NJ, **2000**.

Martin, J. G. P., Junior, M. D. M., Almeida, M. A. A., Santos, T., & Spoto, M. H. F. Avaliação sensorial de bolo com resíduo de casca de abacaxi para suplementação do teor de fibras. *Revista Brasileira de Produtos Agroindustriais*, **2012**, *14*(3), 281–287.

Mata-Alvarez, J., Dosta, J., Romero-Güiza, M. S., Fonoll, X., Peces, M., & Astals, S. A critical review on anaerobic co-digestion achievements between 2010 and 2013. *Renewable and Sustainable Energy Reviews*, **2014**, *36*, 412–427.

Mkhize, S. P., Atkinson, B. W., & Bux, F. Assessment of a biological nutrient removal process for remediation of edible oil effluent. *Water SA*, **2000**, *26*(4), 555–558.

Mohamedin, A. H. Isolation, identification and some cultural conditions of a protease producing thermophilic *Streptomyces* strain grown on chicken feather as a substrate. *International Journal of Biodeterioration and Biodegradation*, **1999**, *43*, 13–21.

Mohapatra, P. K. X. *Textbook of Environmental Biotechnology*. IK International Publishing House, New Delhi, India, **2006**.

Mondal, A. K. Production of single cell protein from fruits waste by using *Saccharomyces cerevisiae*. *Am. J. Food Technol.*, **2006**, *58*, 117–134.

Mondal, A. K., Sengupta, S., & Bhowal, J., Bhattacharya D K. Utilization of fruit wastes in producing single cell protein. *International Journal of Science Environment and Technology*, **2012**, *1*(5), 430–438.

Movahedin, H., Shokoohi, R., Parvaresh, A., Hajia, M., & Jafri, J. A. Evaluating the effect of glucose on phenol removal efficiency and changing the dominant microorganisms in the serial combined biological system. *J. Res. Health Sci.*, **2006**, *6*(1), 8–13.

Munavalli, G. R., & Saler, P. S. Treatment of dairy wastewater by water hyacinth. *Water Sci. Tech.*, **2009**, *59*, 713–722.

Mussatto, S. I., Ballesteros, L. F., Martins, S. L., & Teixeira, J. A. Use of agro-industrial wastes in solid-state fermentation processes. In: Show, K. Y., & Guo, X., (eds.), *Industrial Waste* (pp. 121–140). InTech, Croatia, **2012**.

Mutreja, R., Das, D., Goyal, D., & Goyal, A. Bioconversion of agricultural waste to ethanol by SSF using recombinant cellulase from *Clostridium thermocellum*. *Enzyme Research*, **2011**, 340279, https://doi.org/10.4061/2011/340279 (Accessed on 7 October 2019).

Nallathambi, G. V. Anaerobic digestion of biomass for methane production: A review. *Biomass Bioenergy*, **1997**, *13*, 83–114.

Nasseri, A., Rasoul-Amini, S., Morowvat, M., & Ghasemi, Y. Single cell protein: Production and process. *American Journal of Food Technology*, **2011**, *6*(2), 103–116.

Nathoa, C., Sirisukpoca, U., & Pisutpaisal, N. Production of hydrogen and methane from banana peel by two phase anaerobic fermentation. *Energy Procedia*, **2014**, *50*, 702–710.

Nelson, C., & Cox, M. *Principles of Biochemistry* (pp. 47–50). W. H. Freeman, New York, **2004**.

Nigam, J. Cultivation of *Candida langeronii* in sugar cane bagasse hemicellulosic hydrolyzate for the production of single cell protein. *World Journal of Microbiology and Biotechnology*, **2000**, *16*(4), 367–372.

Nitisinprasert, S., Pornuirum, W., & Keawsompong, S. Characterizations of two bacterial strains showing high keratinase activities and their synergism in feather degradation. *Kasetsart Journal of Natural Science*, **1999**, *33*, 191–199.

Nova, T. D. Utilization of rumen content of slaughterhouse waste fermented with *Trichoderma harzianum* as fed in the ducks. *MS Thesis*, University Northern Sumatera, Medan, **2000**.

Noval, J., & Nickerson, W. J. Decomposition of keratin by *Streptomyces fradiae*. *Journal of Bacteriology*, **1959**, *77*, 251–263.

Nwachukwu, P. C., & Adedokun, O. M. King tuber mushroom: Bioconversion of fluted pumpkin, sawdust and paper. *African Journal of Plant Science*, **2014**, *8*(3), 164–166.

Obodai, M., Cleland-Okine, J., & Johnson, P. N. T. Use of agricultural wastes as substrate for the mushroom *Volvariella volvacea*. *Trop. Sci.*, **2003**, *43*, 121–124.

Ojo, O. A. Petroleum hydrocarbon utilization by native bacterial population from a wastewater canal south west Nigeria. *Afr. J. Biotechnol.*, **2006**, 5333–5337.

Oleskowicz-Popiel, P., Kádár, Z., Heiske, S., Klein-Marcuschamer, D., Simmons, B. A., Blanch, H. W., & Schmidt, J. E. Co-production of ethanol, biogas, protein fodder and natural fertilizer in organic farming – evaluation of a concept for a farm-scale biorefinery. *Bioresour. Technol.*, **2012**, *104*, 440–446.

Onifade, A. A., Al-Sane, N. A., Al-Musallam, A. A., & Al-Zarban, S. A. Potentials for biotechnological applications of keratin-degrading microorganisms and their enzymes for nutritional improvement of feathers and other keratins as livestock feed resources. *Bioresour. Technol.*, **1998**, *66*, 1–11.

Onuoha, C. I., & Ukaulor, B. C. Cultivation of *Pleurotus pulmonarius* (mushroom) using some agrowaste materials. *Agricultural Journal*, **2009**, *4*(2), 109–112.

Pandey, A., Soccol, C. R., & Larroche, C. Current developments in solid-state fermentation, In: Pandey, A., Soccol, C. R., & Larroche, C., (eds.), *Current Developments in Solid-State Fermentation* (pp. 13–25). Springer Science, New York, USA, **2007**.

Pandey, A., Soccol, C. R., & Nigam, P. Biotechnological potential of agro-industrial residues: I–Sugarcane bagasse. *Bioresour. Technol.*, **2000a**, *74*, 69–80.

Pandey, A., Soccol, C. R., & Nigam, P. Biotechnological potential of agro-industrial residues: II – Cassava bagasse. *Bioresour. Technol.*, **2000b**, *74*, 81–87.

Pandey, A., Soccol, C. R., & Nigam, P. Biotechnological potential of coffee pulp and coffee husk for bioprocesses. *Biochem. Eng. J.*, **2000c**, *6*, 153–162.

Pangnakorn, U. Valuable added the agricultural waste for farmers using in organic farming groups in Phitsunlok, Thailand. In: *Proceeding of the Prosperity and Poverty in a Globalized World-Challenges for Agricultural Research* (pp. 11–13, 275–278). Bonn, Germany, **2006**.

Papanikolaou, S., Chevalot, I., Komaitis, M., Aggelis, G., & Marc, I. Kinetic profile of the cellular lipid composition in an oleaginous *Yarrowia lipolytica* capable of producing a cocoa-butter substitute from industrial fats. *Antonie Van Leeuwenhoek*, **2001**, *80*(3/4), 215–224.

Parawira, W., Read, J. S., Mattiasson, B., & Björnsson, L. Energy production from agricultural residues: High methane yields in pilot-scale two-stage anaerobic digestion. *Biomass and Bioenergy*, **2008**, *32*(1), 44–50.

Park, C. H., Keyhan, M., Wielinga, B., Fendorf, S., & Matin, A. Purification to homogeneity and characterization of a novel chromate reductase from *Pseudomonas putida*. *Applied and Environmental Microbiology*, **2000**, *66*, 1788–1792.

Paterson, R. R. M. *Ganoderma*-a therapeutic fungal biofactory. *Phytochem*, **2006**, *67*, 1985–2001.

Perez, S. R., Garcia, O. N., & Bermudez, R. C. Decolorisation of mushroom farm wasterwater by *Pleurotus ostreatus*. *Biodegrad.*, **2007**, *19*(4), 519–526.

Philippoussis, A., Diamantopoulou, P., & Euthimiadou, H. The composition and porosity of lignocellulosic substrates influence mycelium growth and respiration rates of *Lentinus edodes* (Berk.) *Sing Int. J. Med. Mush.*, **2001c**, *3*(2/3), 198.

Philippoussis, A., Diamantopoulou, P., & Zervakis, G. Calcium chloride irrigation influence on yield, calcium content, quality and shelf-life of the white mushroom *Agaricus bisporus*. *J. Sci. Food Agric.*, **2001b**, *81*(15), 1447–1454.

Philippoussis, A., Diamantopoulou, P., & Zervakis, G. Correlation of the properties of several lignocellulosic substrates to the crop performance of the shiitake mushroom *Lentinula edodes*. *World J. Microbiol. Biotechnol.*, **2003**, *19*(6), 551–557.

Philippoussis, A., Diamantopoulou, P., & Zervakis, G. Monitoring of mycelium growth and fructification of *Lentinula edodes* on several agricultural residues. In: Sanchez, J. E., Huerta, G., & Montiel, E., (eds.), *Mushroom Biology and Mushroom Products* (pp. 279–287). UAEM, Cuernavaca, **2002**.

Philippoussis, A., Diamantopoulou, R., Zervakis, G., & Ioannidou, S. Potential for the cultivation of exotic mushroom species by exploitation of Mediterranean agricultural wastes. In: Van Griensven, L. J. L. D., (ed.), *Proceedings of the XVth International Congress on the Science and Cultivation of Edible Fungi* (pp. 523–530). Balkema, Rotterdam, **2000**.

Philippoussis, A., Zervakis, G., & Diamantopoulou, P. Bioconversion of lignocellulosic wastes through the cultivation of the edible mushrooms *Agrocybe aegerita*, *Volvariella volvacea* and *Pleurotus* spp. *World J. Microbiol. Biotechnol.*, **2001a**, *17*(2), 191–200.

Poppe, J. Use of agricultural waste materials in the cultivation of mushrooms. In: Van Griensven, L. J. L. D., (ed.), *Proceedings of the XVth International Congress on the Science and Cultivation of Edible Fungi* (pp. 3–23). Balkema, Rotterdam, **2000**.

Porwal, H. J., Mane, A. V., & Velhal, S. G. Biodegradation of dairy effluent by using microbial isolates obtained from activated sludge. *Water Resources and Industry*, **2015**, *9*, 1–15.

Pozo del, R., Diez, V., & Beltran, S. Anaerobic pre-treatment of slaughterhouse wastewater using fixed-film reactors. *Bioresour. Technol.*, **2000**, *71*, 143–149.

Queiroz, M., & Koetz, P. R. Caracterização do efluente da parboilização do arroz. *Revista Brasileira de Agrociencia*, **1997**, *3*(3), 139–143.

Ragunathan, R., Gurusamy, R., & Palaniswamy, M. Cultivation of *Pleurotus* spp. on various agro residues. *Food Chem.*, **1996**, *55*(2), 139–144.

Raja, R., Hemaiswarya, S., Kumar, N. A., Sridhar, S., & Rengasamy, R. A perspective on the biotechnological potential of microalgae. *Cr. Rev. Microbiol.*, **2008**, *34*, 77–88.

Randive, S. D. Cultivation and study of growth of oyster mushroom on different agricultural waste substrate and its nutrient analysis. *Advances in Applied Science Research*, **2012**, *3*(4), 1938–1949.

Rao, M. N., & Datta, A. K. *Waste Water Treatment: Rational Methods of Design and Industrial Practices* (pp. 254–258). Oxford and IBH, Bombay, **1978**.

Rasoul-Amini, S., Ghasemi, Y., Morowvat, M. H., & Mohagheghzadeh, A. PCR amplification of 18S rRNA, single cell protein production and fatty acid evaluation of some naturally isolated microalgae. *Food Chem.*, **2009**, *116*, 129–136.

Rathna, G. S., Saranya, R., & Kalaiselvam, M. Bioethanol from sawdust using cellulose hydrolysis of *Aspergillus ochraceus* and fermentation by *Saccharomyces cerevisiae*. *Int. J. Curr. Microbiol. Appl. Sci.*, **2014**, *3*, 733–742.

Reddy, L. V., Reddy, O. V. S., & Wee, Y. J. Production of ethanol from mango (*Mangifera indica* L.) peel by *Saccharomyces cerevisiae* CFTRI101. *African Journal of Biotechnology*, **2011**, *10*, 4183–4189.

Reguła, J., & Siwulski, M. Dried shiitake (*Lentinulla edodes*) and oyster (*Pleurotus ostreatus*) mushroom as a good source of nutrient. *Acta. Sci. Pol. Technol. Aliment*, **2007**, *6*(4), 135–142.

Renganathan, P., Eswaran, A., & Balabaskar, P. Effect of various additives to the bed substrate on the sporophore production by *Pleurotus flabellatus* (Berk. and Br.) Sacc. *Mysore J. Agric. Sci.*, **2008**, *42*(1), 132–134.

Rodrigues, A., Xavier, S., & Bernardes, F. Application of photo electrochemical-electro dialysis treatment for the recovery and reuse of water from tannery effluents. *J. Clean. Prod.*, **2007**, *10*, 1–7.

Rodrigues, R., & Koetz, P. Remoçao de nitrogenio de efluente da industria de arroz parboilizado por incorporaçao em biomassa celular de *Candida utilis*-IZ-1840. *Revista Brasileira de Agrociencia*, **1996**, *2*(3), 141–146.

Rodriguez, S., & Toca, J. L. Industrial and biotechnological applications of laccases: A review. *Biotechnol. Adv.*, **2006**, *24*, 500–513.

Rubio-Rodriguez, N., Beltran, S., Jaime, I., De Diego, S. M., Sanz, M. T., & Carballido, J. R. Production of omega-3 polyunsaturated fatty acid concentrates: A review. *Innov. Food Sci. Emerg. Technol.*, **2010**, *11*, 1–12.

Ruiz-Duenas, F. J., & Martinez, A. T. Microbial degradation of lignin: How a bulky recalcitrant polymer is efficiently recycled in nature and how we can take advantage of this. *Microb. Biotechnol.*, **2009**, *2*, 164–177.

Rupani, P. F., Singh, R. P., Ibrahim, H., & Esa, N. Review of current palm oil mill effluent (POME) treatment methods: Vermicomposting as a sustainable practice. *World Applied Sciences Journal*, **2010**, *11*(1), 70–81.

Sabota, C. Strain of Shiitake mushroom *Lentinula edodes* (Berk.) Pegler and wood species affect the yield of Shiitake mushrooms. *Hort. Technol.*, **1996**, *6*(4), 388–393.

Sagagi, B. S., Garba, B., & Usman, N. S. Studies on biogas production from fruits and vegetable waste. *Bayero Journal of Pure and Applied Sciences*, **2009**, *2*(1), 115–118.

Salminen, E., & Rintala, J. Anaerobic digestion of organic solid poultry slaughterhouse waste – a review. *Bioresour. Technol.*, **2002**, *83*, 13–26.

Salminen, E., Rintala, J., Harkonen, J., Kuitunen, M., Hogmander, H., & Oikari, A. Anaerobically digested solid poultry slaughterhouse wastes to be used as fertilizer on agricultural soil. *Bioresour. Technol.*, **2001**, *78*, 81–88.

Samadi, S., Mohammadi, M., & Najafpour, G. D. Production of single cell protein from sugarcane bagasse by *Saccharomyces cerevisiae* in tray bioreactor. *International Journal of Engineering, Transactions B: Applications*, **2016**, *29*(8), 1029–1036.

Sang, L., Li, L., & Zheng, F. Study on the comprehensive utilization of banana by-product. *Heilongjiang Agric. Sci.*, **2006**, *4*, 96–98.

Sangali, S., & Brandelli, A. Feather keratin hydrolysis by a *Vibro* sp. strain kv 2. *Journal of Applied Microbiology*, **2000**, *89*, 735–743.

Saranraj, P., & Stella, D. Bioremediation of sugar mill effluent by immobilized bacterial consortium. *International Journal of Research in Pure and Applied Microbiology*, **2012**, *2*(4), 43–48.

Sawatdeenarunat, C., Surendra, K. C., Takara, D., Oechsner, H., & Khanal, S. K. Anaerobic digestion of lignocellulosic biomass: Challenges and opportunities. *Bioresource Technology*, **2015**, *178*, 178–186.

Schink, B. Energetics of syntrophic cooperation in methanogenic degradation. *Microbiology and Molecular Biology Reviews*, **1997**, *61*(2), 262–280.

Schneid, A. S., Gil de los Santos, J. R., Elias, M., & Gil-Turnes, C. Wastewater of rice parboiling process as substrate for probiotics. In: *Proceedings of the 2ⁿᵈ International Probiotic Conference*, Kosice, Slovakia, **2004**.

Schnurer, A., & Jarvis, A. *Microbiological Handbook for Biogas Plant* (pp. 1–74). Swedish Waste Management, Swedish Gas Centre, Malmö, **2009**.

Seif, H., & Moursy, A. *Sixth International Water Technology Conference*. Alexandria, Egypt, **2001**.

Shah, Z. A., Ashraf, M., & Ishtiaq, C. M. Comparative study on cultivation and yield performance of oyster mushroom (*Pleurotus ostreatus*) on different substrates (wheat straw, leaves, saw dust). *Pakistan J. Nutr.*, **2004**, *3*(3), 158–160.

Sharma, R., Chisti, Y., & Banerjee, U. C. Production, purification, characterization and applications of lipases. *Biotechnol. Adv.*, **2001**, *19*, 627–662.

Shen, F., Zeng, Y., Deng, S., & Liu, R. Bioethanol production from sweet sorghum stalk juice with immobilized yeast. *Procedia Environmental Science*, **2011**, *11*, 782–789.

Shen, K. T., Chen, M. H., Chan, H. Y., Jeng, J. H., & Wang, Y. J. Inhibitory effects of chit oligosaccharides on tumor growth and metastasis. *Food Chem. Toxicol.*, **2009**, *47*, 1864–1871.

Silveira, M. L. L., Furlan, S. A., & Ninow, J. L. Development of an alternative technology for the oyster mushroom production using liquid inoculum. *Cienc. Tecnol. Aliment Campinas*, **2008**, *28*(4), 858–862.

Singh, N. B., Ruchi, S., & Mohammed, M. I. Waste water management in dairy industry: Pollution abatement and preventive attitudes. *International Journal of Science and Environment*, **2014**, *3*(2), 672–683.

Singh, S. N., & Tripathi, R. D. *Environmental Bioremediation Technologies*, Springer-Verlag, Heidelberg, **2007**.

Soccol, C. R., & Vandenberghe, L. P. S. Overview of applied solid-state fermentation in Brazil. *Biochem. Eng. J.*, **2003**, *13*, 205–218.

Sonnenberg, A. S., Baars, J. J. P., Obodai, M., & Asagbra, A. Cultivation of oyster mushrooms on cassava waste. *Proceedings of the 8ᵗʰ International Conference on Mushroom Biology and Mushroom Products* (pp. 286–291). New Delhi, India, **2014**.

Souza, R. A. T., Branco, T. R., Larissa, L. S. C., Alecrim, M. M., Felipe, L., & Teixeira, M. F. S. Nutritional composition of bioproducts generated from semi-solid fermentation of pineapple peel by edible mushrooms. *Afr. J. Biotechnol.*, **2016**, *15*(12), 451–457.

Srirangan, K., Akawi, L., Moo-Young, M., & Chou, C. P. Towards sustainable production of clean energy carriers from biomass resources. *Appl. Energy*, **2012**, *100*, 172–186.

Stamets, P. *Growing Gourmet and Medicinal Mushrooms*. Ten Speed Press, Berkeley, **2000**.

Stover, E. L., & Campana, C. K. Effluents from food processing/on-site processing of waste. In: Caballero, B., Trugo, L. C., & Finglas, P. M., (eds.), *Encyclopedia of Food Sciences and Nutrition* (pp. 1975–1976). Academic Press, London, UK, **2003**.

Subbu, L., & Sornaraj, R. Utilization of wastes for cultivation of the edible mushroom *Pleurotus flabellatus*. *Afr. J. Biotechnol.*, **2014**, *13*(17), 1779–1785.

Sun, Y., & Cheng, J. Hydrolysis of lignocellulosic materials for ethanol production: A review. *Bioresource Technology*, **2002**, *83*(1), 1–11.

Taherzadeh, M. J., & Karimi, K. Acid-based hydrolysis processes for ethanol from lignocellulosic materials: A review. *BioResources*, **2007**, *2*(3), 472–499.

Thakur, I. S., & Srivastava, S. Bioremediation and bioconversion of chromium and Pentachlorophenol in tannery effluent by microorganisms. *International Journal of Technology*, **2011**, *3*, 224–233.

Thompson, A. E., Dierig, D. A., Johnson, E. R., Dahlquist, G. H., & Kleiman, R. Germplasm development of *Vernonia galamensis* as a new industrial oilseed crop. *Industrial Crops and Products*, **1994**, *3*, 185–200.

Tian, S., Zhou, G., Yan, F., Yu, Y., & Yang, X. Yeast strains for ethanol production from lignocellulosic hydrolysates during *in situ* detoxification. *Biotechnology Advances*, **2009**, *27*, 656–660.

Tolfa da Fontoura, J., Rotermund, S., Araujo, A. L., Ramirez, N., Rubleske, M., Farenzena, M., & Gutterres, M. *XXXIII IULTCS Congress*, Novo Hamburgo, Brazil, **2015**.

Toppe, J., Albrektsen, S., Hope, B., & Aksnes, A. Chemical composition, mineral content and amino acid and lipid profiles in bones from various fish species. *Comp. Biochem. Physiol. B.*, **2007**, *146*, 395–401.

Triki-Ellouz, Y., Ghorbel, B., Souissi, N., Kammoun, S., & Nasri, M. Biosynthesis of protease by *Pseudomonas aeruginosa* MN7 grown on fish substrate. *World J. Microbiol. Biotechnol.*, **2003**, *19*, 41–45.

Tsai, G. I., Wu, Z. Y., & Su, W. H. Antibacterial activity of a chit oligosaccharide mixture prepared by cellulose digestion of shrimp chitosan and its application to milk preservation. *J. Food Prot.*, **2000**, *63*, 747–752.

Ugalde, U. O., & Castrillo, J. I. Single cell proteins from fungi and yeasts. *Appl. Mycol. Biotechnol.*, **2002**, *2*, 123–149.

Upadhyay, R. C., Verma, R. N., & Singh, S. K. Effect of organic nitrogen supplementation in *Pleurotus* sp. In: Sanchez, J. E., Huerta, G., & Montiel, E., (eds.), *Mushroom Biology and Mushroom Products* (pp. 225–235). Universidad Autonoma del Estado de Morelos, Mexico, **2002**.

Uquiche, E., Valle, J. M., & Ortiz, J. Supercritical carbon dioxide extraction of red pepper (*Capsicum annuum* L.) oleoresin. *J. Food Eng.*, **2004**, *65*, 55–66.

USDA. *Livestock and Poultry: World Markets and Trade*. United States Department of Agriculture, Washington, D.C., USA, **2014**.

Vazquez, J. A., Docasal, S. F., Miron, J., Gonzalez, M. P., & Murado, M. A. Proteases production by two *Vibrio* species on residuals marine media. *J. Ind. Microbiol. Biotechnol.*, **2006**, *33*, 661–668.

Vazquez, J. A., Docasal, S. F., Prieto, M. A., Gonzalez, M. P., & Murado, M. A. Growth and metabolic features of lactic acid bacteria in media with hydrolyzed fish viscera, an approach to bio-silage of fishing by-products. *Bioresour. Technol.*, **2008**, *99*, 6246–6257.

Velmurugan, B., & Ramanujam, R. A. Anaerobic digestion of vegetable wastes for biogas production in a fed-batch reactor. *Int. J. Emerg. Sci.*, **2011**, *1*(3), 478–486.

Vendruscolo, F., Albuquerque, P. C. M., Streit, F., Esposito, E., & Ninow, J. L. Apple pomace: A versatile substrate for biotechnological applications. *Crit. Rev. Biotechnol.*, **2008**, *28*, 1–12.

Verheijen, L. A. H. M., Weirsema, D., Hwshoff Pol, L. W., & Dewit, J. *Livestock and the Environment: Finding a Balance Management of Waste from Animal Product Processing*. International Agriculture Center, Wageningen, The Netherlands, **1996**.

Vidal, G., Carvalho, A., Mendez, R., & Lema, J. M. Influence of the content in fats and proteins on the anaerobic biodegradability of dairy wastewaters. *Bioresour. Tech.*, **2000**, *74*, 231–239.

Wang, S. L., & Yeh, P. Y. Production of a surfactant and solvent stable alkaliphilic protease by bioconversion of shrimp shell wastes fermented by *Bacillus subtilis* TKU007. *Process Biochem.*, **2006**, *41*, 1545–1552.

Wang, S. L., Shih, I. L., Liang, T. W., & Wang, C. H. Purification and characterization of two antifungal chitinases extracellularly produced by *Bacillus amyloliquefaciens* V656 in a shrimp and crab shell powder medium. *J. Agric. Food Chem.*, **2002**, *50*, 2241–2248.

Wang, X., Yang, G., Feng, Y., Ren, G., & Han, X. Optimizing feeding composition and carbon–nitrogen ratios for improved methane yield during anaerobic co-digestion of dairy, chicken manure and wheat straw. *Bioresource Technology*, **2012**, *120*, 78–83.

Ward, A. J., Lewis, D. M., & Green, F. B. Anaerobic digestion of algae biomass: A review. *Algal Res.*, **2014**, *5*, 204–214.

Ware, S. A. *Single Cell Protein and Other Food Recovery Technologies from Waste.* Municipal Environment Research Laboratory, Office of Research and Development, US Environmental Protection Agency, Cincinnati, Ohio, **1977**.

Wasi, S., Jeelani, G., & Ahmad, M. Biochemical characterization of a multiple heavy metal, pesticides and phenol resistant *Pseudomonas fluorescens* strain. *Chemosphere*, **2008**, *71*, 1348–1355.

Wasser, S. P. Medicinal mushrooms as a source of antitumor and immunomodulating polysaccharides. *Appl. Microbiol. Biotechnol.*, **2002**, *60*, 258–274.

Wasser, S. P., & Weis, A. L. Medicinal properties of substances occurring in higher basidiomycetes mushrooms: Current perspectives. *Int. J. Med. Mush.*, **1999**, *1*, 31–62.

Webb, C., Koutinas, A. A., & Wang, R. Developing a sustainable bioprocessing strategy based on a generic feedstock. *Adv. Biochem. Eng. Biotechnol.*, **2004**, *86*, 195–268.

Weiland, P. Biogas production: Current state and perspectives. *Appl. Microbiol. Biotechnol.*, **2010**, *85*, 849–860.

Wen, C. M., Tseng, C. S., Cheng, C. Y., & Li, Y. K. Purification, characterization and cloning of a chitinase from *Bacillus* sp. NCTU2. *Biotechnol. Appl. Biochem.*, **2002**, *35*, 213–219.

Wildbrett, G. Dairy plant effluent: Nature of pollutants. In: Roginski, H., Fuquay, J. F., & Fox, P. F., (eds.), *Encyclopedia of Dairy Sciences* (pp. 727–733). Elsevier, Oxford, **2002**.

Wu, J. S. B., Chen, H., & Fang, T. Mango juice. In: Nagy, S., Chen, C. S., & Shaw, P. E., (eds.), *Fruit Juice Processing Technology* (pp. 620–655). Agscience, Auburndale, USA, **1993**.

Yabaya, A., & Ado, S. A. Mycelial protein production by *Aspergillus niger* using banana peel. *Sci. World J.*, **2008**, *3*(4), 9–12.

Yalinkilic, M. K., Altun, L., Baysal, E., & Demirci, Z. *Development of Mushroom Cultivation Techniques in Eastern Black Sea Region of Turkey* (pp. 1–287). Project of the Scientific and Technical Research Council of Turkey, **1994**.

Yamamoto, M., Saleh, F., Ohtsuka, A., & Hayashi, K. New fermentation technique to process fish waste. *Anim. Sci. J.*, **2005**, *76*, 245–248.

Yamamura, S., Mosita, Y., Hasan, Q., Yokoyama, K., & Tamiya, E. Keratin degradation: A cooperative action of two enzymes from *Stenotrophomonas* sp. *Biochem. Biophl. Res. Commun.*, **2002**, *294*, 1138–1143.

Yazid, N. A., Barrena, R., & Sánchez, A. Assessment of protease activity in hydrolyzed extracts from SSF of hair waste by and indigenous consortium of microorganisms. *Waste Manag.*, **2016**, *49*, 420–426.

Ye, J., Li, D., Sun, Y., Wang, G., Yuan, Z., Zhen, F., & Wang, Y. Improved biogas production from rice straw by co-digestion with kitchen waste and pig manure. *Waste Management*, **2013**, *33*(12), 2653–2658.

Yousuf, A., Sannino, F., Addorisio, V., & Pirozzi, D. Microbial conversion of olive oil mill wastewaters into lipids suitable for biodiesel production. *J. Agr. Food Chem.*, **2010**, *58*, 8630–8635.

Zampolli, A., Bysted, A., Leth, T., Mortensen, A., De Caterina, R., & Falk, E. Contrasting effect of fish oil supplementation on the development of atherosclerosis in marine models. *Atherosclerosis*, **2006**, *184*, 78–85.

Zervakis, G., & Philippoussis, A. Management of agro-industrial wastes through the cultivation of edible mushrooms. In: *Proceedings of IV European Waste Forum 'Innovation in Waste Management.'* CIPA, Milan, **2000**.

Zhang, Y. H. P. Reviving the carbohydrate economy via multi-product lignocellulose biofineries. *J. Ind. Microbiol. Biotechnol.*, **2008**, *35*, 367–375.

Zhao, G., Zhang, W., & Zhang, G. Production of single cell protein using waste capsicum powder produced during capsanthin extraction. *Letters in Applied Microbiology*, **2010**, *50*, 187–191.

Zheng, Y., Zhao, J., Xu, F., & Li, Y. Pretreatment of lignocellulosic biomass for enhanced biogas production. *Progress in Energy and Combustion Science*, **2014**, *42*, 35–53.

Zhengyun, Z., Rui, X., Huanyun, D., Qiuxia, W., Bin, Y., Jiahong, H., Yage, Y., & Fuxian, W. Biogas yield potential research of the wastes from banana manufacturing process under mesophilic anaerobic fermentation. *Research Journal of Applied Sciences, Engineering and Technology*, **2013**, *5*(19), 4740–4744.

Zubi, W. Production of single cell protein from base hydrolyzed of date extract byproduct by the fungus *Fusarium graminearum*. *MS Thesis*, Garyounis University, Benghazi, **2005**, *19*, 167–225.

CHAPTER 7

SOIL MICROBIAL BIOFILM COMMUNITIES AND THEIR INTERACTIONS

BINDU SADANANDAN, PRIYA ASHRIT, and V. VIJAYALAKSHMI

Department of Biotechnology, MS Ramaiah Institute of Technology, Bengaluru–560054, Karnataka, India

7.1 INTRODUCTION

Microbial populations in the soil are diverse both at the functional and taxonomical levels (Schmidt and Waldron, 2015). This diversity is very critical to maintaining soil quality and also other soil inhabitants. Microbial flora of soil mostly consists of bacteria, fungi, algae, actinomycetes, and protozoa. Bacteria predominate soil microbial diversity and carry out important roles of nutrient cycling/biogeochemical cycling, which form the core of the functional ecosystem (Meliani et al., 2012). Microbial soil communities are essential for organic matter mineralization to maintain plant nutrition, growth, and development of a good ecosystem. Most of the fundamental microbial processes occur in the rhizosphere region. Various plant-microbe as well as microbe-microbe interactions occur in this region. There are various types of bacteria in the rhizosphere region of the soil which associate with plants. Some of them like *Rhizobium* sp., *Agrobacterium* sp. and *Pseudomonas* sp. occur predominantly in the soil and carry out most functions of plants like nitrogen fixation and soil bioremediation (Lakshmanan et al., 2012). Likewise, not only bacteria but fungi are also known to colonize plants and occupy the rhizosphere niche. The common and classic fungal species is mycorrhiza, which associates with plant roots. All these microbial populations, along with many others, are mostly known to occur as planktonic or free-floating organisms in most natural environments. Many bacterial species form structures called biofilm, thus associating and forming various types of interactions with plants and with one another in the rhizosphere region.

7.2   BIOFILM FORMATION BY SOIL MICROBES

7.2.1   WHAT ARE BIOFILMS?

The development of a biofilm structure was observed for the first time by Antony Van Leeuwenhoek in 1708, where he observed tissues colonized by microbial cells (Romling et al., 2014). Biofilms are structurally complex colonies of microorganisms that grow as surface-attached communities (Douterelo et al., 2014). Cluster of microbial cells develops 3D spatial arrangement on biotic/abiotic substrate to form biofilm (De Encreft et al., 2015). A matrix composed of exopolysaccharide (EPS) acts as a protective layer to the organisms residing within the biofilm (Burmolle et al., 2010). Biofilm formation is a natural mechanism for microorganisms to overcome unfavorable conditions (Lambert et al., 2014). Biofilms are ubiquitous and can be found in any environment like food, water, humans, and soil. They are part and parcel of our natural ecosystem involving aggregation of multiple species of microbes. The structure and architecture of biofilm are very complex, and thus it confers numerous advantages to the organism.

7.2.2   HOW ARE BIOFILMS FORMED?

Biofilm systems of the environment consist of diverse species of bacteria that inhabit the soil, which is a perfect platform for multispecies biofilm formation (Ren et al., 2015). Biofilms are formed in stressful and unfavorable conditions. During biofilm formation, individual microbial species can interact through diverse signaling mechanisms and colonize the surface by adhering to it. The signaling mechanism between the species results in the development of a 3D mesh. The signaling mechanism also helps in the production of EPS. There is always a high level of cooperation between the species because EPS produced by one cell is used by others within the biofilm (Xavier and Foster, 2007). EPS mostly acts as a bridge between the organism and the conditioning film (Kokare et al., 2009). The different phases of biofilm formation have been indicated pictorially (Figure 7.1). The first step of biofilm formation starts with the identification and adhesins where individual cells adhere strongly to the surface within 1–2 hours of colonization. Surface associations involve non-specific interactions through the production of cell signaling molecules. The adhered cells undergo growth and maturation. This phase involves the aggregation of adhered cells to form microcolonies through irreversible interactions. The first two stages approximately take 2–3 hours. The last phase of biofilm detachment begins with sloughing, where individual cells disperse from the site of attachment and travel through to form new attachment sites. The presence of multiple species within a biofilm, thus confers added advantages to the resident microbes by offering resistance to biotic and abiotic stresses.

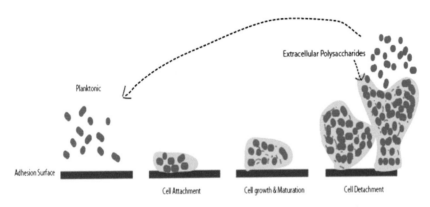

FIGURE 7.1   Pictorial representation of different stages in biofilm formation.

7.2.3   SOIL MICROBIAL BIOFILM FORMERS

Soil microflora is usually predominated by bacteria but is not limited only to them. They also consist of fungi, Blue-Green Algae (BGA), and protozoa. Multispecies bacteria present in the soil can interact among themselves or can associate with plants to form a biofilm. Polymicrobial biofilms/multispecies biofilms that are formed on leaf, root, or any other plant parts are indicated pictorially (Figure 7.2). Biofilms play a vital role in maintaining a stable ecosystem. Soil particles such as clay minerals and metal oxides strongly influence biofilm formation by soil microflora (Ma et al., 2017). Of the vast majority of soil microflora, bacteria like the nitrogen fixer *Rhizobium* sp., *Agrobacterium* sp., *Pseudomonas* sp., and *Bacillus* sp. are known to form a biofilm. Mushrooms like *Pleurotus ostreatus*; fungi like *Penicillium* sp., also form biofilms. The plant-microbial and microbial-microbial associations may result in different types of interactions between them as discussed in the following section.

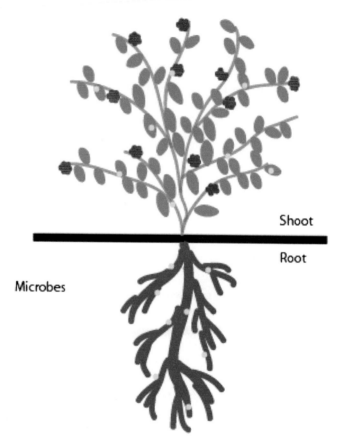

FIGURE 7.2   Different regions of a plant that may be colonized by microbial populations in the soil.

7.3   INTERACTIONS/ASSOCIATIONS OF MICROBES IN SOIL

7.3.1   PLANT-MICROBE BIOFILM ASSOCIATIONS

Microbial cells within the soil release numerous enzymes, Quorum Sensing (QS) molecules, and various growth factors that promote the growth of other cells around them (Foster and Bell, 2012). Microbes can associate with plants at different sites and exhibit different types of the interaction effect. The association of microbes with plant tissues mostly occurs at the rhizosphere region (root), phyllosphere region (leaf), and vasculatures wherein microbes invade the host tissues. The association at these regions can prove to be beneficial, symbiotic, or pathogenic. Some of the bacterial species also exhibit a biocontrol effect and provide protection against other invading pathogens. Some of the organisms invading different parts of plant tissues, and their infection sites have been summarized in Table 7.1 (modified from Lakshmanan et al., 2012). The table consists of Gram-positive bacteria as well as Gram-negative bacteria, both of which equally associate with a wide variety of plants forming biofilm, but vastly differ in their interaction effects.

TABLE 7.1   Some of the Plant-Microbial Biofilms with Their Interaction Effects and Sites of Association

Association Site	Interaction Effect	Plant	Organism
Leaf	Pathogenic	Leafy vegetables	*Escherichia coli*
		Cabbage	*Pseudomonas syringae*
Vasculature	Pathogenic	Apple/Pear	*Erwinia amylovora*
Root	Symbiotic	Leguminous plants	*Rhizobium* sp.
	Beneficial	Rice	*Rhizobium leguminosarum*
		Wheat	*Azospirillum brasilense*
		Wheat	*Klebsiella pneumoniae*
	Biocontrol	Crop plants	*Bacillus subtilis*
			Pseudomonas putida
			Pseudomonas fluorescens

7.3.1.1   INTERACTIONS WITH SYMBIOTIC AND BENEFICIAL EFFECT

7.3.1.1.1   Plant-Rhizobium Interactions

'Molecular Dialogue' is the term used to refer to the exchange of chemical signals during plant-bacterial symbiosis. The most classical case is the infection of root hair of leguminous plants with nitrogen-fixing bacteria

Rhizobium at the rhizosphere. *Rhizobia* may survive under different lifestyles; soil, root hair adherence, and infection of root hair, or it can also grow within the root nodules of legumes where they fix nitrogen (Downie, 2010). Plant Growth Promoting Rhizobacteria (PGPR) employs a range of mechanisms, either alone or in combination, to successfully colonize the root system (Compant et al., 2010). Plant-*Rhizobium* interaction begins with the adherence of single Rhizobium cells to the meristematic cells of the root leading to nodulation. Root exudates from the legumes, mainly flavonoids act as inducing factors for the expression of Nod genes of Rhizobia (Cooper, 2007). Nod factors cause infection of root hair (deformation), root curling, and nodule initiation, which are most important for symbiotic associations. Symbiotic association is dependent on four main types of surface polysaccharides (Fujishige et al., 2006). Adhesion, active nodule formation, and biofilm development is mainly due to the release of EPS and Lipopolysaccharides (LPS) (Sorroche et al., 2012). Flavonoids, along with LPS and EPS, contribute significantly to nodule activity (Cooper, 2007). Biofilm formation is independent of chemical signals from plants. Plant microbial association in the rhizosphere region has been shown pictorially (Figure 7.3).

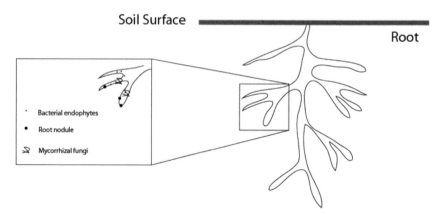

FIGURE 7.3   Plant-microbe association in the rhizosphere region of the soil.

The first step during biofilm formation is the attachment of rhizobia (*Rhizobium, Bradyrhizobium, Mesorhizobium, Sinorhizobium,* and other related genera) to host roots causing infection and nodulation. The second binding step begins with the synthesis of bacterial cellulose fibrils that result in an irreversible binding and formation of bacterial aggregates onto the host surface. Nodulation genes (nodDABC) occur in both alpha and beta types of rhizobia that nodulate legumes (Giraud et al., 2007). Hence biofilm formation is believed to be conserved by evolution with the significant role of a Nod factor gene (Fujishige et al., 2008).

7.3.1.1.2 *Plant-Azospirillum Interactions*

Biofilm lifestyle is often crucial for the survival of bacteria and for the establishment of specific symbiosis with actinorhizal host plants or nonspecific root colonization (Danhorn and Fuqua, 2007; Rodríguez-Navarro et al., 2007). *Azospirillum brasilense* although attaches with wheat roots; it generally does not exhibit species-specificity. Attachment begins with a weak, reversible, and nonspecific binding. The association occurs due to the bacterial surface proteins, capsular polysaccharides, and adhesion component of flagella (Zhu et al., 2002). The first step of attachment and aggregation of *A. brasilense* is due to the outer membrane of bacteria. Biofilm assembly and disassembly is attributed to several QS molecules along with mechanical and nutritional stress (Karatan and Watnick, 2009). QS are vital in regulating bacterial colonization both in the rhizosphere and the rhizoplane (Soto et al., 2006; Compant et al., 2010). Biofilm formation is controlled by the production of *N*-Acyl-homoserine lactones (AHLs). However, there are also reports that AHL's are not always required for plant colonization (Müller et al., 2009; Compant et al., 2010). Bacteria residing within the biofilm function due to intercellular communication facilitated by bacterial products that diffuse from cells (Watnick and Kolter, 2000).

Receptor proteins of cellular pathways are mediated by secondary messenger, cyclic-di-Guanosine Monophosphate (c-di-GMP), which controls flagellar motor speed (Boehm et al., 2010). c-di-GMP, in few species like *Vibrio cholerae* and *Shewanella oneidensis* increases biosynthesis of EPS to facilitate cell adhesion (Beyhan et al., 2006; Thormann et al., 2006; Krasteva et al., 2010).

A. brasilense under aerobic conditions produces considerable amounts of Nitric Oxide (NO), which influences lateral root formation in host plants (Creus et al., 2005; Molina-Favero et al., 2008). NO also stimulates biofilm formation by controlling the levels of c-di-GMP (Plate and Marletta, 2012). Production of NO in *A. brasilense* Sp245 derived from denitrification is a key regulatory step in biofilm formation (Di Palma et al., 2013). NO also triggers the disassembly of *Pseudomonas aeruginosa* biofilms as shown by (Barraud et al., 2006, 2009) through upstream c-di-GMP signaling pathway. Thus NO is a regulatory molecule that can cause biofilm formation or its dispersion.

7.3.1.1.3   Role of Plant Growth Promoting Rhizobacteria (PGPR)

Bacteria that colonize roots and promote plant growth are known as PGPR. PGPR activity is exhibited by bacteria that belong to genera like *Azospirillum*, *Azotobacter*, *Bacillus*, *Clostridium*, *Enterobacter*, *Gluconacetobacter*, *Pseudomonas*, and *Serratia*. Gram-positive bacterium like *Bacillus subtilis* is associated with roots of many crop plants. Studies have demonstrated that *B. subtilis* promotes plant growth through secretion of cytokinins and other volatiles. These volatiles help the plant to overcome salt stress through its effect on high-affinity ion transporter HKT1 (Beauregard et al., 2013). The overall effect of PGPRs on the plant has been represented pictorially (Figure 7.4).

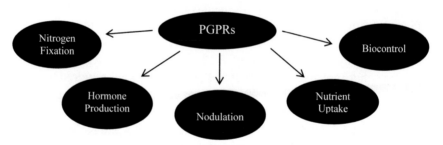

FIGURE 7.4   Different roles of PGPRs in the rhizosphere region.

7.3.1.2   INTERACTIONS WITH BIOCONTROL EFFECT

7.3.1.2.1   Plant-Bacillus Interactions

Bacillus subtilis form biofilm on roots (Danhorn et al., 2007). The extracellular matrix in *Bacillus* is composed of two major components, an EPS and protein tasA that polymerize into amyloid-like fibers. EPS production is regulated by epsA-O operon and tasA under the control of transcriptional regulator Spo0A, whose activity, in turn, depends on its phosphorylation (Beauregard et al., 2013). Biofilm initiation in *B. subtilis* is regulated by Spo0A through various environmental signals. abrB and sinR pathways of anti-repression directly or indirectly control 15-gene eps operon that is required for biosynthesis of EPS, the tapA-sipWtasA operon, bslA gene encoding a hydrophobic biofilm coat protein and also motility genes such as hag, lytA, and lytF (Ma et al., 2017). It is the abrB and sinR pathways that are critical for biofilm formation (Cairns et al., 2014).

7.3.1.2.2 Plant-Pseudomonas Interactions

Pseudomonas fluorescens is a Gram-negative bacterium that colonizes the roots and the endophytic tissues at the root surface level forming biofilms (Kiely et al., 2006). AHLs are important mediators for surface attachment in Gram-negative *P. fluorescens*. Biocontrol effect is exhibited at rhizoplane level where they form discontinued colonies in the grooves of epidermal cells. Root and seed exudates from plants are used effectively by bacteria for their growth. *P. fluorescens* acts non-specifically to protect plants from soil phytopathogens (Couillerot et al., 2009). Competitive inhibition could also serve as a mechanism that can offer an advantage to host plants. Biofilm formation regulated through QS signals in *P. fluorescens* allows the organism to create a niche, where it produces antifungal compounds, mainly phenazine antibiotic, which protects the wheat rhizosphere.

7.3.1.2.3 Plant-Erwinia Interactions

Bacterial adhesins, which may be monomeric or complex protein structures are responsible for cell and surface attachment and include the pili and fimbriae along with multiple appendages (Pizarro-Cerda et al., 2006; Kim et al., 2008). *Erwinia amylovora*, a Gram-negative plant pathogen causes fire blight disease. This organism is highly virulent and rapidly disseminates through vasculatures of rosaceous species like apple and pear trees (Koczan et al., 2011). The most important requisites for biofilm formation by *E. amylovora* are amylovoran (an EPS and pathogenicity factor) and levan (Koczan et al., 2009). The former protects cells from host-elicited immune responses and the latter is a known virulence factor. *E. amylovora* encodes cell surface structures that are necessary for initial cell attachment along with peritrichous flagella and a type III secretion apparatus; however, the role of other surface appendages remains unknown (Koczan et al., 2011).

7.3.2 MICROBE-MICROBE BIOFILM ASSOCIATIONS

Soil biofilm communities are cohabited by diversified ecosystem with high levels of synergy among most species (Ren et al., 2015). Multispecies biofilm showed more protective mechanism than individual microbial biofilm as reported by Lee et al. (2014). This shows that there is a high level of synergy and cooperative social behavior among the mixed communities.

Seeds of peanut are infected by fungus *Aspergillus niger* that causes crown root disease. Biological control of peanut seed from this fungus gained widespread interest owing to environmental concerns due to the use of chemical pesticides. Several studies have also revealed the use of PGPR's in the biological control of pests. One such organism is the bacterium *Paenibacillus polymyxa* found in the rhizosphere of wheat. This plant beneficial bacterium forms biofilm communities around the roots and the biofilm inturn utilizes root exudates secreted by plants. Biocontrol effect could be due to antibiotic compounds produced by the plants (Haggag et al., 2008). Antagonistic effect of *Paenibacillus* can also be exhibited by competing with the pathogen for colonization sites on the roots by making it unavailable for the pathogen (Timmusk et al., 2005).

Biofilm communities are important in an agricultural setting (Powers et al., 2015). Interspecific interactions in a mixed-species biofilm are mediated via AHL's that act as costimulatory signals. Biofilm communities of *B. subtilis* and *P. protegens* show antagonistic interactions as demonstrated by Powers et al. (2015). The active molecule involved in the inhibitory mechanism is 2-4-diacetylphloroglucinol (DAPG) produced by *Pseudomonas*. DAPG inhibits both sporulation as well as biofilm formation by *B. subtilis*.

B. subtilis also acts as a biocontrol agent by preventing plant infection by bacterial pathogens through the secretion of AiiA enzyme, a lactonase enzyme. This enzyme functions as an inactivator of acylhomoserine-lactone molecules released by plant pathogens (Chen et al., 2013). *B. subtilis* acts as a bio controller by secreting surfactins that act as an antimicrobial agent against the plant pathogen *Pseudomonas syringae*. *B. subtilis* also provides protection against other plant pathogens like *Erwinia* and *Xanthomonas* (Bais et al., 2006). It also enhances plant protection by inducing systemic resistance (SR) in plants for protection against an array of pathogens. Fungal-bacterial interactions play a dynamic role depending on the species, strain, and environment, but they can also demonstrate endosymbiotic, synergistic or antagonistic behavior. For example, the plant fungal pathogen *Rhizopus microsporus* exhibits endosymbiotic relationship with Gram-negative bacteria, *Burkholderia rhizoxinica* and *Burkholderia endofungorum*. The fungus, in turn, uses bacteria to produce rhizoxin, which causes rice seedling blight (Dixon and Hall, 2015).

Thus the commercial use of PGPR's as replacements to chemical pesticides and supplements helps to control various plant diseases (Shivasakthi et al., 2014) and is a rapidly expanding area of research. Late blight disease of tomato is one such example where *Phytopthora infestans* infects tomato crops. In a study by Kumar et al. (2015), a combination of *Trichoderma harzianum* OTPB3, *B. subtilis* OTPB1, and *P. putida* OTf1 were used to induce SR. These organisms control the disease through the production of Indole Acetic Acid (IAA) and Gibberellins (G3) (Chowdappa et al., 2013).

7.4   BIOREMEDIATION

In today's world, the main reason for environmental pollution is the non-specific and disproportionate utilization of biocides (Sanchez-Vizuete et al., 2015). "Remediate" means to solve a problem, and "bio-remediate" means to use biological organisms to solve an environmental problem by eliminating contaminants from the environment. The use of microorganisms as agents for *in situ* remediations is a well-known subject. Factors like chemical nature and the concentration of pollutants, microbial access to these pollutants, and also physicochemical characteristics of the prevailing environment influence this process (Mrozik, 2010). As reported by El Fantroussi et al. (2005), process efficiency can be improved by biostimulation (addition of excess nutrients) and bioaugmentation (increasing specific microbial load). Bioaugmentation improves bioremediation only if the microbial populations being used possess the following features: fast growth, easy *in vitro* culturing and also their ability to withstand and survive in extreme environmental conditions with high concentrations of xenobiotic compounds (Singer et al., 2005; Thompson et al., 2005). Studies on the use of mixed cultures of aromatic-degrading microbial strains for the clean-up of the environment has been positive and are being encouraged (Goux et al., 2003; Ghazali et al., 2004).

The property of microorganisms to adhere to surfaces and form biofilm communities is a well-known concept now. Microbes in biofilm communities show differences in both morphological and physiological characteristics. The ability of microbes to form biofilms has been exploited to clean up environmental pollutants as safer alternatives. Microbial biofilms present better bioremediation than their planktonic counterparts. They are better adapted to overcome stress, nutrient deficiency, extremes of pH, temperature, and other harsh environmental conditions as they are protected within a matrix made up of EPS. The survival ability of biofilms under such conditions has been attributed to gene transfer mechanisms (Singh et al., 2006). *In vitro* studies by Goris et al. (2003) through a bioreactor system, showed the transfer of pC1 of *Delftia acidovorans,* which was tagged with a mini-Tn5 transposon encoding the gene for oxidative deamination of 3-chloroaniline. This tagged plasmid was later transferred to *Pseudomonas putida* where 3-chloroaniline was mineralized. Springael et al. (2002) through their bioreactor system studies reported that *P. putida* BN210 contains a transposon called *clc*-element which carries out degradation of 3-clorobenzoate. Table 7.2 lists the organisms involved in the bioremediation of recalcitrants.

The success of bioremediation using biofilm formers largely depends upon the understanding of microbial interaction within biofilms and also the recalcitrants in the surrounding environment. Microorganisms that form biofilms on hydrocarbon surfaces secrete polymers. These are best suited for recalcitrant treatment of slow degrading compounds because of their ability to immobilize compound by processes like biosorption, bioaccumulation, and biomineralization (Singh et al., 2006). Table 7.3 shows different bioremediation methods for mineralization of heavy metals. Every microbial group is versatile and capable of degrading a wide spectrum of substrates. Thus bioremediation is not universally restricted to any particular microbial population.

7.5   CONCLUSION

Microbial soil communities play a fundamental role in mineralization of soil organic matter, a key process for plant nutrition, growth, and production in agricultural ecosystems. A number of microbial processes take place in the rhizosphere and rhizoplane region which serves as "hotspot" for all biological and physico-chemical activity, transfers, and biomass production. The knowledge of rhizosphere processes is, however, still scanty, especially regarding the interactions between physico-chemical processes and interactions occurring between plant and soil microbes and also between soil microorganisms. Thus there is lot of diversity and abundance of soil-borne microbes that may be strongly influenced by abiotic and biotic factors. Microbes exhibit a natural tendency to interact with surfaces and form complex structures called biofilms in a natural ecosystem. These structured microbial cell communities can be composed of either a single or multiple species. The key to the formation of mixed-species biofilm is cell-cell signaling. These mechanisms of signaling occur during plant microbial interactions and also during inter microbial interactions.

TABLE 7.2 Pollutants Studied *In Vitro* Along With Organisms Involved in the Bioremediation Process

Pollutants	Organism(s)	Bioreactor System
Chlorophenols		
4-Chlorophenol	Bacterial consortium from rhizosphere of *Phragmites australis*	Granular activated-carbon biofilm reactor
2,4-Dichlorophenol	*Pseudomonas putida*	Rotating perforated tube biofilm reactor
2,4,6-Trichlorophenol	*Pseudomonas* sp., *Rhodococcus* sp.	Fluidized bed biofilm reactor
2,4,6-Tetrachlorophenol	*Pseudomonas* sp., *Rhodococcus* sp.	Fluidized bed biofilm reactor
Herbicides		
2-(2-methyl-4-chlorophenoxy) propionic acid (MCPP)	Mixed culture of herbicide-degrading bacteria	Granular activated-carbon biofilm reactor
2, 4 Dichlorophenoxy acetic acid	Mixed culture of herbicide-degrading bacteria	Granular activated-carbon biofilm reactor

Source: Modified from Singh et al., 2006.

TABLE 7.3 Methods of Heavy Metal Bioremediation

Heavy metals	Method of bioremediation in a bioreactor system
Zn, Cd, Ni	Biosorption
Cu, Zn, Ni, Co	Biosorption, bioprecipitation
Co	Immobilization
Cd, Cu, Zn, Ni	Adsorption
Cd, Zn, Cu, Pb, Y, Co, Ni, Pd, Ge	Bioprecipitation

Adapted and modified from Singh et al., 2006.

A diverse array of bacteria associates with root, vasculature, stem, and leaf of plants to form a biofilm. The interaction effect of plants with these bacteria can have many responses such as symbiosis, beneficial, pathogenic, or biocontrol effect. Biofilm formation of soil microbiota also depends on components of soil like clay, minerals, and metal oxides. However, the underlying mechanism and the effect of soil on its formation of biofilm are poorly understood. Apart from the associated microbes and plants, biofilms also affect the compounds within their vicinity and decide their fate. The role of biofilm communities in the soil can also be extended to their use in the process of bioremediation. The complex architecture of biofilm communities accompanied by the diverse interplay of intercellular genetic exchange, QS signals, and metabolites help in the massive diffusion of nutrients to surviving microbes. Effective bioremediation and recalcitrant treatment would require a better understanding of soil biofilms to develop better bio-degradative processes to improve *in situ* remediations of the environment. In spite of all this knowledge about biofilm, the research is still very challenging and draws attention to develop better techniques to understand and exploit the advantages offered by the microbes residing as biofilm communities.

KEYWORDS

- acyl-homoserine lactones
- *Aspergillus niger*
- *Azospirillum brasilense*
- *Bacillus subtilis*
- biocontrol effect
- biofilm
- bioremediation
- *Erwinia amylovora*
- exopolysaccharide
- nitric oxide
- polymicrobial biofilm
- *Pseudomonas fluorescens*
- quorum sensing
- *Rhizobium*
- rhizosphere
- symbiotic effect

REFERENCES

Bais, H. P., Weir, T. L., Perry, L. G., Gilroy, S., & Vivanco, J. M. The role of root exudates in rhizosphere interactions with plants and other organisms. *Annu. Rev. Plant Biol.*, **2006**, *57*, 233–266.

Barraud, N., Hassett, D. J., Hwang, S. H., Rice, S. A., Kjelleberg, S., & Webb, J. S. Involvement of nitric oxide in biofilm dispersal of *Pseudomonas aeruginosa*. *J. Bacteriol.*, **2006**, *188*, 7344–7353.

Barraud, N., Schleheck, D., Klebensberger, J., Webb, J. S., Hassett, D. J., Rice, S. A., & Kjelleberg, S. Nitric oxide signaling in *Pseudomonas aeruginosa* biofilms mediates phosphodiesterase activity, decreased cyclic di-GMP levels, and enhanced dispersal. *J. Bacteriol.*, **2009**, 7333–7342.

Beauregard, P. B., Chai, Y., Vlamakis, H., Losick, R., & Kolter, R. *Bacillus subtilis* biofilm induction by plant polysaccharides. *Proc. Natl. Acad. Sci. USA*, **2013**, 1621–1630.

Beyhan, S., Tischler, A. D., Camilli, A., & Yildiz, F. H. Transcriptome and phenotypic responses of *Vibrio cholerae* to increased cyclic di-GMP level. *J. Bacteriol.*, **2006**, *188*, 3600–3613.

Boehm, A., Kaiser, M., Li, H., Spangler, C., Kasper, C. A., Ackermann, M., Kaever, V., Sourjik, V., Roth, V., & Jenal, U. Second messenger-mediated adjustment of bacterial swimming velocity. *Cell*, **2010**, *14*, 107–116.

Burmolle, M., Thomsen, T. R., Fazli, M., Dige, I., Christensen, L., & Homoe, P. Biofilms in chronic infections – a matter of opportunity-monospecies biofilms in multispecies infections. *FEMS Immunol. Med. Microbiol.*, 2010, *59*, 324–336.

Cairns, L. S., Hobley, L., & Stanley-Wall, N. R. Biofilm formation by *Bacillus subtilis*: New insights into regulatory strategies and assembly mechanisms: Regulation and assembly of *Bacillus subtilis* biofilms. *Mol. Microbiol.*, **2014**, *93*, 587–598.

Chen, F., Gao, Y., Chen, X., Yu, Z., & Li, X. Quorum quenching enzymes and their application in degrading signal molecules to block quorum sensing-dependent infection. *Int. J. Mol. Sci.*, **2013**, *14*, 17477–17500.

Chowdappa, P., Kumar, S., Lakshmi, M., & Upreti, K. K. Growth stimulation and induction of systemic resistance in tomato against early and late blight by *Bacillus subtilis* OTPB1 or *Trichoderma harzianum* OTPB3. *Biol. Cont.*, **2013**, *65*, 109–117.

Compant, S., Clément, C., & Sessitsch, A. Plant growth-promoting bacteria in the rhizo- and endosphere of plants: their role, colonization, mechanisms involved and prospects for utilization. *Soil Biol. Biochem.*, **2010**, *42*, 669–678.

Cooper, J. E. Early interactions between legumes and *Rhizobia*: disclosing complexity in a molecular dialogue. *J. Appl. Microbiol.*, **2007**, *103*, 1355–1365.

Couillerot, O., Prigent-Combaret, C., Caballero-Mellado, J., & Moenne-Loccoz, Y. *Pseudomonas fluorescens* and closely-related fluorescent pseudomonads as biocontrol agents of soil-borne phytopathogens. *Lett. Appl. Microbiol.*, **2009**, *48*, 505–512.

Creus, C. M., Graziano, M., Casanovas, E. M., Pereyra, M. A., Simontacchi, M., Puntarulo, S., Barassi, C. A., & Lamattina, L. Nitric oxide is involved in the *Azospirillum brasilense*-induced lateral root formation in tomato. *Planta*, **2005**, *221*, 297–303.

d'Enfert, C., & Janbon, G. Biofilm formation in *Candida glabrata*: What have we learnt from functional genomics approaches? *FEMS Yeast Res.*, **2016**, *16*(1), 111, doi: 10.1093/femsyr/fov111.

Danhorn, T., & Fuqua, C. Biofilm formation by plant-associated bacteria. *Annu. Rev. Microbiol.*, **2007**, *61*, 401–422.

Di Palma, A. A., Pereyra, C., Moreno, R. L., Vázquez, X., María, L., Baca, B. E., Pereyra, M. A., Lamattina, L., & Creus, C. M. Denitrification-derived nitric oxide modulates biofilm formation in *Azospirillum brasilense*. *FEMS Microbiol. Lett.*, **2013**, *338*, 77–85.

Dixon, E. R., & Hall, R. A. Noisy neighborhood: Quorum sensing in fungal-polymicrobial infections. *Cell Microbiol.*, **2015**, *17*, 1431–1441.

Douterelo, I., Sharpe, R., & Boxall, J. Bacterial community dynamics during the early stages of biofilm formation in a chlorinated experimental drinking water distribution system: Implications for drinking water discoloration. *J. Appl. Microbiol.*, **2014**, *177*, 286–301.

Downie, A. J. The roles of extracellular proteins, polysaccharides, and signals in the interactions of *Rhizobia* with legume roots. *FEMS Microbiol. Rev.*, **2010**, *34*, 150–170.

El Fantroussi, S., & Agathos, S. N. Is bioaugmentation a feasible strategy for pollutant removal and site remediation? *Curr. Opin. Microbiol.*, 2005, *8*, 268–275.

Foster, K. R., & Bell, T. Competition, not cooperation, dominates interactions among culturable microbial species. *Curr. Biol.*, **2012**, *22*, 1845–1850.

Fujishige, N. A., Kapadia, N. N., De Hoff, P. L., & Hirsch, A. M. Investigations of *Rhizobium* biofilm formation. *FEMS Microbiology Ecol.*, **2006**, *56*, 195–206.

Fujishige, N. A., Lum, M. R., De Hoff, P. L., Whitelegge, J. P., Faull, K. F., & Hirsch, A. M. *Rhizobium* common nod genes are required for biofilm formation. *Mol. Microbiol.*, **2008**, *67*, 504–515.

Ghazali, F. M., Rahman, R. N. Z. A., Salleh, A. B., & Basri, M. Biodegradation of hydrocarbons in soil by microbial consortium. *Int. Biodeterior. Biodegrad.*, **2004**, *54*, 61–70.

Giraud, E., Moulin, L., Vallenet, D., Barbe, V., Cytryn, E., Avarre, J. C., Jaubert, M., Simon, D., Cartieaux, F., Prin, Y., & Bena, G. Legume symbioses: Absence of *nod* genes in photosynthetic *Bradyrhizobia*. *Science*, **2007**, *316*, 1307–1312.

Goris, J., Boon, N., Lebbe, L., Verstraete, W., De Vos, P. Diversity of activated sludge bacteria receiving the 3-chloroaniline-degradative plasmid pC1gfp. *FEMS Microbiol. Ecol.*, **2003**, *46*, 221–230.

Goux, S., Shapir, N., El Fantroussi, S., Lelong, S., Agathos, S. N., & Pussemier, L. Long term maintenance of rapid atrazine degradation in soils inoculated with atrazine degraders. *Water Air Soil Pollut. Focus*, **2003**, *3*, 131–142.

Haggag, W. M., & Timmusk, S. Colonization of peanut roots by biofilm forming *Paenibacillus polymyxa* initiates biocontrol against crown rot disease. *J. Appl. Microbiol.*, **2008**, *104*, 961–969.

Karatan, E., & Watnick, P. Signals, regulatory networks, and materials that build and break biofilm formation. *Microbiol. Mol. Biol. Rev.*, **2009**, *73*, 310–347.

Kiely, P. D., Haynes, J. M., Higgins, C. H., Franks, A., Mark, G. L., Morrissey, J. P., & O'Gara, F. Exploiting new systems-based strategies to elucidate plant-bacterial interactions in the rhizosphere. *Microb. Ecol.*, **2006**, *51*, 257–266.

Kim, T., Young, B. M., & Young, G. M. Effect of flagellar mutations on *Yersinia enterocolitica* biofilm formation. *Appl. Environ. Microbiol.*, **2008**, *74*, 5466–5474.

Koczan, J. M., Lennemenn, B. K., McGrath, M. J., & Sundin, G. W. Cell Surface attachment structures contribute to biofilm formation and xylem colonization by *Erwinia amylovora*. *Appl. Environ. Microbiol.*, **2011**, *77*, 7031–7039.

Koczan, J. M., McGrath, M. J., Zhao, Y., & Sundin, G. W. Contribution of *Erwinia amylovora* exopolysaccharides amylovoran and levan to biofilm formation: implications in pathogenicity. *Phytopathol.*, **2009**, *99*, 1237–1244.

Kokare, C. R., Chakraborty, S., Khopade, A. N., & Mahadik, K. R. Biofilm: Importance and applications. *Indian J. Biotechnol.*, **2009**, *8*, 159–168.

Krasteva, P. V., Fong, J. C., Shikuma, N. J., Beyhan, S., Navarro, M. V., Yildiz, F. H., & Sondermann, H. *Vibrio cholerae* VpsT regulates matrix production and motility by directly sensing cyclic di-GMP. *Science*, **2010**, *327*, 866–868.

Kumar, S. P. M., Chowdappa, P., & Krishna, V. Development of seed coating formulation using consortium of *Bacillus subtilis* OTPB1 and *Trichoderma harzianum* OTPB3 for plant growth promotion and induction of systemic resistance in field and horticultural crops. *Indian Phytopath.*, **2015**, *68*, 25–31.

Lambert, G., Bergman, A., Zhang, Q., Bortz, D., & Austin, R. Physics of biofilms: The initial stages of biofilm formation and dynamics. *New J. Phy.*, **2014**, *16*, 045005, doi:10.1088/1367-2630/16/4/045005.

Laxmanan, V., Kumar, A. S., & Bais, H. P. The ecological significance of plant-associated biofilms. In: Lear, G., & Leis, G. D., (eds.), *Microbial Biofilms: Current Research and Applications* (pp. 43–60). Caister Academic Press, Auckland, **2012**.

Lee, K. W. K., Periasamy, S., Mukherjee, M., Xie, C., Kjelleberg, S., & Rice, S. A. Biofilm development and enhanced stress resistance of a model, mixed species community biofilm. *ISME J.*, **2014**, *8*, 894–907.

Ma, W., Peng, D., Walker, S. L., Cao, B., Gao, C., Huang, Q., & Cai, P. *Bacillus subtilis* biofilm development in the presence of soil clay minerals and iron oxides. *Biofilms and Microbiomes*, **2017**, 3, 4.

Meliani, A., Bensoltane, A., & Mederbel, K. Microbial diversity and abundance in soil: Related to plant and soil type. *Am. J. Plant. Nutri. Fert. Technol.*, **2012**, *2*, 10–18.

Molina-Favero, C., Creus, C. M., Simontacchi, M., Puntarulo, S., & Lamattina, L. Aerobic nitric oxide production by *Azospirillum brasilense* Sp245 and its influence on root architecture in tomato. *Mol. Plant Microbe Interact*, **2008**, *21*, 1001–1009.

Mrozik, A., & Piotrowska-Seget, Z. Bioaugmentation as a strategy for cleaning up of soils contaminated with aromatic compounds. *Microbiol. Res.*, **2010**, *165*, 363–375.

Müller, H., Westendorf, C., Leitner, E., Chernin, L., Riedel, K., Schmidt, S., Eberl, L., & Berg, G. Quorum-sensing effects in the antagonistic rhizosphere bacterium *Serratia plymuthica* HRO-C48. *FEMS Microbiol. Ecol.*, **2009**, *67*, 468–478.

Pizarro-Cerdá, J., & Cossart, P. Bacterial adhesion and entry into host cells. *Cell*, **2006**, *124*, 715–727.

Plate, L., & Marletta, M. A. Nitric oxide modulates bacterial biofilm formation through a multicomponent cyclic-di-GMP signaling network. *Mol. Cell*, **2012**, *46*, 449–460.

Powers, M. J., Sanabria-Valentin, E., Bowers, A. A., & Shank, E. A. Inhibition of cell differentiation in *Bacillus subtilis* by *Pseudomonas protegens*. *J. Bacteriol.*, **2015**, *197*, 2129–2138.

Ren, D., Madsen, J. S., Sorensen, S. J., & Burmello, M. High prevalence of biofilm synergy among bacterial soil isolates in cocultures indicates bacterial interspecific cooperation. *The ISME J.*, **2015**, *9*, 81–89.

Rodriguez, S. J., & Bishop, P. L. Three-dimensional quantification of soil biofilms using image analysis. *Environ. Eng. Sci.*, **2007**, *24*, 96–103.

Rodríguez-Navarro, D. N., Dardanelli, M. S., & Ruíz-Saínz, J. E. Attachment of bacteria to the roots of higher plants. *FEMS Microbiol. Lett.*, **2007**, *272*, 127–136.

Romling, U., Kjelleberg, S., Normark, S., Nyman, L., Uhlin, B. E., & Akerlund, B. Microbial biofilm formation: A need to act. *J. Intern. Medicine*, 2014, *276*, 98–110.

Sanchez-Vizuete, P., Orgaz, B., Aymerich, S., Le Coq, D., & Briandet, R. Pathogens protection against the action of disinfectants in multispecies biofilms. *Front Microbiol.*, **2015**, *6*, 705.

Schmidt, T. M., & Waldron, C. Microbial diversity in soils of agricultural landscapes and its relation to ecosystem function. In: Hamilton, S. E., Doll, J. E., & Robertson, G. P., (eds.), *The Ecology of Agricultural Landscapes: Long-Term Research on the Path to Sustainability* (pp. 135–157). Oxford University Press, New York, **2015**.

Shivasakthi, S., Usharani, G., & Saranraj, P. Biocontrol potentiality of plant growth promoting bacteria (PGPR): *Pseudomonas fluroscens* and *Bacillus subtilis*: A review. *App. J. Agric. Res.*, **2014**, *7*, 1265–1277.

Singer, A. C., Van Der Gast, C. J., & Thompson, I. P. Perspectives and vision for strain selection in bioaugmentation. *Trends Biotechnol.*, 2005, *23*, 74–77.

Singh, R., Paul, D., & Jain, R. K. Biofilms: Implications in bioremediation. *Trends Microbiol.*, **2006**, *14*(9), 389–397.

Sorroche, F. G., Spesia, M. B., Zorreguieta, Á., & Giordano, W. A positive correlation between bacterial autoaggregation and biofilm formation in native *Sinorhizobium meliloti* isolates from Argentina. *Appl. Environ. Microbiol.*, **2012**, *78*, 4092–4101.

Soto, M. J., Sanjuan, J., & Olivares, J. *Rhizobia* and plant-pathogenic bacteria: Common infection weapons. *Microbiol.*, **2006**, *152*, 3167–3174.

Springael, D., Peys, K., Ryngaert, A., Roy, S. V., Hooyberghs, L., Ravatn, R., Heyndrickx, M., Meer, J. R. V. D., Vandecasteele, C., Mergeay, M., & Diels, L. Community shifts in a seeded 3-chlorobenzoate degrading membrane biofilm reactor: Indications for involvement of *in situ* horizontal transfer of the *clc*-element from inoculum to contaminant bacteria. *Environ. Microbiol.*, **2002**, *4*, 70–80.

Thompson, I. P., Van Der Gast, C. J., Ciric, L., & Singer, A. C. Bioaugmentation for bioremediation: The challenge of strain selection. *Environ. Microbiol.*, **2005**, *7*, 909–915.

Thormann, K. M., Duttler, S., Saville, R. M., Hyodo, M., Shukla, S., Hayakawa, Y., & Spormann, A. M. Control of formation and cellular detachment from *Shewanella oneidensis* MR-1 biofilms by cyclic di-GMP. *J. Bacteriol.*, **2006**, 2681–2691.

Timmusk, S., Grantcharova, N., & Wagner, E. G. H. *Paenibacillus polymyxa* invades plant roots and forms biofilms. *Appl. Environ. Microbiol.*, **2005**, *11*, 7292–7300.

Watnick, P., & Kolter, R. Biofilm, city of microbes: mini review. *Journal of Bacteriology*, **2000**, *182*(10), 2675–2679.

Xavier, B. J., & Foster, R. K. Cooperation and conflict in microbial biofilms. *Proc. Natl. Acad. Sci. USA*, **2007**, *104*, 876–881.

Zhu G-Y, Dobbelaere, S., & Vanderleyden, J. Use of green fluorescent protein to visualize rice root colonization by *Azospirillum irakense* and *A. brasilense*. *Funct. Plant Biol.*, **2002**, *29*, 1279–1285.

CHAPTER 8

BACTERIOLOGICAL REMOVAL OF AZO DYES: AN ECO-FRIENDLY APPROACH

SHANMUGAPRIYA SARAVANABHAVAN,[1]
MANIVANNAN GOVINDASAMY,[1] SIVAKUMAR NATESAN,[2] and
SELVAKUMAR GOPAL[3]

[1]Department of Microbiology and Biotechnology, NMSS Vellaichamy Nadar College, Madurai–625019, Tamil Nadu, India

[2]Department of Molecular Microbiology, School of Biotechnology, Madurai Kamaraj University, Madurai–625021, Tamil Nadu, India

[3]Department of Microbiology, Allagappa University, Karaikudi–630003, Tamil Nadu, India

8.1 INTRODUCTION

Colors and dyes are commonly used in textile, paper, food, cosmetics, and pharmaceutical industries. There are above 1,00,000 different human-made synthetic dyes available on the market, and worldwide its production has around 7,00,000 tons/year (Hao et al., 2000). Most of the synthetic dyes are lethal to living organisms due to their toxic and carcinogenic properties. Most of the commonly used dyes in textile industries belong to a class of compounds called azo dyes, having the functional group R-N=N-R' (where R and R' can be either aryl or alkyl). Wastewater from textile industries carries 10% of the dyestuffs, which has been a significant cause of environmental pollution and has been released into different water bodies. Azo dyes in the effluent give a strong color, high pH, higher chemical oxygen demand (COD), and lower dissolved oxygen (DO). The removal of dyes from industrial wastewaters could be very important due to their toxicity and carcinogenicity. The structural complexity of azo dyes makes effluent treatment difficult by conventional physicochemical methods due to their high cost and low effectiveness (Latif, 2010). Among the various effluent treatment practices, the biological method of remediation has significantly removed the dyes. Both fungal and bacterial bioremediation process is effective, which removes the dyes by either anaerobic or aerobic metabolism. Microbial consortia become the best alternative for the single-microbial treatment process. This chapter describes a brief review on methods of azo dye bioremediation.

8.2   DYES

Dye is a natural or synthetic colored substance that is used to impart color the substrate to which it binds and becomes an integral part. It is composed of chromophore and auxochromes groups. Chromophore group ($-N=N$, $-C=O$, $-NO_2$, CH_4, and quinoid group) is responsible for the color of dye while auxochromes group ($-COOH$, $-NH_2$, $-SO_3H$ and $-OH$) intensify the color of the chromophore (Sapna Kochher and Sandeep Kumar, 2012). Dye is used for the coloring of paper, cotton, polyester, nylon, silk, leather, plastics, and hair. Pigments are also used for coloring the substances. Both dyes and pigments can absorb a particular wavelength of light. Most of the pigments are insoluble in water, so they do not have an affinity to the substrate. Dyes must have the characters such as (1) color the substrate, (2) soluble in water/solvents, (3) ability to absorb and retained by fiber (or) chemically combined with the substrate, and (4) ability to withstand washing or dry cleaning and resist to exposure to light.

8.2.1   CLASSIFICATION OF DYES

Each class of dye has a very unique structure, chemistry, and bonding. For example, some can make strong bonds when it reacts chemically with the substrates, but others can be held by physical forces. Dyes are classified in numerous ways. Based on its origin, dyes are classified into (i) natural and (ii) synthetic dyes. Natural dyes are mainly obtained from plant resources such as roots, bark, leaves, berries, fruits or flower, wood, algae, and lichens. At the Neolithic period, people used these natural dye materials for coloring the textiles. Human inspired to search alternative coloring materials due to the scarcity and the limited color availabilities.

Human-made synthetic dyes were discovered in the late 19th century, which are prepared from petroleum by-products and earth minerals.

Besides, dyes are broadly classified based on their chemical structure particularly the presence of chromophore group and method of application. According to their chromophore groups, dyes are classified into six groups, such as (1) Nitro and Nitroso, (2) Azo, (3) Triaryl methane, (4) Anthraquinone, (5) Indigo, and (6) Sulfur. According to their application dyes are classified into (1) Acid, (2) Basic, (3) Direct, (4) Reactive, (5) Disperse, (6) Vat, (7) Mordant, and (8) Sulfur (Corbmann, 1983; Hao et al., 2000). Among the various dyestuffs, azo compound is effectively used for dyeing the textile products like fibers and fabrics.

8.2.2 AZO DYES

Azo dye is a class of synthetic organic dyes contains the functional group R-N=N-R,' R and R' can be either alkyl or aryl group (Figure 8.1). The term azo comes from 'azote' (Latin name meaning nitrogen). The azo group is linked with naphthalene or benzene groups that may contain different functional groups such as methyl ($-CH_3$), nitro ($-NO_2$), chloro (-Cl), hydroxyl (-OH), amino ($-NH_2$), SO_3H, NR_2 and carboxyl (-COOH) leads to formation of different types of azo dyes (Zollinger, 1991). Azobenzene (Figure 8.2) is the chromophore group of azo dyes, the intensity and color of the molecule can be altered by changing the auxochromes group (Figure 8.2). The polar auxochromes make the dye either water-soluble or insoluble, and dye binds with fabric by forming a chemical bond.

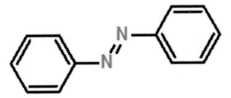

FIGURE 8.1   Structure of azobenzene.

Based on the number of azo groups, azo dyes are classified into mono azo (examples, orange-II, acid orange-12, aniline yellow), diazo (examples, oil red, congo red, direct blue-1, sudanblack-B), triazole (example, direct blue-71), tetra-azo, and polyazole dyes (Figure 8.3).

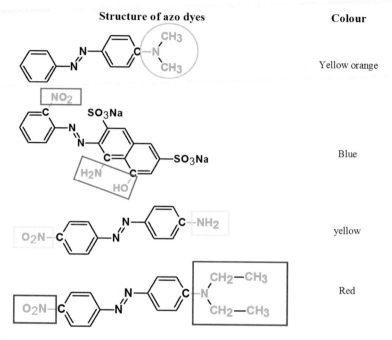

Structure of azo dyes	Colour
	Yellow orange
	Blue
	yellow
	Red

FIGURE 8.2 Auxochrome group (green color) influences the color of dye.

Azo dyes absorb particular wavelength of light in the visible spectrum because of their chemical structure (Chang et al., 2000), give bright, high-intensity colors and have fair to good fastness properties. Aryl azo compound has vivid color includes orange, yellow, and red as a consequence of electron delocalization. Azo dyes are used in dyeing of cloth, natural, and synthetic materials, ink, food, medicine, cosmetics, and paints (Telke et al., 2008). Azo dyes account for approximately 60–70% of all dyes used in food and textile industries because of economic feasibility, ease of the synthesis, stability, and the availability of different colors of azo dyes compared to other types of dyes (Chang et al., 2004).

8.3 HAZARDOUS NATURE OF AZO DYE

In the market, there are several thousand numbers of azo dyes are used for various purposes, of these 500, and more azo dyes contain carcinogenic aromatic amines such as benzidine, 4-aminodiphenyl, 3,3'-dichlorobenzidine and 2-naphthylamine in their chemical formulations. It has been estimated that textile industries consumed about two-third of the total dye produced annually (Stolz, 2001). The textile industry effluents are complex, having a variety of synthetic dyes and other products, such as acids, bases, dispersants, salts, detergents, humectants, oxidants, and so on. Textile dye in water can absorb and reflect the sunlight entering in aquatic ecosystem thereby reduce photosynthetic activity of algae and oxygenation ability of water. It can also affect germination rates and biomass of different plant species that provide habitat for living organisms and organic matter essential for soil fertility (Ghodake et al., 2009a).

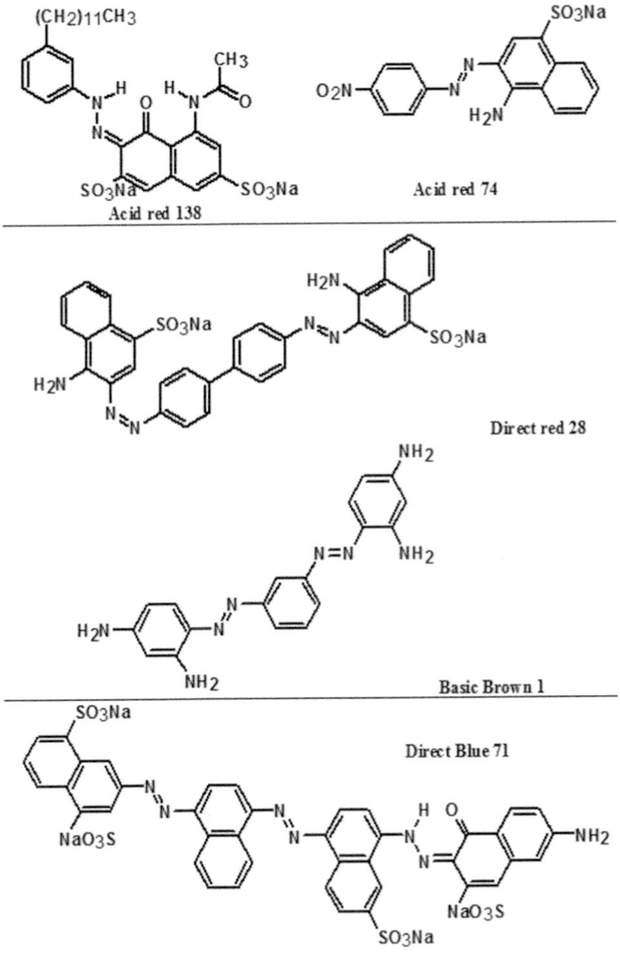

FIGURE 8.3   Chemical structure of azo dyes.

Toxic compounds from textile effluent enter into aquatic life pass via the food chain and finally reach the human being and cause physiological disorders such as sporadic fever, renal damage, hypertension, and cramps. Azo dyes and its degradation products are toxic, carcinogenic, and mutagenic. It causes human bladder cancer, hepatocarcinoma, and splenic sarcoma and induces chromosomal aberrations in experimental animal cells (Puvaneswari et al., 2006). Genotoxicity potential of the azo dye is based on its binding capability on double-stranded DNA, while mutagenicity has been associated with the formation of free radicals (Wang et al., 2012).

Synthetic azo compounds and its derivatives 3,3-dichlorobenzidine, 2-amino-4-nitrotoluene and phthalocyanine have been used in tattoo colorants (Baumler et al., 2000). Normal microflora of human skin does not produce the harmful effect by their presence under normal circumstances. When azo dye exposed on the skin by tattoo ink, the topical use of skin colorants, and wearing colored textiles, the normal microflora of the skin react with the dye and produces potentially toxic aromatic amines (Levine, 1991; Chung et al., 1992; Platzek et al., 1999). If azo dyes entering into the human body, they are metabolized by azoreductase enzyme present in the gastrointestinal tract and liver that produces aromatic amines, and it create adverse effects (Platzek et al., 1999). For example, anaerobic metabolism of 1-Amino-2-Naphthol based azo dyes (Sudan dyes) by human intestinal microflora leads to the formation of aromatic amines, 1-Amino-2-naphthol, 1,4-phenylene-diamine, and 2,5-diaminotoluene, which cause harmful effects (Figure 8.4) (Xu et al., 2008).

When effluent containing azo dyes are entering into the water bodies, environmental microorganisms can readily reduce the dyes to produce carcinogenic amines under anaerobic conditions. The following three mechanisms convert azo dyes to the carcinogenic aromatic amine, such as (1) Aromatic amines are produced by reduction and cleavage of the azo bonds, (2) Oxidation of azo dye with a structure containing free aromatic amine group without reduction of azo linkage, and (3) Direct oxidation of azo linkage leads to activation of azo dyes to highly reactive electrophilic diazonium salts (Brown and Devito, 1993). Therefore, there is a need to remove/treat the textile effluents before releasing into the environments. Hence the government legislature is imposing textile industries to treat the effluent before discharging to the environment. Currently, several physical, chemical, and biological methods have been used to treat the effluents discharged from the dye industries (Figure 8.5).

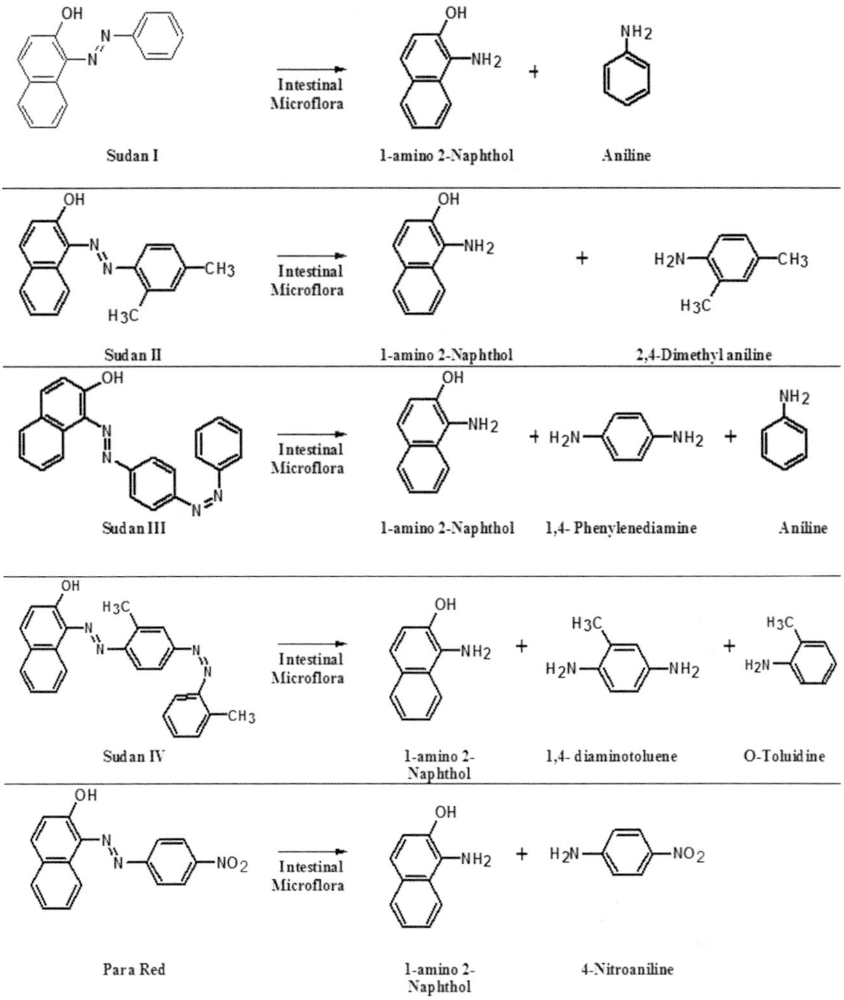

FIGURE 8.4   Anaerobic metabolism of 1-amino-2-naphthol-based azo dyes (Sudan dyes).

8.4   METHODS OF TREATMENT

8.4.1   PHYSICAL TREATMENT METHODS

Many different physical methods including membrane filtration, reverse osmosis, electrolysis, nanofiltration, and various absorption techniques are used to remove the dyes. Among the different methods, coagulation, and flocculation of dyes using Fe^{2+}, Al^{2+}, and Mg^{2+} salts is effective, but this method mainly remove disperse and sulfur dyes, and it showed very low capacity for other dyes (Kharub, 2012). Adsorption methods have been considered to the superior of other physical methods because of their efficacy for the removal of a wide variety of dyes, the simplicity of design, flexibility, and intensive to toxic pollutants. Adsorptions of dye rely on many physicochemical factors such as dye-adsorbents interaction, the surface area of adsorbent, temperature, contact time and particle size (Guptha and Suhas, 2009). Activated carbon is a very effective amongst all adsorbent, but it is not used due to its higher cost (Robinson et al., 2001). Peat fly ash, bentonite clay, polymeric resins, corn/maize cobs, maize, and wheat bran and straw are also used for the removal of the color of dye in wastewater (Ramakrishna and Viraraghavan, 1997; Kobya, 2004; Sulak, 2007; Kharub, 2012). In addition, filtration methods such as ultrafiltration and nanofiltration have been used to clarify and concentrate dyes from the wastewater. The subsequent process of filtration after coagulation-flocculation revealed good color removal. However, membranes used for filtration have some limitations including high cost, periodic replacement due to clogging of pores and the production of secondary waste (Robinson et al., 2001; Dos Santos et al., 2007) (Figure 8.6).

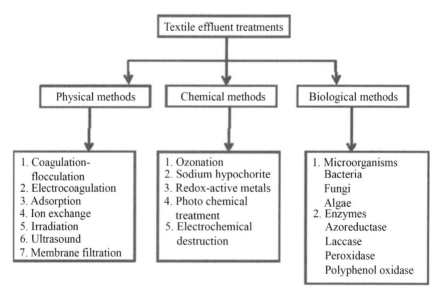

FIGURE 8.5   Textile effluent treatment techniques.

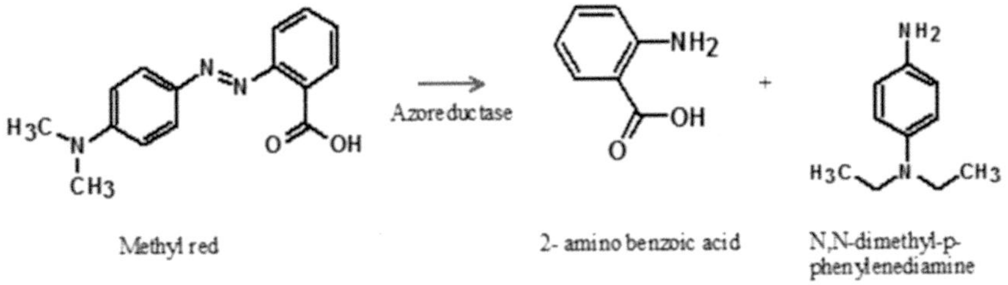

FIGURE 8.6   Methyl red degradation by *Staphylococcus aureus*.

8.4.2   CHEMICAL TREATMENT METHODS

Chemical oxidation is the common method used for the removal of colors from the effluents. These techniques use different oxidizing agents, such as hydrogen peroxide (H_2O_2), ozone (O_3), and permanganate (MnO_4) to destruct or decomposition of dye molecules. In the presence of oxidizing agent, transformation of a group or modification of the chemical composition of a compound could takes place, which makes the dye susceptible to degradation (Metcalf, 2003). In this view, Ozone has been used to remove colors effectively from textile wastewater containing several azo dyes (Alaton et al., 2002). But, ozone shows ineffectiveness on removal of disperse dyes and also incompetent to remove COD because of its short half-life period (20 minutes) and the higher cost practical application of ozonation method has been restricted (Anjaneyulu et al., 2005). However, advanced oxidation processes have been successfully used for the removal of recalcitrant dyes present in the textile effluents. The effectiveness of this process is based on the formation of very powerful oxidation molecule like hydroxyl radical (OH⁻) that destructs the hazardous dye molecules. Most commonly, the Fenton reaction method has been used to remove both soluble as well as insoluble dyes. Although it is a cheap and efficient color removal method, which generates high sludge that limits the usage of this process (Robinson et al., 2001). Electrochemical oxidation is a new technique to effectively destroy organic pollutant and produce nontoxic end products. However, little or no consumption of chemicals and the high cost of the electricity requirement limit its application. Physical/chemical methods of dye removal have several drawbacks, such as (i) cost, (ii) unable to remove xenobiotic azo dyes and their degradative products, (iii) chemical and photolytic stability of dyes and (iv) generation of a large amount of sludge (Anjaneyulu et al., 2005).

8.4.3   BIOLOGICAL TREATMENT

In the present scenario, biological decolorization methods are gaining importance to remove toxic waste from textile effluents. In this approach, microorganisms are widely studied because they acclimatize to a wide range of toxic compounds and develop new strains as expected, which can transform toxic compounds into a harmless form. Commonly, microbial decolorization of dyes may take place in two ways: (i) degradation of the chemical structure of dyes by microbial cells, and (ii) adsorption of dyes on living and dead cell biomass. Microbial degradation has been attracted in several ways because of (i) environmentally friendly, (ii) economically feasible, (iii) complete mineralization or nontoxic end product, (iv) less sludge production, and (v) less water consumption. However, the imperfection of bioremediation process limits its use at the field level (Bafana et al., 2011).

8.5   ROLE OF MICROORGANISMS IN AZO DYE DEGRADATION

8.5.1   FUNGI

In general, fungi is a saprophytic organism, which can rapidly use different nature of nutritional sources, because it can produce a significant number of intra and extracellular enzymes that are needed to degrade several complex organic pollutants such as dyestuffs, polyaromatic compounds, organic waste and steroids (Gadd, 2001; Humnabadkar et al., 2008). Last few decades, the fungal system have been utilized in the treatment of colored and metallic textile effluents (Ezeronye and Okerentugba, 1999) because they can produce nonspecific enzymes such as lignin peroxidase (LiP), manganese peroxidase (MnP) and laccase (Christian et al., 2005) that can mineralize dyes. Mostly, the fungi such as *Phanerochaete chrysosporium*, *Trametes versicolor*, *Coriolus versicolor*, *Bjerkan deraadusta*, *Aspergillus niger*, *Geotrichum candidum*, *Pleurotus ostreatus* and *Cunninghamella elegans* have shown effective dye decolorization (Ventura-Camargo and Marin-Moralas, 2013). In the last few years, azo dye decolorization by fungi has been reported by many researchers (Table 8.1).

TABLE 8.1   Decolorization of Azo Dyes by Fungi

Fungi Name	Dye	Mechanism	References
Aspergillus niger	Acid blue 29	Adsorption	Fu and Viraraghavan, 1999
Phanerochaete chrysosporium	Congo red	Lignin degradation	Tatarko and Bumpus, 1998
Phanerochaete chrysosporium	Acid yellow 9, orange II	Peroxidase	Pasti Grigsby et al., 1992
Trametes versicolor	Acid green 27	Laccase	Wong and Yu, 1999
Ganoderma sp.	Orange II	Adsorption	Mou et al., 1991
Geotrichum fuci	Reactive black 5	Adsorption	Polman and Breckenridge, 1996
Cyathus bulleri	Malachite green	Laccase	Vasdev et al., 1995
Trametes versicolor	Rhodamine B	Laccase	Khammuang and Sarnthima, 2009
Thelephora sp.	Congo red, Orange G, Amido black 10B	Adsorption	Selvam et al., 2003
Armillaria sp. F022	Reactive Black 5	Laccase	Hadibarata et al., 2012
Cerrena unicolor	Acid Red 27	Laccase	Michniewicz et al., 2008
Coprinopsis cineria	Methyl Orange	Laccase	Tian et al., 2014
Ganoderma sp.	Methyl Orange	Laccase	Zhao et al., 2011
Geobacillus catenulatus MS5	Congo Red	Laccase	Verma and Shirkot, 2014
Lentinus polychrous	Congo Red	Laccase	Suwannawong et al., 2010
Pleurotus ostreatus	Remazol Brilliant Blue R	Laccase	Palmieri et al., 2005
Pleurotus ostreatus	Synazol Red HF6BN	Laccase	Ilyas et al., 2012
Pycnoporus sanguineus	Trypan Blue	Laccase	Annuar et al., 2009
Thelephora sp.	Orange G	Laccase	Selvam et al., 2003
Trametes versicolor, Ganoderma lucidum, Irpex lacteus	Black Dycem	Laccase	Baccar et al., 2011
Trametes versicolor	Reactive Black 5	Laccase	Bibi and Bhatti, 2012
Providencia rettgeri HSL1	Reactive Blue 172	Laccase, Azoreductase, NADH-DCIP reductase	Harshad Lade et al., 2015

TABLE 8.1   *(Continued)*

Fungi Name	Dye	Mechanism	References
Aspergillus flavus	R Navy, Blue M3R, R Red M8B, R Green HE4B, R Orange M2R, R RedM5B, Dt Orange, RS, Dt Black BT, Dt Blue GLL, and Dt Sky Blue FF	Lignin peroxidase> laccase> manganese peroxidase> tyrosinase	Laxmi and Nikam, 2015

In addition to enzymatic biodecolorization, fungi remove dyes from the effluent by adsorption mechanism. Both living and dead fungal biomass adsorb dye from wastewater by physicochemical interactions such as adsorption, deposition, and ion-exchange. However, adsorption of dye by wet fungal biomass is very fast (Mou et al., 1991), and adsorbed dye are gradually degraded within a week by fungal cells depending on the type of molecule. It is stated that compared with living cells, dead cells decolorize dye solution by adsorption process effectively because of an increased surface area that is due to cell rupture upon death (Mou et al., 1991). It is advantageous that dead cells may be used or stored for a longer period over living cells that need nutrient supply as well as cultural maintenance. The demerits of bioabsorption are the dye remains unaltered form after adsorptions and will create secondary pollution.

8.5.2   ALGAE

Algae can decolorize azo dyes with the aid of an induced form of an azo reductase enzyme. It removes the color by three different mechanisms: (i) Assimilation of chromophores for the synthesis of algal biomass, carbon dioxide, and water, (ii) Transformation of colored to non-colored molecules, and (iii) Adsorption mechanism. Some species of algae such as *Chlorella* and *Oscillatoria* have the capability to reduce azo dye into corresponding aromatic amines that subsequently metabolize to simpler compounds or carbon dioxide (Acuner and Dilek, 2004). It has been found that *Chlorella pyrenoidosa*, *Chlorella vulgaris*, and *Oscillateria tenuis* decomposed more than 30 azo dyes into simpler aromatic amines (Yan and Pan, 2004). In the last few years, azo dye decolorization by algae has been well documented (Table 8.2).

TABLE 8.2   Algae in Azo Dyes Decolorization

Algae	Dye	References
Phormidium valderianum	Acid red 119, Direct black 115	Vishal et al., 2001
Spirogyra rhizopus	Acid red 247	Ozer et al., 2006
Cosmarium sp.	Triphenylmethane dye, Malachite green	Daneshvar et al., 2007
Chlorella vulgaris	G-Red	El-Sheekh et al., 2009
Nostoc lincki	Methyl red	El-Sheekh et al., 2009
Oscillatoria formosa	Amido black dye	Mubarak Ali et al., 2011
Enteromorpha sp.	Basic Red 46	Khataee et al., 2013
Nostoc muscorum	RGB-Red	Surbhi et al., 2015
Haematococcus sp.	Congo red	Mahalakshmi et al., 2015
Spirogyra sp., *Cladophora* sp.	Reactive Blue	Waqas et al., 2015

8.5.3   ACTINOMYCETES

Actinomycetes, especially *Streptomyces*, can degrade lignin and azo dyes by producing extracellular enzymes laccases and peroxidases. Halotolerant laccase enzyme produced from *S. ipomoea* in the presence of a redox mediator could detoxify 90% of the azo dye Orange II (Molina-Guijarro et al., 2009). A peroxidase enzyme has been identified from *S. chromofuscu,* which has the ability to degrade sulfonated dyes (Goszczynski et al., 1994). Combined actions of different enzymes such as lignin peroxidase, NADH-dichloroindophenol oxidoreductase (DCIP), and Methylenetetrahydrofolate reductases (MR) from *S. krainskii* degraded reactive blue 59 completely within 24 hours (Mane et al., 2008). Laccase from *S. psammoticus* showed extensive decolorization of remazol brilliant blue R and to a less significant range of methyl orange, Bismarck brown, and acid orange (Niladevi and Prema, 2008) degradation. Table 8.3 shows a few important azo dyes degrading actinomycetes and their substrates.

8.5.4   BACTERIA

Many researchers have found the role of various groups of bacteria in decolorization and degradation of azo dyes. Since bacteriological decolorization is, (i) inexpensive, (ii) higher degree of biodegradation and mineralization, (iii) applicable to a wide variety of azo dyes, (iv)

TABLE 8.3   Azo Dyes Degradation by Actinomycetes

Strains	Type of Enzyme	Dyes Degraded	References
Streptomyces chromofuscus	Peroxidase	3,5-dimethyl-4-hydroxy-aobenzene-4'-sulfonic acid (I), 3-methoxy-4-hydroxyazobenzene-4'-sulfonamide (II)	Goszczvnski et al., 1994
Streptomyces psammoticus	Laccase	Remazol brilliant blueR	Niladevi and Prema, 2008
Streptomyces skrainskii	Ligninperoxidase, NADH-DCIP reductase, MR- reductases	Reactive blue 59	Mane et al., 2008
Streptomyces ipomoea	Laccase oxidoreductase	Orange II	Molina-Guijarro et al., 2009
Nocardia sp. KN5	-	Congo red	Bhoodevi et al., 2015
Saccharothrix aerocolonigenes	-	Reactive Red 1, Reactive Orange (RY107) and Reactive black 5	Rizwana and Uma, 2015
Streptomyces spp.	-	Azo blue and azo orange	Jai et al., 2015
Micromonospora sp., *Streptomyces* sp., *Micropolyspora* sp.	-	Amido black	Mohamed et al., 2016

faster mineralization than fungi, (v) eco-friendliness and (vi) less sludge production (Verma and Madamwar, 2003; Rai et al., 2005; Khehra et al., 2006; Saratale et al., 2009c). Several members of the genus including *Micrococcus, Aeromonas, Escherichia, Acinetobacter, Pseudomonas, Enterococcus, Klebsiella, Desulfovibrio, Rhodopseudomonas, Rhizobium, Acinetobacter, Alcaligenes, Bacillus, Brevibacterium, Proteus, Microbacterium, Sphingomonas* sp., *Staphylococcus,* and *Gracilibacillus* are degrading various azo dyes by their unique metabolic process.

Bacteria have the ability to decolorize a broad spectrum of dyes. The bacterial decolorization of azo dyes can be mainly by either aerobic or anaerobic metabolism. However, a wide variety of azo dyes degradation takes place under both anaerobic and aerobic conditions, the preliminary step of bacterial decolorization is the reductive cleavage of -N=N linkage with the help of azoreductase enzyme under anaerobic conditions, which forms colorless potentially toxic aromatic amines (Chang and Kuo, 2000; Vander Zee and Villaverde, 2005). Further, aromatic amines are broken down aerobically or anaerobically (Joshi et al., 2008). It has been reported that mixed bacterial culture can degrade dyes more successfully than pure bacterial strains (Nigam et al., 1996) due to synergistic metabolic activities of the microbial consortia. The pure culture of bacteria such as *Proteus mirabilis, Pseudomonas luteola,* and *Pseudomonas* sp. degrade azo dyes under axenic conditions (Chen et al., 1999; Yu et al., 2001; Kalyani et al., 2008). Moreover, decolorization by pure culture method helps to understand the mechanisms of biodegradation in-depth in the course of biochemical and molecular studies; this information may useful to produce modified strains with higher enzyme activities. However, a pure culture of bacteria cannot degrade azo dye completely and also produce a toxic intermediate which requires further decomposition to become non-toxic (Joshi et al., 2008). It has been suggested that a pure culture of bacteria requires a long-term adaptation process for efficient decolorization of azo dyes (Saratale et al., 2011). However, in a mixed bacterial culture system, aromatic amines that are formed by reductive cleavage of the azo bond are further degraded by co-existing organisms. Therefore, the mixed culture system is effective because dye molecules are attacked at a different positions by various bacteria or formed intermediate may be further decomposed by complementary organisms. The reported azo dye decolorizing bacteria are tabulated in Table 8.4.

TABLE 8.4   Decolorization of Azo Dyes by Bacteria

Organism	Dye Degraded	Type of Enzyme	References
Pseudomonas luteola	Red G	Azoreductase	Hu, 1994
Klebsiella pneumoniae	Methyl Red	Reductive	Wong and Yuen, 1996
Rhodopseudomonas palustris	Reactive Brilliant Red	Reductive	Wong and Yuen, 1996
Sphingomonas sp. BN6	Amaranth	Azoreductase	Kudlich et al., 1997
Desulfovibrio desulfuricans	Reactive Orange 96, Reactive Red 120	Reductive	Yoo et al., 2000
Bacillus strain SF	Reactive Black 5, Mordant black 9	Reductive	Maier et al., 2004
Pseudomonas aeruginosa	Navitan fast blue S5R, Amaranth, Orange G	Azoreductase	Nachiyar and Rajkumar, 2005
Staphylococcus aureus	Orange II, Ponceau BS, Ponceau S	Azoreductase	Chen et al., 2005
Bacillus cereus	Indigo caramine, Ruby red, Flame orange	Azoreductase	Pricelius et al., 2007
Gracilibacillus sp.	Acid red B	Azoreductase	Uddin et al., 2007
Bacillus velezensis AB	Congo red	Azoreductase	Bafana et al., 2008
Rhizobium radiobacter	Reactive Red 141	Oxidative and reductive	Telke et al., 2008
Enterococcus faecalis	Methyl red	Azoreductase	Punj and John, 2009
Micrococcus glutamicus	Reactive Green 19A	Oxidative and reductive	Saratale et al., 2009c
Aeromonas hydrophila	Reactive Red 198, Reactive Black 5	Reductive	Hsueh et al., 2009
Escherichia coli	Direct Blue 71	Reductive	Jin et al., 2009
Acinetobacter calcoaceticus	Direct Brown MR	Oxidative and reductive	Ghodake et al., 2009a
Enterococcus gallinarum	Direct Black 38	Reductive	Bafana et al., 2009
Mutant *Bacillus* sp. ACT2	Congo Red	Reductive	Gopinath et al., 2009
Escherichia coli JM109 (pGEX-AZR)	Direct Blue 71	Reductive	Jin et al., 2009

TABLE 8.4 *(Continued)*

Organism	Dye Degraded	Type of Enzyme	References
Enterococcus gallinarum	Direct Black 38	Reductive	Bafana et al., 2009
Enterococcus gallinarum	Direct Black 38	Reductive	Bafana et al., 2009
Mutant *Bacillus* sp. ACT2	Congo Red	Reductive	Gopinath et al., 2009
Pseudomonas aeruginosa	Remazol Orange	Reductive	Sarayu and Sandhya, 2010
Pseudomonas aeruginosa	Remazol Orange	Reductive	Sarayu and Sandhya, 2010
Pseudomonas aeruginosa	Remazol Orange	Reductive	Sarayu and Sandhya, 2010
Acinetobacter radioresistens	Acid Red	Reductive	Ramya et al., 2010
Bacillus megaterium	Red 2G	Reductive	Khan, 2011
Bacillus subtilis ORB7106	Methyl Red	Reductive	Leelakriangsak and Borisut, 2012
Brevibacterium sp. strain VN-15	RY107	Reductive	Franciscon et al., 2012
Proteus sp.	Congo Red	Reductive	Perumal et al., 2012
Alcaligenes sp. AA09	Reactive Red BL	Reductive	Pandey and Dubey, 2012
Brevibacterium sp. strain VN-15	Reactive Yellow 107	Tyrosinase activity	Elisangela Franciscon et al., 2012
Bacillus lentus BI377	Reactive Red 141	Reductive	Oturkar et al., 2013
Pseudomonas spp.	Reactive Violet 5	Laccase	Shah, 2014
Pseudomonas stutzeri	Disperse Yellow (D4), Disperse Blue (R16), Reactive Red, Synozol (R4)	-	Amr Fouda et al., 2016
Microbacterium sp. B12 Mutant	Reactive Blue 160	Reductive	Chetana et al., 2016
Pseudomonas entomophila	Reactive Black 5	Azoreductase	Sana and Abdul, 2016

8.5.5   AEROBIC BACTERIAL DECOLORIZATION

Bacteria are capable of decolorizing azo dyes in the presence of oxygen by a reductive mechanism with the aid of aerobic azoreductase enzyme. In general, azo reductases require NADH/NADPH for their redox reaction. Perhaps azo dyes are resisting to bacterial decolorization under aerobic conditions, because aerobic respiration may take over to utilize NADH, thus inhibiting the electron transfer from NADH to azo bonds. Thus organisms with specific azo dye reducing enzymes can only degrade azo dye. Bacteria such as *Bacillus, Rhodobacter spheroids, Enterococcus, Shigella flexneri, E. coli, Xenophilus azovorans, Pseudomonas aeruginosa,* and *Pigmentiphaga kullae* (Bafana and Chakrabarti, 2008) decolorize azo dyes aerobically with the help of aerobic azoreductase enzyme which reduces the azo group to aromatic amines (Bafana et al., 2009). For example, azoreductase from *Staphylococcus aureus* cleaved methyl red into 2-aminobenoic acid (Carcinogenic) and N,N-dimethyl-*p*-phenylenediamine (Figure 8.7) (Chen et al., 2005).

Some bacteria utilize azo dyes for its carbon and energy requirements. They can slowly degrade the complex dyestuffs and to minerals in aerobic conditions without production of carcinogenic aromatic amines. Example, *Xenophilus azovorans* KF 46 and *Pigmentiphaga kullae* K24 can grow on media with azo compounds carboxy orange 1 and carboxy orange II, as the sole carbon source (McMullan et al., 2001), but *Sphingomonas* sp. degrades azo dye by reductases enzymes for its carbon and energy requirements (Coughlin et al., 1999). Normally, these bacteria reduce -N=N- bonds and utilize amines as carbon and energy source for their growth through aerobic metabolism.

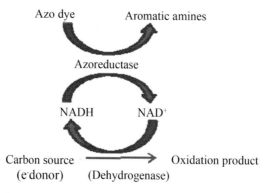

FIGURE 8.7   Direct enzymatic method of azo dye degradation.

8.5.6 ANAEROBIC BACTERIAL DECOLORIZATION

One such approach used to degrade the dyes by the anaerobic method is electron acceleration, in which the transfer of an electron to the dye increases the degradation efficiency in an anaerobic environment. Different types of bacteria such as *Eubacterium*, *Proteus*, *Bacteroids*, *Streptococcus*, and *Clostridium* could reduce azo compounds anaerobically (Stolz, 2001). In anaerobic condition, the azo bond is cleaved by nonspecific reduction with the aid of reduced flavin (electron carrier). Cytoplasmic flavin reductases assist the transfer of an electron to the azo dye through reduced electron carriers in an anaerobic environment. But, azo dye substituted with a sulfonate group cannot be reduced by cytoplasmic enzyme because the permeability of the plasma membrane is low for sulfonated dyes (Wuhrmann et al., 1980). They can be reduced extracellularly with the assistance of redox mediator that transports electrons between cell and the dye. For example, *Sphingomonas xenophaga* produces redox mediator itself during reduction of azo dye (Keck et al., 2002). Some anaerobic bacteria use their metabolic end products such as Fe^{2+} and H_2S as a redox mediator to reduce azo dyes (Kim et al., 2007). The redox mediators (RMs) such as riboflavin and quinones are increasing the efficiency of degradation. It exhibits intermediate reduction potentials and simplifies the electron transfer from the substrate (donor) to the dye (acceptor); thereby speed up the decolorization process. RMs are very effective in azo dye reduction process because of the unstable nature of the azo bond, which can readily receive electron from riboflavin and quinines. Redox mediator dependent anaerobic process is accomplished in two steps (Rau et al., 2002), such as (i) RM reduction by the reducing equivalents produced by substrate metabolism, and (ii) Azo dye reduction by the reduced form of RM (RMreduc). This type of anaerobic process has been widely used for dye removal and reduction of COD in textile effluents. It has more advantageous, because of less sludge production, minimum energy requirements, and used to produce energy (methane-biofuel) (Diego et al., 2014).

However, anaerobic bacteria efficiently decolorize azo dye by nonspecific azo reductase enzyme activity; the formation of aromatic amines would become a problem. For example, metabolism of a dye direct black 38 by human intestinal microflora was carried out in a semi-continuous culture system that mimics the lumen of human large intestine. After 7 days of incubation, compounds such as benzidine, 4-aminobiphenyl, mono acetylbenzidine, and acetylaminobiphenyl have produced as the end products which were identified by GC-MS analysis (Manning et al., 1985). The aromatic amines generated by the reduction of the azo bond are not degraded further under anaerobic conditions. But, they are mineralized under aerobic conditions by nonspecific enzymes via hydroxylation and ring-opening (Feigel et al., 1993). It is proposed that the combined action of anaerobic and aerobic bacteria remove recalcitrant azo dye from the environment. This has been proved by the decolorization of reactive azo dyes such as remazol brilliant violet 5R, remazol black B, and remazol brilliant orange 3R was occurred effectively by mixed culture of bacteria in an anaerobic-aerobic treatment process (Supaka et al., 2004; Popli and Patel, 2015).

8.6   ENZYMES IN AZO DYES DECOLORIZATION

The azo bond (N=N) is the most labile region of azo dyes, which may readily undergo cleavage either by enzymatic biodegradation (i.e., metabolic cleavage) or abiotic (thermal and photochemical) degradation (Weber and Adams, 1995; Ollgaard et al., 1998). Biological systems contain enzymes, azoreductases, laccases, peroxidase, and polyphenol oxidase that involved in degradation of azo dyes. The azo bond of water-insoluble azo dyes is usually not available for the intracellular enzymatic breakdown. However, the possible degradation of azo dyes may occur through specific pathways (Engel et al., 2010). However, in sulfonated azo dyes, it is restricted to the release of aromatic amines. Several oxidoreductases from microorganisms have effectively doing the breakdown process either intracellularly or specific pathways.

8.6.1   AZOREDUCTASES

Azoreductases (EC 1.7.1.6), otherwise known as azobenzene reductase, are a major group of enzymes produced by bacteria and fungi. It can decolorize azo dye by reducing the azo group (-N=N-) into their corresponding amines (Pandey et al., 2007), which involve the breakdown of azo bond leads to degradation of dye. Azoreductases belongs to the family of oxidoreductases, and it catalyzes the reduction reaction only in the presence of coenzymes like NADH, NADPH, and $FADH_2$. The catalytic reaction of this enzyme proceeds via a ping-pong mechanism by using two molecules of NAD(P)H to reduce one molecule of the azo compound, a substrate. Mainly bacterial decolorization of azo dye is attributed by the enzyme azoreductase (Figure 8.7). In bacteria, the reduction reaction takes place either intracellularly or extracellularly at the cell membrane (Ram Lakhan Singh et al., 2015). Several azoreductases have been identified from bacteria and are tabulated in Table 8.4.

However, in recent years, intracellular azo dye reduction has been suspected because of high polarities, high molecular weight, and sulfonate group substitution on azo dyes (Joshni et al., 2011). Hence, a mechanism other than reduced flavin-dependent azoreductases must exist for sulfonated azo dye reduction in bacteria. Moreover, these reductants may link between the outer membrane of bacterial cells with the intracellular electron transport system and a complex dye molecule. The mediator compounds may be either metabolic products of certain substrates utilized by bacteria or added externally. For example, riboflavin significantly increased mordant yellow 10 reductions by anaerobic granular sludge (Field and Brady, 2003); the addition of anthraquinone-2,6-disulfonate, and a synthetic electron carrier could greatly enhance the reduction of many azo dye (Vander Zee et al., 2001b). Cell-free extract of *Spingomonas* sp. strain BN6 that was grown in the presence of 2-naphthyl sulfonate under aerobic condition showed 10–20 times better decolorization rate of amaranth in anaerobic condition (Keck et al., 1997). Kudlich et al. (1997) suggested that the membrane-bound and the cytoplasmic azoreductases not be the same enzyme systems because the activity of membrane-bound azoreductase is dependent on redox mediator (Figure 8.8).

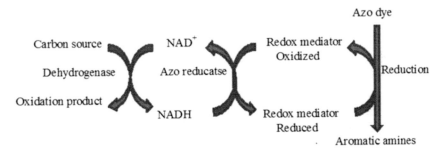

FIGURE 8.8   Decolorization of azo dye by redox mediator assisted method.

Azoreductases can be classified based on their oxygen requirements, structure, and function. On the basis of oxygen requirement, it is classified either active in the presence or absence of oxygen (Bafana and Chakrabarti, 2008). There are two types of oxygen-insensitive azoreductases identified in bacteria: one is monomeric flavin-free enzymes containing a putative NAD(P)H binding motif, and the other is polymeric flavin-dependent enzymes (Chen, 2006). Monomeric flavin-free azoreductases also show very narrow substrate specificity, but polymeric flavin-dependent azoreductase families can catalyze a variety of substrates that differ in size and complexity. Biochemical characteristics and the protein structures of several bacterial FMN-dependent azoreductases have been determined (Wang et al., 2007; Chen et al., 2010). Flavin dependent reductases are further categorized into three groups based on their requirement of NADH, NADPH, or both coenzymes (Punj and John, 2009) as an electron donor. Two monomeric flavin-free azoreductases (Azo A and Azo B) have been described in *Pigmentiphaga kullae* K24 (Blumel and Stolz, 2003). AzoB is very efficient in reducing Orange I compared to AzoA from *P. kullae* K24. Functionally AzoB is higher active than AzoA. AzoB requires two molecules of NADPH (4 electron donor) for the complete reductive cleavage of Orange I to sulfanilic acid and 1-amino-4-naphthaol. The active site of the AzoB has two binding sites that allow the binding of the substrate and NADPH that implies an appropriate position for direct hydride transfer rather than a proton-relay catalytic reaction. Also, the enzyme AzoB utilizes either NADH or NADPH as reductants for the reaction. Chen et al. (2010) revealed that there is a clear correlation between structure, cofactor requirement, and substrate specificity in azoreductases.

8.6.2   LACCASES

Laccases (EC 1.10.3.2) are multicopper phenol oxidases and were first identified from the sap of the *Rhusvernicifera*, Japanese lacquer tree (Ram Lakhan Singh et al., 2015). Laccases act on phenol containing compounds and similar other molecules and perform one-electron oxidations. It decolorizes azo dyes and produces phenolic compounds rather than toxic aromatic amines while decolorizing azo dyes through absorbing a highly nonspecific free radical mechanism (Chivukula and Renganathan, 1995; Wong and Yu, 1999). It shows less substrate specificity and has the capability to degrade different types of xenobiotic substances, including dyes (De Souza et al., 2006). Therefore, it has been attaining great importance in the bioremediation of colored textile wastewater. Laccases so far reported were mainly from fungi and plant and also from a small number of bacteria (Gianfreda et al., 1999).

Laccases use O_2 as an electron acceptor to catalyze the oxidation of aromatic amines, phenols, polyphenols, and various nonphenolic compounds (Kiiskinen et al., 2002; Viswanath et al., 2008). Industrially relevant laccase enzyme is produced by bacteria, insects, higher plants, and fungi (Delanoy et al., 2005). Laccase has been described from a large number of bacteria like *Azospirillum lipoferum*, *Escherichia coli*, *Bacillus licheniformis*, *Bacillus halodurans*, *Streptomyces coelicolor* and *Thermus thermophillus* (Sharma et al., 2007; Singh et al., 2007; Koschorreck et al., 2009). Fungal and bacterial laccases have a similar structure, but their amino acid sequences are quite different (Claus, 2004), and bacterial laccases often occurred in the monomeric form, but some fungal laccases are isozymes/isoenzymes that form a multimeric complex by oligomerization. Laccase uses low molecular weight substances like 2,2-azino-bis-3-ethylbenzothiazoline-6-sulfonic acid (ABTS) as a redox mediator in the actual electron transfer steps (Wong and Yu, 1999). However, on the other hand, laccase enzymes that purified from mushroom *Hypsizygus ulmarius* decolorized methyl orange without using redox mediator (Ravikumar et al., 2013). It has been suggested that in the presence of redox mediators, dye decolorization could be improved considerably (Reyes et al., 1999; Abadulla et al., 2000; Soares et al., 2001).

Laccases are an oxidative enzyme oxidizes phenol compounds by transfer of one electron to generate a phenoxy radical which is further oxidized by the same enzyme to produce a carbonium ion in which the charge is restricted on the phenolic ring of carbon bearing the azo linkage. Water makes a nucleophilic attack on the carbonium ion and produces 4-sulfophenyldiazene and a benzoquinone. 4-Sulfophenyldiazene is an apparently unstable compound in the presence of oxygen, which oxidizes it to analogous phenyl diazene radical. Then it readily loses molecular nitrogen and to produce a sulfophenyl radical, which is then scavenged by O_2 to yield 4-sulfophenylhydroperoxide (Figure 8.9). 4-sulfophenylhydroperoxide is uncommon peroxide and is known to be formed only oxidation of sulfonated azo dyes by peroxidases, whereas organic peroxides are unstable in presence of metal ions.

8.6.3 PEROXIDASES

Peroxidases (EC 1.11.1.7) are oxidoreductases that catalyze reactions like reduction of peroxides such as hydrogen peroxide (H_2O_2) and oxidation of a wide variety of organic and inorganic compounds. It is a hemoprotein that catalyzes the oxidations of lignin and other phenolic compounds under acidic pH (pH 3.0 to pH 4.5) and in the presence of hydrogen peroxide (electron acceptor) (Duran et al., 2002). It contains iron (III) protoporphyrin IX as the prosthetic group. It has the potential to reduce environmental pollution by bioremediation of wastewater containing phenols, cresols, chlorinated phenols, and synthetic textile azo dyes (Bansal and Kanwar, 2013). It is well known in fungi to mineralize a wide variety of recalcitrant toxic azo dyes. This character is attributed to their ability to produce exo-enzymes such as lignin peroxidases, manganese peroxidases, and polyphenol oxidases (PPOs). Predominantly, white-rot fungus secretes lignin peroxidases (LiP) under aerobic conditions as a secondary metabolite in the stationary phase. Lignin and phenolic compounds are degraded by (LiP) in the presence of H_2O_2 (cosubstrate) and veratryl alcohol (mediator). In this degradation, H_2O_2 is reduced to H_2O by accepting an electron from LiP (which can oxidize itself). The oxidized LiP returns to its native form (reduced) by gaining an electron from veratryl alcohol, thereby veratryl aldehyde is formed. Veratryl aldehyde gets reduced back to veratryl alcohol by accepting an electron from the substrate. Lignin peroxidase enzyme production has been reported from organisms such as *Candida krusei, Phanerochaete chrsosporium, Pleurotus streatus, Citrobacter freundii,* and *Pseudomaonas desmolyticum* (Bansal and Kanwar, 2013).

Lignin-degrading Basidiomycetes fungus produces manganese peroxidases (MnP) extracellularly. MnP oxidizes Mn^{2+} to Mn^{3+} that act as a mediator for the oxidation of many phenolic compounds (ten Have and Teunissen, 2001). *Phanerochaete crysosporium, P. sordid, C. subvermispora, P. radiate, D. squalens,* and *P. rivulosu* can produce MnP extracellularly (Bansal and Kanwar, 2013). Microbial production of MnP is dependent on nutrient on which it grows. For example, in the presence of glutamic acid *Trametes trogii* showed higher laccase and MnP activities, causing decolorization of several azo dyes effectively (Levin et al., 2010). For example, *Brevibacterium casei* removes azo dye Acid Orange7 (AO7) and chromate Cr(VI) under nutrient-limiting conditions. Cr(VI) is reduced by the reduction enzyme of *B. casei* with the help of AO7 that acts as an electron donor. The reduced chromate Cr(III) form complex with the oxidized AO7 produced a purple intermediate (Ng et al., 2010). Kalyani et al. (2011) purified 86 kDa peroxidase enzymes from *Pseudomonas* sp. SUK1, which oxidized various lignin-related phenols and decolorized many textile dyes under the optimum pH (3.0) and temperature (40°C). Compared to fungal peroxidases, the bacterial peroxidases are advantageous, because of bacterial species are more suitable for protein engineering to improve their catalytic properties successfully (Bugg et al., 2011).

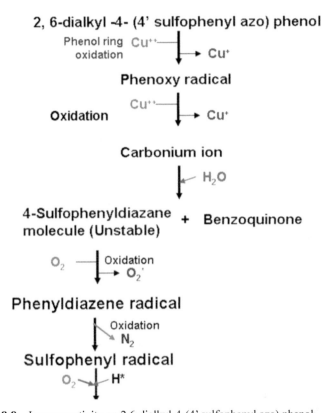

2, 6-dialkyl -4- (4' sulfophenyl azo) phenol

Phenol ring oxidation Cu^{++} → Cu^+

Phenoxy radical

Oxidation Cu^{++} → Cu^+

Carbonium ion

H_2O

4-Sulfophenyldiazane molecule (Unstable) + **Benzoquinone**

O_2 — Oxidation → O_2'

Phenyldiazene radical

Oxidation N_2

Sulfophenyl radical

O_2 — H^*

FIGURE 8.9   Laccase activity on 2,6-dialkyl-4-(4' sulfophenyl azo) phenol.

8.6.4 POLYPHENOL OXIDASES (PPOS)

PPOs (EC 1.14.18.1) is referred to as tyrosinase or monophenol mono-oxygenase. It is a tetrameric protein, contains four copper atoms and has binding sites for oxygen and two aromatic compounds. It uses molecular oxygen as an oxidant, thereby reduces the cost of applying the technology (Wu et al., 2001). It has been reported that tyrosinase enzyme from *Bacillus thuringiensis* has to be used for decontamination of wastewater having phenols (EL-Shora and Metwally, 2008). It catalyzes phenol oxidation in two phases. In the first step, it catalyzes the hydroxylation of monophenols to o-diphenols. In the second step, o-diphenols is further oxidized to o-quinones. This process has been demonstrated with the aid of tyrosinase enzyme of *Pseudomonas desmolyticum* NCIM 2112 to degrade Direct blue-6 (Kalme et al., 2007) and mixed culture of *Galactomyces geotrichum* and *Bacillus* sp. VUS to degrade Disperse dye brown 3REL (Jadhav et al., 2008b). Husain and Jan (2000) stated that PPOs are a nonspecific enzyme that can act on a broad range of substrate and have the ability to remove the pollutant from contaminated sites at very low concentrations.

8.7   BIOREACTOR SYSTEM FOR INDUSTRIAL EFFLUENT TREATMENT

Industrial effluent treatment plays a key role in the development of a sustainable environment. Classical approaches have several limitations to produce reusable water. Moreover, the preparation of reusable wastewater is a necessity to reduce water utility. Last few decades industries have adopted several methods for the effluent treatment process, but their efficiency is too low. Recently, biotechnologist has developed several reactor based strategies to remove toxic substances from the industrial effluents which have several advantages than the classical methods. High COD containing effluents are easily treated by anaerobic reactors, which help to produce energy generation and low sludge production. However, in realistic applications, anaerobic treatment experiences the low growth rate of the microorganisms, a little settling rate, unstable end products, and the need for post-treatment of the toxic anaerobic effluent, which often contains hydrogen sulfide (HS$-$) and ammonium ion (NH$_4^+$). The systems such as a stirred tank reactor, air-lift, and bubble column reactor, fixed-bed bioreactor, rotating disk reactor, and silicone-membrane reactor have been widely used. Following are the several reactor systems employed for bioremediation of textile effluent using microorganisms (Figure 8.10) such as, (i) up-flow anaerobic sludge blank reactor (UASB), (ii) Rotating drum bioreactor (RDBR), (iii) up-flow column reactor (UFCR), (iv) Fluidized bed reactor (FBR), (v) Anaerobic baffled reactor (ABR), (vi) Continuous packed bed bioreactor, (vii) Pulsed packed bed bioreactor, (viii) Airlift reactor, and (ix) Tubular photobioreactor for cyanobacteria. The choice of reactor depends on the organism that is used for the treatment and process of interest (Padmanaban et al., 2013). Decolorization of textile industrial effluents is a continuous process, which required continuous flow bioreactors. For example, continuous with fixed film bioreactors remove disperse dye (Red-553) up to 80% within 20 days (Yang et al., 2003).

Darwesh et al. (2015) designed prototype bioreactor with dual oxygenation (anoxic and aerobic) level for the decolorization of azo dye Reactive Blue by *Pseudomonas aeruginosa* strain OS4. This bioreactor consists of two compartments, up-flow fixed film column (UFC), and continuously stirred aerobic (CSA) container. The bioreactor could be operated at a flow rate of 50ml per hour with a hydraulic retention time (HRT) of 20 hours using the immobilized bacterial cell. This system showed 99% decolorization of Reactive blue in sequential anoxic and aerobic treatment and found that decolorization of dye is due to the formation of lignin peroxidase in both conditions. They prepared and modified surface properties of lignin peroxidase magnetic nanoparticle and tested it for bioremediation of reactive blue.

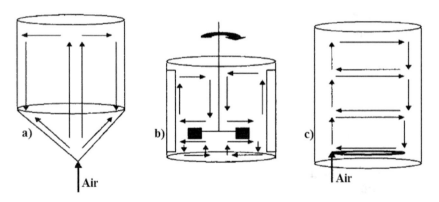

FIGURE 8.10 (a) Conical up-flows system, (b) Stirred reactor, and (c) Bubble column reactor.

Andleeb et al. (2012) studied the anthraquinone dye biodegradation by immobilized *Aspergillus flavus* in fluidized bed bioreactor. The reactor functions continuous flow mode, at room temperature (25–30°C). The reactor was filled with sand-immobilized fungal mass, and simulated textile effluent (STE) medium with 50 mg l^{-1} of dye concentration was fed into the reactor. The reactor effectively removes overall biological oxygen demand (BOD), COD, and color up to 85.57%, 84.70%, and

71.3%, respectively. Buitron et al. (2004), studied the application of sequencing batch reactor (SBR) pilot system biofilter filled with porous volcanic rock for the aerobic degradation of the azo dye acid red 151, which removes the color up to 99% using an initial concentration of 50 mg AR151.

A trickling filter system was made with small river rock of 2.5 to 7.5 cm for the treatment of textile effluent. It reduces the BOD level in wastewater from 600 to 100 mg/L (Shanooba et al., 2011). Efficient decolorization of Reactive Red 198 (>90%) was possible in anaerobic/aerobic SBR system. The color removal above 90% was attained under anaerobic phase, and concentration of Reactive Red 198 at 20 and 50 mg l^{-1} did not create an adverse effect on the activity of the microorganisms (Kocyigit and Ugurlu, 2015). Sandhya et al. (2005) reported about 94% azo dye color removal in microaerophilic-aerobic condition. Hosseini Koupaie et al. (2013) conducted biodegradation of Acid Red 18 in integrated anaerobic-aerobic fixed bed sequencing batch biofilm reactor and observed approximately 90% removal of color, COD, and intermediate compounds. They found polyethylene as a medium for immobilization of microorganisms for better azo dye removal.

Upflow Column Bioreactor was used for mineralization and detoxification of the carcinogenic azo dye Congo red. In this system, the textile effluent is treated by a Polyurethane Foam Immobilized Microbial Consortium. A bacterial consortium consisting of *Providencia rettgeri* strain HSL1 and *Pseudomonas* sp. SUK1 reduce azo dye Congo red under microaerophilic incubation conditions and yield toxic aromatic amines, biphenyl-4,4'-diamine and sodium-3,4-diaminonaphthalene-1-sulfonate. Biphenyl-4,4'-diamine and sodium-3,4-diaminonaphthalene-1-sulfonate converted into biphenyl and naphthalene respectively in a consequent aerobic process due to the activities of oxidative enzyme, laccase (Figure 8.11).

Textile effluents composition is obviously variable, comprising organics, nutrients, sulfur compounds, salts, and toxic chemical substances because of usage of wide varieties of synthetic dye that have the different chemical structure (Chen et al., 2003; Pearce et al., 2003; Ghodake et al., 2009b). Therefore, bacterial decolorization is directly influenced by several factors includes oxygen concentration, the level of agitation, pH, temperature, the structure of dye, dye concentration, carbon, and nitrogen source availability, electron donor, and redox mediator (Saratale et al., 2011).

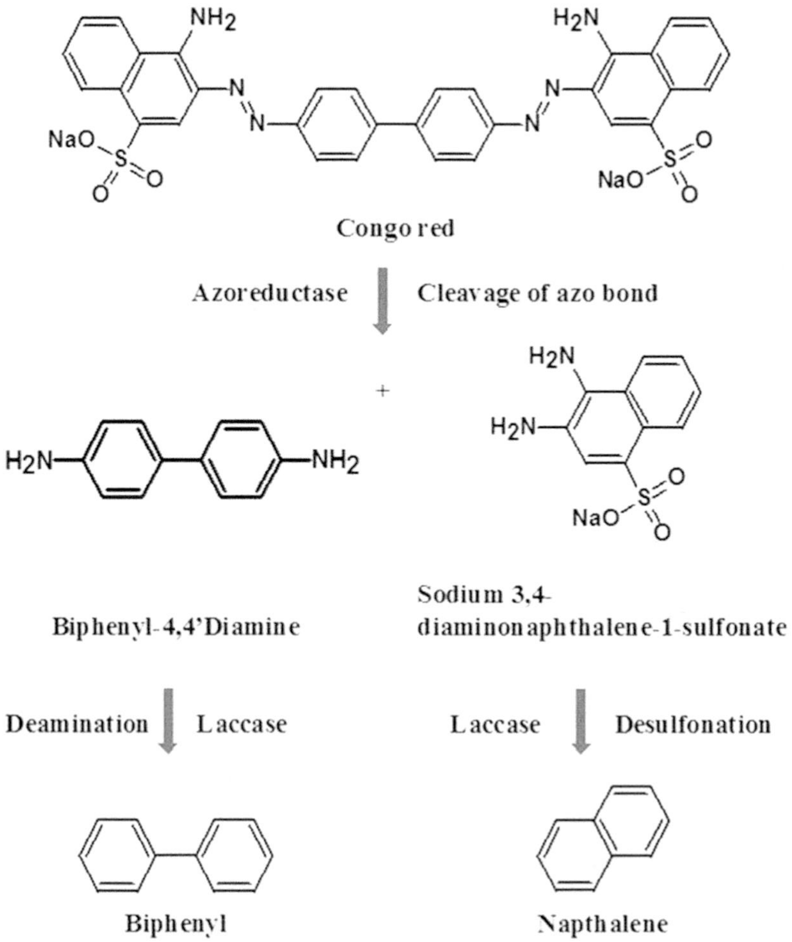

Congo red

Azoreductase | Cleavage of azo bond

+

Biphenyl-4,4'Diamine

Sodium 3,4-
diaminonaphthalene-1-sulfonate

Deamination | Laccase   Laccase | Desulfonation

Biphenyl   Napthalene

FIGURE 8.11   Biodegradation of Congo red.

8.8   CONCLUSION

Azo compounds constitute the most diverse group of human-made synthetic dyes and are widely used in several industries such as textiles, food, cosmetics, and paper printing. Azo dyes are resistant to biodegradation due to their recalcitrant nature. However, microbes are highly adaptable in extreme environments and synthesis of different enzymes for the decolorization and mineralization of dyes under suitable environmental conditions. Several genera of bacteria have been reported for azo dyes degradation. Anaerobic or aerobic azo dye degradation by several pure and mixed bacterial cultures has been reported. Different mechanisms, including enzymatic as well as RMs, have been used for this non-specific reductive cleavage. However, few aerobic bacteria that can utilize azo dye as their growth substrates. Through biotechnological advancements, various reactor systems are devised and are employed for decolorization and degradation of textile dyes using microorganisms. The systems such as UASB, ABR, FBR, and RDBR are suitable for bacterial operation. The selection of an appropriate type of reactor system facilitated with a suitable condition can efficiently remove the dyes from the textile industrial effluents.

KEYWORDS

- aerobic degradation
- anaerobic degradation
- azo dyes
- azo-reductases
- bacterial decolorization
- bioreactor based decolorization
- bioremediation
- laccases
- oxidoreductases
- physical methods
- polyphenol oxidases

REFERENCES

Abadulla, E., Tzonov, T., Costa, S., Robra, K. H., & Cavaco, P. A. Decolorization and detoxification of textile dyes with a laccase from *Trametes hirsute*. *Appl. Environ. Microbiol.*, **2000**, *66*, 3357–3362.

Acuner, E., & Dilek, F. B. Treatment of tectilon yellow 2G by *Chlorella vulgaris*. *Process Biochem.*, **2004**, *39*, 623.

Alaton, I. A., Balcioglu, I. A., & Bahnemann, D. W. Advanced oxidation of a reactive dye bath effluent: Comparison of O_3, H_2O_2/UV-C, and TiO_2/UV-A process. *Water Res.*, **2002**, *36*, 1143–1154.

Amr, F., Saad, E. H., Mohamed, S. A., & Ebrahim, S. Decolorization of different azo dyes and detoxification of dyeing wastewater by *Pseudomonas stutzeri* (SB13) isolated from textile dyes effluent. *Br. Biotechnol. J.*, **2016**, *15*(4), 1–18.

Andleeb, S., Naima, A., Robson, G. D., & Safia, A. An investigation of anthraquinone dye biodegradation by immobilized *Aspergillus flavus* in fluidized bed bioreactor. *Environ. Sci. Pollut. Res.*, **2012**, *19*(5), 1728–1737.

Anjaneyulu, Y., Sreedhara, C. N., & Raj, S. S. D. Decolorization of industrial effluents – available methods and emerging technologies – a review. *Rev. Environ. Sci. Biotechnol.*, **2005**, *4*, 245.

Annuar, M. S. M., Adnan, S., Vikneswari, S., & Chisti, Y. Kinetics and energetic of dye decolorization by *Pycnoporus sanguine*. *Water Air Soil Pollut.*, **2009**, *202*, 179–188.

Baccar, R., Blanquez, P., Bouzid, J., Feki, M., Attiya, H., & Sarra, M. Decolorization of a tannery dye: from fungal screening to bioreactor application. *J. Biochem. Eng.*, **2011**, *56*, 184–189.

Bafana, A., & Chakrabarti, T. Lateral gene transfer in phylogeny of azoreductase system. *Comput. Biol. Chem.*, **2008**, *32*, 191–197.

Bafana, A., Chakrabarti, T., & Devi, S. S. Azoreductase and dye detoxification activities of *Bacillus velezensis* strain, A. B. *Appl. Microbiol. Biotechnol.*, **2008**, *77*, 1139–1144.

Bafana, A., Chakrabarti, T., Muthal, P., & Kanade, G. Detoxification of benzidine-based azo dye by *Enterococcus gallinarum*: Time course study. *Ecotoxicol. Environ. Saf.*, **2009**, *72*, 960–964.

Bafana, A., Saravana, D. S., & Chakrabarti, T. Azo dyes: Past, present and the future. *Environ. Rev.*, **2011**, *19*, 350–370.

Bansal, N., & Kanwar, S. S. Peroxidase(s) in environment protection. *The Scientific World Journal*, **2013**, 714639, doi: 10.1155/2013/714639.

Baumler, W., Eibler, E. T., Hohenleutner, U., Sens, B., Sauer, J., & Landthaler, M. Q-switched laser and tattoo pigment: First results of the chemical and photophysical analysis of 41 compounds. *Lasers. Surg. Med.*, **2000**, *26*, 13–21.

Bibi, I., & Bhatti, H. N. Biodecolorization of reactive black 5 by laccase mediator system. *Afr. J. Biotechnol.*, **2012**, *11*, 7464–7471.

Blumel, S., & Stolz, A. Cloning and characterization of the gene coding for the aerobic azoreductase from *Pigmentiphaga kullae* K24. *Appl. Microbiol. Biotechnol.*, **2003**, *62*(2/3), 186–190.

Brown, M. A., & De Vito, S. C. Predicting azo dye toxicity. *Crit. Rev. Environ. Sci. Technol.*, **1993**, *23*, 249.

Bugg, T. D. H., Ahmad, M., Hardiman, E. M., & Singh, R. The emerging role for bacteria in lignin degradation and bio-product formation. *Curr. Opin. Biotechnol.*, **2011**, *22*, 394–400.

Buitron, G., Quezada, M., & Moreno, G. Aerobic degradation of the azo dye acid red 151 in a sequencing batch biofilter. *Biores. Technol.*, **2004**, *92*, 143–149.

Chang, J. S., & Kuo, T. S. Kinetics of bacterial decolorization of azo dye with *Escherichia coli* NO3. *Bioresour. Technol.*, **2000**, *75*, 107.

Chang, J. S., Chen, B. Y., & Lin, Y. S. Stimulation of bacterial decolorization of an azo dye by extracellular metabolites from *Escherichia coli* strain NO_3. *Bioresour. Technol.*, **2004**, *91*, 243.

Chang, J. S., Kuo, T. S., Chao, Y. P., Ho, J. Y., & Lin, P. J. Azo dye decolorization with a mutant *Escherichia coli* strain. *Biotechnol. Lett.*, **2000**, *22*, 807.

Chen, H. Recent advances in azo dye degrading enzyme research. *Curr. Protein Pept. Sci.*, **2006**, *7*(2), 101–111.

Chen, H., Feng, J., Kweon, O., Xu, H., & Cerniglia, C. E. Identification and molecular characterization of a novel flavin-free NADPH preferred azoreductase encoded by azoB in *Pigmentiphaga kullae* K24. *BMC Biochem.*, **2010**, *11*, 13, doi: 10.1186/1471-2091-11-13.

Chen, H., Hopper, S. L., & Cerniglia, C. E. Biochemical and molecular characterization of an azoreductase from *Staphylococcus aureus* a tetrameric NADPH-dependent flavoprotein. *Microbiology*, **2005**, *151*, 1433–1441.

Chen, K. C., Huang, W. T., Wu, J. Y., & Houng, J. Y. Microbial decolorization of azo dyes by *Proteus mirabilis*. *J. Ind. Microbiol. Biotechnol.*, **1999**, *23*, 686–690.

Chen, K. C., Wu, J. Y., Liou, D. J., & Hwang, S. C. J. Decolorization of the textile dyes by newly isolated bacterial strains. *J. Biotechnol.*, **2003**, *101*, 57.

Chetana, R., Avinash, K., Tallika, P., & Shailesh, D. Biodegradation of diazo dye, reactive blue 160 by isolate *Microbacterium* sp. B12 mutant: Identification of intermediates by LC-MS. *Int. J. Curr. Microbiol. App. Sci.*, **2016**, *5*(3), 534–547.

Chivukula, M., & Renganathan, V. Phenolic azo dye oxidation by laccase from *Pycularia oryzae*. *Appl. Environ. Microbiol.*, **1995**, *1*(12), 4374–4377.

Christian, V., Shrivastava, V., Shukla, D., Modi, H. A., & Vyas, B. R. M. Degradation of xenobiotic compounds by lignin-degrading white-rot fungi: Enzymology and mechanisms involved. *Ind. J. Exp. Biol.*, **2005**, *43*, 301.

Chung, K. T., Stevens, S. E., & Cerniglia, C. E. The production of azo dyes by the intestinal microflora. *Crit. Rev. Microbiol.*, **1992**, *18*, 175–190.

Claus, H. Laccases: Structure, reactions, distribution. *Micron.*, **2004**, *35*, 93–96.

Corbman, B. P. *Textiles: Fiber to Fabric* (pp. 201–222). McGraw-Hill, New York, **1983**.

Coughlin, M. F., Kinkle, B. K., & Bishop, P. L. Degradation of azo dyes containing amino naphthol by *Sphingomonas* sp. strain ICX. *J. Ind. Microbiol. Biotechnol.*, **1999**, *23*, 341–346.

Daneshvar, N., Ayazloo, M., Khataee, A. R., & Pourhassan, M. Biological decolorization of dye solution containing malachite green by microalgae *Cosmarium* sp. *Bioresour. Technol.*, **2007**, *98*, 1176–1182.

Darwesh, M. O., Moawad, H., & Abd El-Rahim, W. M. Application of nanostructured microbial enzyme for bioremediation of industrial wastewater. *Intl. J. Water Resources Arid. Environ.*, **2015**, *4*(1), 37–52.

De'Souza, D. T., Tiwari, R., Sah, A. K., & Raghukumara, C. Enhanced production of laccase by a marine fungus during treatment of colored effluents and synthetic dyes. *Enzyme Microb. Technol.*, **2006**, *38*, 504–511.

Delanoy, G., Li, Q., & Yu, J. Activity and stability of laccase in conjugation with chitosan. *Int. J. Biol. Macromol.*, **2005**, *35*(1/2), 89–95.

Diego, R. S. L., Bruno, E. L. B., Gilmare, A. S., Silvana, Q. S., & Sergio, F. A. Use of multivariate experimental designs for optimizing the reductive degradation of an azo dye in the presence of redox mediators. *Quim. Nova*, **2014**, *37*(5), 827–832.

Dos Santos, A. B., Cervantes, F. J., & Van Lier, J. B. Review paper on current technologies for decolorization of textile wastewaters: Perspectives for anaerobic biotechnology. *Bioresour. Technol.*, **2007**, *98*, 2369.

Duran, N., Rosa MA, D'Annibale, A., & Gianfreda, L. Applications of laccases and tyrosinases (phenoloxidases) immobilized on different supports: A review. *Enzyme Microb. Technol.*, **2002**, *31*, 907–931.

El-Sheekh, M. M., Gharieb, M. M., & Abou-El-Souod, G. W. Biodegradation of dyes by some green algae and cyanobacteria. *Int. Biodeterior. Biodegradation*, **2009**, *63*, 699–704.

El-Shora, H. M., & Metwally, M. Use of tyrosinase enzyme from *Bacillus thuringiensis* for the decontamination of water polluted with phenols. *Biotechnology*, **2008**, *7*(2), 305–310.

Engel, E., Vasold, R., Santarelli, F., Maisch, T., Gopee, N. V., Howard, P. C., Landthaler, M., & Baumler, W. Tattooing of skin results in transportation and light-induced decomposition of tattoo pigments – a first quantification *in vivo* using a mouse model. *Exp. Dermatol.*, **2010**, *19*, 54–60.

Ezeronye, O. U., & Okerentugba, P. O. Performance and efficiency of a yeast biofilter for the treatment of a Nigerian fertilizer plant effluent. *World J. Microbiol. Biotechnol.*, **1999**, *15*, 515.

Feigel, B. J., & Knackmuss, H. J. Syntropic interactions during degradation of 4-aminobenzenesulfonic acid by a two species bacterial culture. *Arch. Microbiol.*, **1993**, *159*, 124–130.

Field, J. A., & Brady, J. Riboflavin as a redox mediator accelerating the reduction of the azo dye mordant yellow 10 by anaerobic granular sludge. *Water Sci. Technol.*, **2003**, *48*, 187.

Franciscon, E., Grossman, M. J., Rizzato, P. J. A., Reyes, R. F. G., & Durrant, L. R. Decolorization and biodegradation of reactive sulfonated azo dyes by a newly isolated *Brevibacterium* sp. strain VN-15. *Springer Plus*, **2012**, *1*(37), doi.org/10.1186/2193–1801–1–37.

Fu, Y., & Viraraghavan, T. Removal of acid blue 29 from an aqueous solution by fungus *Aspergillus niger*. In: Niko-Laidis, N., Erkey, C., & Smets, B. F., (eds.), *Proceedings of the 31ˢᵗ Mid-Atlantic Industrial and Hazardous Waste Conference* (pp. 510–519). Storrs, Conneticut, USA.

Gadd, G. M. *Fungi in Bioremediation*. Cambridge University Press, Cambridge, UK, **2001**.

Ghodake, G. S., Telke, A. A., Jadhav, J. P., & Govindwar, S. P. Potential of *Brassica juncea* in order to treat textile effluent contaminated sites. *Int. J. Phytorem.*, **2009a**, *11*, 297–312.

Ghodake, G., Jadhav, S., Dawkar, V., & Govindwar, V. Biodegradation of diazo dye direct brown MR by *Acinetobacter calcoaceticus* NCIM, 2890. *Int. Biodeter. Biodegr.*, **2009b**, *63*, 433.

Gianfreda, L., Xu, F., & Bollag, J. M. Laccases: A useful group of oxidoreductive enzymes. *Bioremediation J.*, **1999**, *3*, 1–25.

Gopinath, K. P., Murugesan, S., Abraham, J., & Muthukumar, K. *Bacillus* sp. mutant for improved biodegradation of Congo red: random mutagenesis approach. *Bioresour. Technol.*, **2009**, *100*, 6295–6300.

Goszczynski, S., Paszczynski, A., Pasti-Grigsby, M. B., Crawford, R. L., & Crawford, D. L. New pathway for degradation of sulfonated azo dyes by microbial peroxidases of *Phanerochaete chrysosporium* and *Streptomyces chromofuscus*. *J. Bacteriol.*, **1994**, *176*, 1339–1347.

Gupta, V. K., Suhas. Application of low cost adsorbents for dye removal – a review. *J. Environ. Manage*, **2009**, *90*(8), 2313–2342.

Hadibarata, T., Yusoff, A. R. M., Aris, A., Hidayat, T., & Kristanti, R. A. Decolorization of azo, triphenylmethane and anthraquinone dyes by laccase of a newly isolated *Armillaria* sp. F022. *Water Air Soil Pollut.*, **2012**, *223*, 1045–1054.

Hao, O. J., Kim, H., & Chiang, P. C. Decolorization of wastewater. *Crit. Rev. Env. Sci. Tec.*, **2000**, *30*, 449–505.

Harshad, L., Sanjay, G., & Diby, P. Low-cost biodegradation and detoxification of textile azo dye C.I. reactive blue 172 by *Providencia rettgeri* Strain HSL1. *J. Chem.*, **2015**, 1–10.

Hosseini, K. E., Alavi Moghaddam, M. R., & Hashemi, S. H. Investigation of decolorization kinetics and biodegradation of azo dye acid red 18 using sequential process of anaerobic sequencing batch reactor/moving bed sequencing batch biofilm reactor. *Int. Biodeterior. Biodegrad.*, **2012**, *71*, 43–49.

Hsueh, C. C., Chen, B. Y., & Yen, C. Y. Understanding effects of chemical structure on azo dye decolorization characteristics by *Aeromonas hydrophila*. *J. Hazard Mater.*, **2009**, *167*, 995.

Hu, T. L. Decolorization of reactive azo dyes by transformation with *Pseudomonas luteola*. *Bioresour. Technol.*, **1994**, *49*, 47–51.

Humnabadkar, R. P., Saratale, G. D., & Govindwar, S. P. Decolorization of purple 2R by *Aspergillus ochraceus* (NCIM-1146). *Asian J. Microbiol. Biotechnol. Environ. Sci.*, **2008**, *10*, 693.

Husain, Q., & Jan, U. Detoxification of phenol and aromatic amines from polluted waste water by using phenol oxidases. *J. Sci. Ind. Res.*, **2000**, *59*, 286–293.

Ilyas, S., Sultan, S., & Rehman, A. Decolorization and degradation of azo dye synozol red HF6BN by *Pleurotus ostreatus*. *Afr. J. Bioechnol.*, **2012**, *11*, 15422–15429.

Jadhav, S. U., Jadhav, U. U., Dawkar, V. V., & Govindwar, S. P. Biodegradation of disperse dye brown 3REL by microbial consortium of *Galactomyces geotrichum* MTCC, 1360 and *Bacillus* sp. VUS. *Biotechnol. Bioprocess Eng.*, **2008b**, *13*, 232.

Jai, S., Pillai, H. P., Girish, K., & Dayanand, A. Isolation, characterization and screening of actinomycetes from textile industry effluent for dye degradation. *Int. J. Curr. Microbiol. App. Sci.*, **2014**, *3*(11), 105–115.

Jin, R., Yang, H., Zhang, A., Wang, J., & Liu, G. Bioaugmentation on decolorization of C.I. Direct blue 71 by using genetically engineered strain *Escherichia coli* JM109(PGEX-AZR). *J. Hazard Mater.*, **2009**, *163*, 1123–1128.

Joshi, T., Iyengar, L., Singh, K., & Garg, S. Isolation, identification and application of novel bacterial consortium TJ-1 for the decolorization of structurally different azo dyes. *Bioresour. Technol.*, **2008**, *99*, 7115.

Joshni, T. C., & Kalidass, S. Enzymatic degradation of azo dyes – a review. *International Journal of Environmental Sciences*, **2011**, *1*(6), 1250–1260.

Kalme, S. D., Parshetti, G. K., Jadhav, S. U., & Govindwar, S. P. Biodegradation of benzidine based dye direct blue-6 by *Pseudomonas desmolyticum* NCIM 2112. *Bioresour. Technol.*, **2007**, *98*, 1405.

Kalyani, D. C., Phugare, S. S., & Shedbalkar, U. U. Purification and characterization of a bacterial peroxidase from the isolated strain *Pseudomonas* sp. SUK1 and its application for textile dye decolorization. *Ann. Microbiol.*, **2011**, *61*, 483.

Kalyani, D. C., Telke, A. A., Dhanve, R. S., & Jadhav, J. P. Ecofriendly biodegradation and detoxification of reactive red 2 textile dye by newly isolated *Pseudomonas* sp. SUK1. *J. Hazard Mater.*, **2008**, *163*, 735.

Keck, A., Klein, J., Kudlich, M., Stolz, A., Knackmuss, H. J., & Mattes, R. Reduction of azo dyes by redox mediators originating in the naphthalene sulfonic acid degradation pathway of *Sphingomonas* sp. strain BN6. *Appl. Environ. Microbiol.*, **1997**, *63*, 3684–3690.

Keck, A., Rau, J., Reemtsma, T., Mattes, R., Stolz, A., & Klein, J. Identification of quinoide redox mediators that are formed during the degradation of naphthalene-2-sulfonate by *Sphingomonas xenophaga* BN6. *Appl. Environ. Microbiol.*, **2002**, *68*(9), 4341–4349.

Khammuang, S., & Sarnthima, R. Mediator-assisted rhodamine B decolorization by *Trametes versicolor* laccase. *Pakistan J. Biol. Sci.*, **2009**, *18*, 616–623.

Khan, J. A. Biodegradation of azo dye by moderately halotolerant *Bacillus megaterium* and study of enzyme azoreductase involved in degradation. *Adv. Biotechnol.*, **2011**, *10*, 21–27.

Kharub, M. Use of various technologies, methods and adsorbents for removal of dye. *J. Environ. Res. Develop.*, **2012**, *6*, 879–883.

Khataee, A., Dehghan, G., Zarei, M., Fallaha, S., Niaeia, G., & Atazadeh, I. Degradation of an azo dye using the green macroalga *Enteromorpha* sp. *Chem. Ecol.*, **2013**, *29*(3), 221–233.

Khehra, M. S., Saini, H. S., Sharma, D. K., Chadha, B. S., & Chimni, S. S. Biodegradation of azo dye C.I. Acid red 88 by an anoxic–aerobic sequential bioreactor. *Dyes Pigments*, **2006**, *70*(1), 1–7.

Kiiskinen, L. L., Viikari, L., & Kruus, K. Purification and characterization of novel laccase from the Ascomycete *Melanarpus albomyces*. *Appl. Microbiol. Biotechnol.*, **2002**, *59*(2/3), 198–204.

Kim, S. Y., An, J. Y., & Kim, B. W. Improvement of the decolorization of azo dye by anaerobic sludge bioaugmented with *Desulfovibrio desulfuricans*. *Biotechnol. Bioprocess Eng.*, **2007**, *12*(3), 222–227.

Kobya, M. Removal of Cr(VI) from aqueous solutions by adsorption on to hazel nut shell activated carbon: Kinetic and equilibrium studies. *Bioresour. Technol.*, **2004**, *91*, 317–321.

Kocyigit, H., & Ugurlu, A. Biological decolorization of reactive azo dye by anaerobic/ aerobic sequencing batch reactor system. *Global NEST J.*, **2015**, *17*(1), 210–219.

Koschorreck, K., Schmid, R. D., & Urlacher, V. B. Improving the functional expression of a *Bacillus licheniformis* laccase by random and site-directed mutagenesis. *BMC Biotechnol.*, **2009**, *9*, 12, https://doi.org/10.1186/1472-6750-9-12 (Accessed on 7 October 2019).

Kudlich, M., Keck, A., Klein, J., & Stolz, A. Localization of the enzyme system involved in anaerobic reduction of azo dyes by *Sphingomonas* sp. strain BN6 and effect of artificial redox mediators on the rate of azo dye reduction. *Appl. Environ. Microbiol.*, **1997**, *63*, 3691–3694.

Latif, A., Noor, Sharif, Q. M., & Najeebullah, M. Different techniques recently used for the treatment of textile dyeing effluents: A review. *J. Chem. Soc. Pakistan*, **2010**, *32*, 115–124.

Laxmi, S., & Nikam, T. D. Decolorisation and detoxification of widely used azo dyes by fungal species isolated from textile dye contaminated site. *Int. J. Curr. Microbiol. App. Sci.*, **2015**, *4*(4), 813–834.

Leelakriangsak, M., & Borisut, S. Characterization of the decolorizing activity of azo dyes by *Bacillus subtilis* azoreductase AzoR1. *Songklankrin. J. Sci. Technol.*, **2012**, *34*, 509–516.

Levin, L., Melignani, E., & Ramos, A. M. Effect of nitrogen sources and vitamins on ligninolytic enzyme production by some white-rot fungi dye decolorization by selected culture filtrates. *Bioresour. Technol.*, **2010**, *101*(12), 4554–4563.

Levine, W. G. Metabolism of azo dyes: Implication for detoxification and activation. *Drug Metab. Rev.*, **1991**, *23*, 253–309.

Mahalakshmi, S., Lakshmi, D., & Menaga, U. Biodegradation of different concentration of dye (Congo red dye) by using green and blue green algae. *Int. J. Environ. Res.*, **2015**, *9*(2), 735–744.

Maier, J., Kandelbauer, A., Erlacher, A., Cavaco-Paulo, A., & Gubitz, G. M. A new alkali thermostable azoreductase from *Bacillus* sp. strain, S. F. *Appl. Environ. Microbiol.*, **2004**, *70*, 837–844.

Mane, U. V., Gurav, P. N., Deshmukh, A. M., & Govindwar, S. P. Degradation of textile dye reactive navy-blue Rx (reactive blue-59) by an isolated Actinomycete *Streptomyces krainskii* SUK-5. *Malays J. Microbiol.*, **2008**, *4*, 1–5.

Manning, W. B., Cerniglia, E. C., & Thomas, W. F. Metabolism of the benzidine-based azo dye direct black 38 by human intestinal microbiota. *Applied and Environmental Microbiology*, **1985**, *50*(1), 10–15.

McMullan, G., Meehan, C., Conneely, A., Kirby, N., Robinson, T., Nigam, P., Banat, I. M., & Smyth, W. F. Microbial decolorization and degradation of textile dyes. *Appl. Microbiol. Biotechnol.*, **2001**, *56*, 81.

Metcalf, E. *Wastewater Engineering: Treatment and Reuse.* McGraw-Hill, New York, USA, **2003**.

Michniewicz, A., Ledakowicz, S., Ullrich, R., & Hofrichter, M. Kinetics of the enzymatic decolorization of textile dyes by laccase from *Cerrena unicolor*. *Dyes Pigments*, **2008**, *77*, 295–302.

Mohamed, M., Raja, M., Mohamed, S. S., & Gajalakshmi, P. Studies on effect of marine actinomycetes on amido black (azo dye) decolorization. *J. Chem. Pharm. Res.*, **2016**, *8*(8), 640–644.

Molina-Guijarro, J. M., Perez, J., Munoz-Dorado, J., Guillen, F., Moya, R., Hernandez, M., & Arias, M. E. Detoxification of azo dyes by a novel pH-versatile, salt-resistant laccase from *Streptomyces ipomoea*. *Int. Microbiol.*, **2009**, *12*, 13–21.

Mou, D. G., Lim, L. L., & Shen, H. P. Microbial agents for decolorization of dye wastewater. *Biotechnol. Adv.*, **1991**, *9*, 613–622.

Mubarak Ali, D., Suresh, A., Kumar, P. R., Gunasekaran, M., & Thajuddin, N. Efficiency of textile dye decolurization by marine cyanobacteria *Oscillatoria formosa* NTDM02. *Afr. J. Basic Appl. Sci.*, **2011**, *3*(1), 9–13.

Nachiyar, C. V., & Rajakumar, S. Degradation of a tannery and textile dye, Navitan fast blue S5R by *Pseudomonas aeruginosa*. *World J. Microbiol. Biotechnol.*, **2005**, *19*, 609–614.

Ng, T. W., Cai, Q., Wong, C., Chow, A. T., & Wong, P. Simultaneous chromate reduction and azo dye decolorization by *Brevibacterium casei*: Azo dye as electron donor for chromate reduction. *J. Hazard Mater.*, **2010**, *182*, 792–800.

Nigam, P., Banat, I. M., Singh, D., & Marchant, R. Microbial process for the decolorization of textile effluent containing azo, diazo and reactive dyes. *Process Biochem.*, **1996**, *31*, 435–442.

Niladevi, K. N., & Prema, P. Effect of inducers and process parameters on laccase production by *Streptomyces psammoticus* and its application in dye decolorization. *Bioresour. Technol.*, **2008**, *99*, 4583–4589.

Ollgaard, H., Frost, L., Galster, J., & Hasen, O. C. *Survey of Azo Colorants in Denmark: Consumption, Use, Health and Environmental Aspects.* Ministry of Environment and Energy, Government of Denmark, Denmark, **1998**.

Oturkar, C. C., Patole, M. S., Gawai, K. R., & Madamwar, D. Enzyme based cleavage strategy of *Bacillus lentus* BI377 in response to metabolism of azoic recalcitrant. *Bioresource Technol.*, **2013**, *130*, 360–365.

Ozer, A., Akkaya, G., & Turabik, M. The removal of acid red 274 from wastewater combined biosorption and biocoagulation with *Spirogyra rhizopus*. *Dyes Pigment*, **2006**, *71*, 83–89.

Padmanaban, V. C., Sandra, J., & Catherine, R. Reactor systems for the degradation of textile dyes. *Int. J. Environ. Sci.*, **2013**, *3*(6), 1868–1873.

Palmieri, G., Cennamo, G., & Sania, G. Remazol brilliant blue R decolorization by the fungus *Pleurotus ostreatus* and its oxidative enzyme system. *Enzyme Microb. Technol.*, **2005**, *36*, 17–34.

Pandey, A. K., & Dubey, V. Biodegradation of azo dye reactive red BL by *Alcaligenes* sp. AA09. *Int. J. Eng. Sci.*, **2012**, *1*, 54–60.

Pandey, A., Singh, P., & Iyengar, L. Bacterial decolorization and degradation of azo dyes. *Int. Biodeter. Biodegrad.*, **2007**, *59*, 73.

Pasti-Grigsby, M. B., Paszcczynski, A., Goszczynski, S., Crawford, D. L., & Crawford, R. L. Influence of aromatic substitution patterns on azo dye degradability by *Streptomyces* sp. and *Phanerochaete chrysosporium*. *Appl. Environ. Microbiol.*, **1992**, *58*(11), 3605–3613.

Pearce, C. I., Lloyd, J. R., & Guthriea, J. T. The removal of color from textile wastewater using whole bacterial cells: A review. *Dyes Pigments*, **2003**, *58*, 179.

Perumal, K., Malleswari, R. B., Catherin, A., & Sambanda, M. T. A. Decolorization of congo red dye by bacterial consortium isolated from dye contaminated soil, Paramakudi, Tamil Nadu. *J. Microbiol. Biotechnol. Res.*, **2012**, *2*, 475–480.

Platzek, T., Lang, C., Grohmann, G., Gi, U. S., & Baltes, W. Formation of a carcinogenic aromatic amine from an azo dye by human skin bacteria. *Hum. Exp. Toxicol.*, **1999**, *18*, 552–559.

Polman, J. L., & Breckenridge, C. R. Biomass-mediated binding and recovery of textile dyes from waste effluents. *Text Chem. Color,* **1996**, *28*(4), 31–35.

Popli, S., & Patel, D. U. Destruction of azo dyes by anaerobic-aerobic sequential biological treatment: A review. *Int. J. Environ. Sci. Technol.*, **2015**, *12*, 405–420.

Pricelius, S., Held, C., Murkovic, M., Bozic, M., Kokol, V., Cavaco-Paulo, A., & Guebitz, G. M. Enzymatic reduction of azo and indigoid compounds. *Appl. Microbiol. Biotechnol.*, **2007**, *77*, 321–327.

Punj, S., & John, G. H. Purification and identification of an FMN-dependent NAD(P)H azoreductase from *Enterococcus faecalis*. *Curr. Issues Mol. Biol.*, **2009**, *11*, 59–66.

Puvaneswari, M. J., & Gunasekaran, P. Toxicity assessment and microbial degradation of azo dyes. *Indian J. Exp. Biol.*, **2006**, *44*, 618–626.

Rai, H., Bhattacharya, M., Singh, J., Bansal, T. K., Vats, P., & Banerjee, U. C. Removal of dyes from the effluent of textile and dyestuff manufacturing industry: A review of emerging techniques with reference to biological treatment. *Crit. Rev. Env. Sci. Tech.*, **2005**, *35*, 219–238.

Ram Lakhan, S., Pradeep, K. S., & Rajat, P. S. Enzymatic decolorization and degradation of azo dyes – a review. *Int. Biodeterior. Biodegradation*, **2015**, *104*, 21–31.

Ramakrishna, K. R., & Viraraghavan, T. Dye removal using low cost adsorbents. *Water Sci. Technol.*, **1997**, *36*, 189.

Ramya, M., Iyappan, S., Manju, A., & Jiffe, J. S. Biodegradation and decolorization of acid red by *Acinetobacter radioresistens*. *Appl. Environ. Microbiol.*, **2010**, *70*, 837–844.

Rau, J., Knackmuss, H. J., & Stolz, A. Effects of different quinoid redox mediators on the anaerobic reduction of azo dyes by bacteria. *Environ Sci. Technol.*, **2002**, *36*, **1497**.

Ravikumar, G., Kalaiselvi, M., Gomathi, D., Vidhya, B., Devaki, K., & Uma, C. Effect of laccase from *Hypsizygus ulmarius* in decolorization of different dyes. *J. Pharm. Sci.*, **2013**, *3*, 150–152.

Reyes, P., Pickard, M. A., & Vazquez-Duhalt, R. Hydroxybenzotriazole increase the range of textile dyes decolorized by immobilized laccase. *Biotechnol. Lett.*, **1999**, *21*, 875–880.

Rizwana, P. S., & Uma, M. D. P. Decolorization and detoxification of reactive azo dyes by *Saccharothrix aerocolonigenes* TE5. *Appl. Environ. Microbiol.*, **2015**, *3*(2), 58–62.

Robinson, T., McMullan, G., Marchant, R., & Nigam, P. Remediation of dyes in textile effluent: A critical review on current treatment technologies with a proposed alternative. *Bioresour. Technol.*, **2001**, *77*, 247.

Sana, K., & Abdul, M. Degradation of reactive black 5 dye by a newly isolated bacterium *Pseudomonas entomophila* BS1. *Can. J. Microbiol.*, **2016**, *62*(3), 220–232.

Sandhya, S., Padmavathy, S., Swaminathan, K., Subrahmanyam, Y. V., & Kaul, S. N. Microaerophilic-aerobic sequential batch reactor for treatment of azo dyes containing simulated wastewater. *Process Biochem.*, **2005**, *40*, 885–890.

Sapna, K., & Sandeep, K. Screening for potential textile dye decolorizing bacteria. *IJSETT*, **2012**, *2*(1), 36–48.

Saratale, R. G., Saratale, G. D., *Chang, J.* S., & Govindwar, S. P. Bacterial decolorization and degradation of azo dyes: A review. *J. Taiwan Inst. Chem. Eng.*, **2011**, *42*(1), 138–157.

Saratale, R. G., Saratale, G. D., Chang, J. S., & Govindwar, S. P. Ecofriendly decolorization and degradation of reactive green 19A using *Micrococcus glutamicus* NCIM-2168. *Bioresour. Technol.*, **2009c**, *110*, 3897.

Sarayu, K., & Sandhya, S. Aerobic biodegradation pathway for remazol orange by *Pseudomonas aeruginosa*. *Appl. Biochem. Biotechnol.*, **2010**, *160*, 1241–1253.

Selvam, K., Swaminathan, K., & Chae, K. S. Decolorization of azo dyes and a dye industry effluent by a white rot fungus *Thelephera* sp. *Bioresour. Technol.*, **2003**, *88*, 115–119.

Shah, M. P. Microbial degradation of azo dye by *Pseudomonas* spp. 2413 isolated from activated sludge of common effluent treatment plant. *International Journal of Environmental Bioremediation and Biodegradation*, **2014**, *2*(3), 133–138.

Shanooba, P., Dhiraj, P., & Yatin, P. Microbial degradation of textile industrial effluents. *Afr. J. Biotechnol.*, **2011**, *10*(59), 12657–12661.

Sharma, P., Goel, R., & Caplash, N. Bacterial laccases. *World J. Microbiol. Biotechnol.*, **2007**, *23*, 823–832.

Singh, G., Capalash, N., Goel, R., & Sharma, P. A pH-stable laccase from alkalitolerant γ-proteobacterium JB: Purification, characterization and indigo carmine degradation. *Enzyme Microb. Technol.*, **2007**, *41*(6/7), 794–799.

Soares, G. M., Pessoad, A., & Costa-Ferreira, M. Use of laccase together with redox mediators to decolorize remazol brilliant blue R. *J. Biotechnol.*, **2001**, *89*, 123–129.

Stolz, A. Basic and applied aspects in the microbial degradation of azo dyes. *Appl. Microbiol. Biotechnol.*, **2001**, *56*, 69.

Sulak, M. T., Demirbas, E., & Kobya, M. Removal of astrazon yellow 7GL from aqueous solutions by adsorption onto wheat bran. *Bioresour. Technol.*, **2007**, *98*, 2590–2598.

Supaka, N., Juntongjin, K., Damronglerd, S., Delia, M. L., & Strehaiano, P. Microbial decolorization of reactive azo dyes in a sequential anaerobic-aerobic system. *Chem. Eng. J.*, **2004**, *99*, 169–176.

Surbhi, S., Subhasha, N., & Rachana, S. Potential of *Nostoc muscorum* for the decolorization of textile dye RGB-Red. *Int. J. Pharm. Bio. Sci.*, **2015**, *6*(3), 1092–1100.

Suwannawong, P., Khammuang, S., & Sarnthima, R. Decolorization of rhodamine B and congo red by partial purified laccase from *Lentinus polychrous* Lev. *J. Biochem. Technol*, **2010**, *3*, 182–186.

Tatarko, M., & Bumpus, J. A. Biodegradation of congo red by *Phanerochaete chrysosporium*. *Water Res.*, **1998**, *32*(5), 1713–1717.

Telke, A., Kalyani, D., Jadhav, J., & Govindwar, S. Kinetics and mechanism of reactive red 141 degradation by a bacterial isolate *Rhizobium radiobacter* MTCC, 8161. *Acta. Chim. Slov.*, **2008**, *55*, 320.

Ten Have, R., & Teunissen, P. J. M. Oxidative mechanisms involved in lignin degradation by white-rot fungi. *Chem. Rev.*, **2001**, *101*, 3397–3413.

Tian, Y. S., Xu, H., Peng, R. H., Yao, Q. H., & Wong, R. T. Heterologous expression and characterization laccase 2 from *Coprinopsis cinerea* capable of decolorizing different recalcitrant dyes. *Biotechnol. Equip.*, **2014**, *28*, 248–258.

Uddin, M. S., Zhou, J., Qu, Y., Guo, J., Wang, P., & Zhao, L. H. Biodecolorization of azo dye acid red B under high salinity condition. *Bull. Environ. Contam. Toxicol.*, **2007**, *79*, 440–444.

Vasdev, L., Luhad, R. C., & Saxena, R. L. Decolorization of triphenylmethane dyes by the bird's nest fungus *Cyathus bulleri*. *Current Microbiol.*, **1995**, *30*, 269–272.

Verma, A., & Shirkot, P. Purification and characterization of thermostable laccase from thermophilic *Geobacillus thermocatenulatus* MS5 and its application in removal of textiles dyes. *Sch. Acad. J. Biosci.*, **2013**, *2*, 479–485.

Verma, P., & Madamwar, D. Decolorization of synthetic dyes by a newly isolated strain of *Serratia maerascens*. *World J. Microbiol. Biotechnol.*, **2003**, *19*, 615.

Vishal, S., Garg, N., & Datta, M. An integrated process of textile dye removal and hydrogen evolution using Cyanobacterium, *Phormidium valderianum*. *World J. Microbiol. Biotechnol.*, **2001**, *17*, 499–504.

Viswanath, B., Chandra, M. S., & Reddy, B. R. Screening and assessment of laccase producing fungi isolated from different environment samples. *Afr. J. Biotechnol.*, **2008**, *7*(8), 1129–1133.

Von Lehmann, G., & Pierchalla, P. *Tatowierungsfarbstoffe*. Derm Beruf Umwelt, **1988**, *36*, 152–156.

Wang, C. J., Hagemeier, C., Rahman, N., Lowe, E., Noble, M., Coughtrie, M., Sim, E., & Westwood, I. Molecular cloning, characterization and ligand-bound structure of an azoreductase from *Pseudomonas aeruginosa*. *J. Mol. Biol.*, **2007**, *373*(5), 1213–1228.

Wang, M., Si, T., & Zhao, H. Biocatalyst development by directed evolution. *Bioresour. Technol.*, **2012**, *115*, 117–125.

Waqas, R., Arshad, M., Asghar, H. N., & Asghar, M. Optimization of factors for enhanced phycoremediation of reactive blue azo dye. *Int. J. Agric. Biol.*, **2015**, *17*, 803–808.

Weber, E. J., & Adams, R. L. Chemical and sediment mediated reduction of the azo dye disperse blue 79. *Environ. Sci. Technol.*, **1995**, *29*, 1163–1170.

Wong, P. K., & Yuen, P. Y. Decolorization and biodegradation of methyl red by *Klebsiella pneumonia* RS-13. *Water Res.*, **1996**, *30*, 1736.

Wong, Y., & Yu, J. Laccase catalyzed decolorization of synthetic dyes. *Water Res.*, **1999**, *33*, 3512–3520.

Wu, F. C., Tseng, R. L., & Juang, R. S. Enhanced abilities of highly swollen chitosan beads for color removal and tyrosinase immobilization. *J. Hazard Mater.*, **2001**, *81*, 167.

Wuhrmann, K., Mechsner, K., & Kappeler, T. Investigations on rate determining factors in the microbial reduction of azo dyes. *European J. Appl. Microbiol. Biotechnol.*, **1980**, *9*, 325.

Xu, H., Heinze, T. M., Donald, D., Paine, D. D., Cerniglia, C. E., & Chen, H. Sudan azo dyes and para red degradation by prevalent bacteria of the human gastrointestinal tract. *Anaerobe*, **2010**, *16*(2), 114–119.

Xu, H., Heinze, T., Chen, S., & Chen, H. Anaerobic metabolism of 1-amino-2-naphthol-based azo dyes (Sudan Dyes) by human intestinal microflora. *Appl. Environ. Microbiol.*, **2008**, *73*(23), 7759–7762.

Yan, H., & Pan, G. Increase in biodegradation of dimethyl phthalate by *Closterium lunula* using inorganic carbon. *Chemosphere*, **2004**, *55*, **1281**.

Yang, Q., Yang, M., Pritsch, K., Yediler, A., Hagn, A., Schloter, M., & Kettrup, A. Decolorization of synthetic dyes and production of manganese-dependent peroxidase by new fungal isolates. *Biotechnol. Lett.*, **2003**, *25*, 709–713.

Yoo, E. S., Libra, J., & Wiesmannn, U. Reduction of azo dyes by *Desulfovibrio desulfuricans*. *Water Sci. Technol.*, **2000**, *41*, 15–22.

Yu, J., Wang, X., & Yue, P. L. Optimal decolorization and kinetic modeling of synthetic dyes by *Pseudomonas* strains. *Water Res.*, **2001**, *35*, 3579–3586.

Zhao, R., Ma, L., Fan, F., Gong, Y., Wan, X., Jian, M., Zhang, X., & Yang, Y. Decolorization of different dyes by a newly isolated white-rot fungal strain *Ganoderma* sp. En3 and cloning and functional analysis of its laccase gene. *J. Hazard Mater.*, **2011**, *192*, 855–873.

Zollinger, H. *Color Chemistry: Synthesis, Properties and Applications of Organic Dyes and Pigments* (pp. 1–187). VCH Publishers, Weinheim, Germany, **1991**.

CHAPTER 9

IMPACT OF SOIL, PLANT-MICROBE INTERACTION IN METAL CONTAMINATED SOILS

NEETU SHARMA, ABHINASHI SINGH, and NAVNEET BATRA

Department of Biotechnology, GGDSD College, Sector-32-C, Chandigarh–160030, India

9.1 INTRODUCTION

With the rapid progress in the field of science leading to new innovations and growing industrialization, the markets are flooded with novel products having better shelf life and quality. The production process of such compounds leads to the production of advanced effluents which when discharged into the environment leads to water, air, and soil pollution. The degradation pathway of naturally occurring compounds is available, but these modified residues are not known to occur in the natural environment; hence their degradation pathways are not present and thus lead to accumulation of such products in the environment. Such man-made compounds are known as xenobiotic compounds. Their toxicity profile depends on many factors like the type of parent compound, degree of complexity that is alkane, alkyne, alkene, aromatics, and others. They are further categorized into persistent and recalcitrant types. The persistent compounds are one that accumulates in the environment but are degradable, although their degradation rate is slow and hence pose a toxic impact on the environment. On the other hand, those persistent compounds which do not get degraded over a long period of time and gets accumulated in the environment are known as recalcitrant compounds and are far more toxic and have an alarming impact on the environment (Thakur, 2011).

Several approaches are available for the removal of these toxic wastes depending on their nature (organic, inorganic) and their form (solid, liquid, or gaseous). Some of the commonly used techniques are physical, chemical, and biological approaches. The commonly used conventional approaches are adsorption, absorption, filtration, convection, ozonation, activated charcoal treatment, UV treatment, volatilization, chemical treatment, chlorination, air stripping, ion exchange methods, centrifugation, gravitation, oxidation ditches and rotating biological contactors (RBC). The liquid waste undergoes specific treatment steps: primary, secondary, and tertiary treatment (Peters, 2015). All the steps are crucial and lead to a drop in the BOD, COD loads at each step. But these techniques also suffer from the major drawbacks of failure to remove a certain class of modified chemical compounds. The biological approaches commonly used by the industries are divided into two categories depending on the nature of microbes: aerobic and anaerobic. The former involves activated sludge treatment, trickling filters. The latter type involves anaerobic reactors, an up-flow anaerobic sludge blanket reactor (UASB), and others. The above mentioned physical and chemical techniques mostly suffer from the main disadvantage of the production of chemical residues, which further contaminate the environment. The biological methods suffer from the poor removal rate as the microbial consortia sometimes are inhibited by the toxic effects of the chemicals (Peters, 2015).

The most promising approach nowadays is the exclusive use of specific organisms for the removal of contaminants from the environment is called bioremediation. There are many definitions available for bioremediation. The most commonly used is given by EPA, which suggests "treatment that uses naturally occurring organisms to break down hazardous substances into less toxic or nontoxic substances." Bioremediation is a broad term and involves many techniques like compositing, co-compositing, phytoremediation, phytovolatilization, rhizoremediation, rhizofilteration, biostimulation, bioaugmentation, bioleaching. Figure 9.1 shows the commonly used bioremediation methods. Bioremediation can also be defined as the breakdown of complex waste of plant, animal, and human origin into simpler compounds by the action of microbes (Thakur, 2011). The process of bioremediation involving natural microflora can be broadly categorized into two types: intrinsic bioremediation and biostimulation. The former involves the action of native bacteria, which utilizes the available energy sources to carry out the degradation of contaminants while the latter depends on the external energy sources to carry out the remediation process. One such study carried out by the US Army Corps of Engineers reported an increase in the rate of bioremediation following the techniques of windrowing and aeration in the sites contaminated with petroleum (Shah et al., 2013a). The amount of bioavailability of metal to be degraded is another important factor affecting the efficiency of Bioremediation. In the absence of regular sources of energy, microbes can efficiently utilize the alternative energy source. In one of the studies carried out by Shah et al. (2013b), the microbes successfully utilize and degrade the nitrogenous organic chemicals in the nitrogen-deficient soils. The soils with high adsorption capacity limit the bioavailability of contaminants and make them persist in the environment (Shah et al., 2013c).

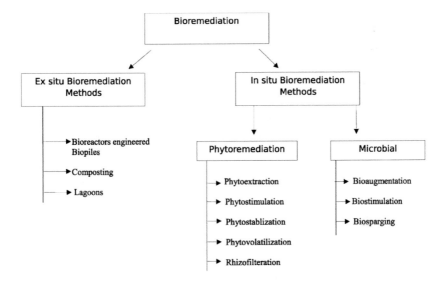

FIGURE 9.1   Bioremediation approaches for treatment of environmental contaminants.

Microorganisms are the most studied and are the key players in bioremediation. From billions of years, they have played a key role in the cycling of nutrients across the globe by the breakdown of waste, which owes to the vast diversity of enzymes produced by them both intracellular and extracellular (Prakash et al., 2010). Apart from microbes, the other biological sources were successfully reported to carry out the removal of contaminants from the environment. One such case was the use of bone char of fish to remove cadmium and lead as a result of adsorption (Shah et al., 2013b; Shah, 2014). Microalgae were successfully employed to remove various contaminants like chromium and nitrate from tannery effluent (Gosavi et al., 2004).

The process of biomagnification poses a major threat to the health of humans and animals as it results in an increase in the concentration of the contaminant at every trophic level. The situation can be controlled by the use of phytoremediating agents. There are different types of phytoremediation, such as phytovolatilization, rhizoremediation, and phytoaccumulation, as of this phytoaccumulation was reported to be an efficient method of removal of contaminants by allowing the accumulation in harvested part of the plant including fruit, stem or flowers (Yerima et al., 2012). It is easy to extract contaminants from harvested part either by concentrating it by incineration or by other method and use it further for industrial use (Singh et al., 2012). In contrast, the native microflora, which was being exposed to petroleum hydrocarbons efficiently degraded aromatic hydrocarbons as reported by Sims (2006). The microbes showed different mechanisms for the removal of contaminants depending upon the novelty of a particular compound. The compound understudy can be directly taken by the cell and utilized or can be transformed prior to uptake. In order to design efficient bioremediation protocols, there is a need to understand the existing metabolic pathways in microbes and to further utilize the knowledge to develop novel pathways to carry out the degradation of persistent and recalcitrant class of xenobiotic compounds.

Recent advancements have also proven successful via the addition of matched microbe strains to the medium to enhance the resident microbe populations' ability to break down contaminants. The elimination of a wide range of pollutants and wastes from the environment requires increasing our understanding of the relative importance of different pathways and regulatory networks to carbon flux in particular environments and for particular compounds, and they will certainly accelerate the development of bioremediation technologies and biotransformation processes (O'Loughlin et al., 2000).

Among all the pollutants, contamination of soils by various heavy metals is a crucial area of research. The persistent rate of heavy metals is reported to be quite high as compared to other pollutants. They exist in the soil in different forms such as free ions, reactive metal ions, soluble form, nonsoluble form, reactive, and non-reactive complex forms that is carbonates, oxides, phosphates, and others (Leyval et al., 1997). Though some metals are required in trace amounts for the growth of plants, beyond a certain limit, they are known to cause negative impacts on the growth of plants and also affect the microbial diversity of that region. Some of the heavy metals like zinc, copper, nickel, iron act as cofactors for essential enzymes required for completing important metabolic pathways in plants as well as microbes. On the other hand, some metals like chromium, selenium, arsenic are highly toxic even at ppb levels and lead to the disruption of growth of plants and microbes by inhibiting the principle enzyme or competing with the crucial substrate (Panda and Chaudhary, 2005). Both natural and anthropogenic sources are responsible for the addition of heavy metals in the soil like weathering of rocks, volcanic eruptions, industrial sources, power plants, mining runoff, floods, municipal waste, refining, and burning of fossil fuels (Wang et al., 2003).

Whether microorganisms will be successful in destroying man-made contaminants in the subsurface depends on three factors: the type of organisms, the type of contaminant, and the geological and chemical conditions at the contaminated site. This chapter explains how these three factors influence the outcome of a subsurface bioremediation project. It also evaluates the role of plant-microbes interaction in the removal of contaminants. It reviews how microorganisms destroy contaminants, their interaction with plants and soil. The chapter will also focus on the impact of the common contaminants on plants and humans and the need for reforms in current practices of bioremediation.

9.2   EFFECT OF METAL CONTAMINATED SOIL ON PLANTS

The presence of heavy metals such as cadmium, chromium, lead, arsenic in an excessive amount not only affects the physical and chemical properties of soil but also poses a great threat to the plant diversity growing in that particular environment. Plants can uptake those heavy metals from soil resulting in the alteration of their cellular functioning, which can ultimately risk the human health consuming those contaminated crops. The toxic effects of some common heavy metals in the soil and plants growing in those contaminated soil are discussed below.

9.2.1   EFFECT OF CHROMIUM ON PLANTS

Chromium is a toxic metal that affects plants in many ways. Chromium is toxic in two forms Cr^{3+} and Cr^{6+}. Chromium in its oxidation state of Cr^{6+} gets easily transported across the cell membrane and affects functions of the cell. Chromium gets easily reduced inside the cell and results in the generation of reactive oxygen species (ROS) such as hydroxyl radicals (OH^-)

and damage the DNA and proteins (Stohs and Bagchi, 1995; Jaishankar et al., 2014). Plants mostly get exposed to chromium through contaminated soils receiving industrial effluents from tanneries, metallurgical industries, or pigment industries (Ghani, 2011). In addition to this, new agricultural methods are also one of the reasons for contaminating the soil and vegetation growing in it through the release of chromium dust or residues (Duan et al., 2010). Chromium toxicity has harmful effects on plant growth and development. It causes a reduction in the growth of roots, chlorosis, and necrosis of leaves, lesser biomass (Ghani, 2011). Several studies have proven detrimental effects of chromium on plant growth. For example, Rout et al. (2000) reported that chromium at a concentration of 200 µM reduces seed germination by 25% in *Echinochloa colona*. In another study, Jain et al. (2000) reported a reduction of 32–57% with chromium at a concentration of 80 ppm in bud germination of sugarcane. Chromium toxicity also affects different enzymes in plant cells including catalase, cytochrome oxidase, amylase, protease, and many more (Nagajyoti et al., 2010; Jaishankar et al., 2014). For example, Zeid (2001) observed a reduction in seed germination is linked towards the decreased activity of amylase, which helps the supply of sugars to the developing seed.

9.2.2   EFFECT OF CADMIUM ON PLANTS

Cadmium is a rare earth element. It exists in two oxidation states Cd^{+1} and Cd^{+2} and is highly toxic in solution form. Soils contaminated with cadmium not only have altered physical and chemical properties. Depending upon their concentration, different metals starts accumulating in the plants from where it enters the food chain and ultimately affects the humans causing severe diseases. Plants growing in cadmium contaminated soils have a great effect on their normal growth and developmental processes. Cadmium toxicity in soil cause modifications in the photosynthetic activities, metabolic processes, and also in the salt and mineral uptake processes (Tran and Popova, 2013). In soil, it occurs mostly in the oxidation state of Cd^{+2}. Uptake of cadmium causes several changes; for example, it causes chlorosis in leaves and shortening of shoots. The harmful effect of cadmium was very well studied by Rascio et al. (2008) in rice. Treatment of rice seedlings with cadmium results in modification of the root system, showing how much cadmium is toxic to plants.

Cadmium also interferes with metabolic processes going in the plant cells by targeting the enzyme system of the cell and thus causing plant damage and sometimes results in plant death (Assche and Clijsters, 1990; Chibuike and Obiora, 2014). Cadmium metal toxicity inhibits the functioning of photosynthetic enzymes like RubisCo by modifying its structure and lowering its binding to its substrate and lowering the enzyme activity (Siedlecka et al., 1998). The concentration of these metals in soil plays a significant role in their affect against plants, animals, and humans. For example, as reported by Ahmad et al. (2012), cadmium at a concentration of 5 mg/L affects the wheat crop and results in less growth of both roots and shoots. In another study, cadmium at a concentration of 250 μM has a very toxic and rapid effect on the cellular division (Fusconi et al., 2007; Tran and Popova, 2013).

9.2.3 EFFECT OF ARSENIC ON PLANTS

Arsenic is a common soil contaminant which, like other metals, affects the growth and development of plants growing in it. Arsenic is mostly toxic in two forms As^{5+} and As^{3+} as these are easily taken up by the plant root system (Finnegan and Chen, 2012). Arsenic is more toxic in its reduced form of As^{3+}. Over the years, a number of toxicity studies related to arsenic with different plants have been done. Shri et al. (2009) studied the effect of arsenic on rice seedlings and how arsenic affects the physiological and biochemical processes of the plant. It was observed that the developmental processes in the rice seedlings were greatly affected, and there was a marked decrease in root and shoot elongation, and also plant biomass was less than normal, and both the forms As^{3+} and As^{5+} accumulated in significant amount. In another study, arsenic toxicity was studied in beans.

9.2.4 EFFECT OF LEAD ON PLANTS

Lead, like other heavy metals discussed before, is one of the most toxic metals present in the environment. It presence in the ecosystem can affects all life forms including plants, animals, microorganisms, and humans. It affects an organism at all the three levels, morphological, physiological, and biochemical (Pourrut et al., 2011). It is an industrially important metal and is used to manufacture batteries, home paints, pipes, and solders (Martin and Griswold, 2009). It is also produced during the combustion automobile fuel. All these sources pollute the soil in one way or the other. Plants generally uptake lead into their system from contaminated soil. Lead is mostly taken up by the roots from where it gets accumulated in the plant. It affects major metabolic processes like photosynthesis by damaging chlorophyll and lipid membrane of the cell (Jaishankar et al., 2014; Najeeb et al., 2017). It also produces ROS which have detrimental effects on cellular DNA. These ROS creates imbalance of antioxidants and free radicals in the plant cell. Lead causes decrease in the concentration of antioxidants such as glutathione which leads to a condition of oxidative stress arises in the cell and causes cellular damage (Jaishankar et al., 2014). Metals like magnesium, zinc, calcium all having an important role in completing different biochemical and metabolic processes in the plant cell, but lead replaces these metals and disturbs processes like cell signaling and enzyme activity.

9.3   EFFECT OF HEAVY METAL TOXICITY ON THE SOIL MICROORGANISMS

Soil is home to a large number of organisms which grows, flourish, and helps in maintaining the structure of the whole ecosystem. Microorganisms are one of the most important parts of soil biological community and play a critical role in maintaining the physical and chemical properties of soil and also regulate the organic matter and nutrient cycles so that other organisms and plants can grow in it. They also act as a nutrient source for other biological communities growing in soil. A part from these microorganisms also causes biotransformation and biodegradation of pollutants, whether organic or inorganic, thus cleaning and protecting the environment (Kavamura and Esposito, 2010; Chodak et al., 2013). Heavy metal contamination in the soil, whether by natural or anthropogenic sources is detrimental to the microbial diversity of soil, resulting in the imbalance of the ecological cycle (Wang et al., 2010). It has been very well reported that a high concentration of heavy metals in the soil leads to a decrease in microbial diversity (Gans et al., 2005; Chodak et al., 2013).

9.4   EFFECT OF HEAVY METAL TOXICITY ON HUMANS

Contamination of soil with heavy metals including chromium, cadmium, arsenic, lead have a direct impact on the health of human beings as the vegetation growing in these contaminated soils contains accumulated heavy metals in them and which will be further consumed by humans leading to different health risks. Heavy metal accumulation also takes place in animals and aquatic organisms such as fishes, which are further eaten up by humans leading to metal toxicity in them (Martin and Griswold, 2009). All heavy metals are toxic to humans at high concentrations and affect cellular processes. These metals tend to accumulate inside the cells and affects or inhibits metabolic reactions by targeting different enzymes and cofactors (Singh et al., 2011). Contaminants can enter the human body from metal-polluted soil by three ways: ingestion, skin, and air.

9.4.1   EFFECT OF CHROMIUM ON HUMANS

Chromium is one of the most abundant, naturally occurring elements on earth. It exists in several oxidation states between Cr^{2+} to Cr^{3+} (Rodriguez et al., 2007). However, they are most stable in the environment in two oxidation states, Cr^{3+} and Cr^{6+}. Both forms are very toxic to animals, plants, and humans (Mohanty and Patra, 2013). It enters the water and soil system mostly through industrial sources, including tannery industries, pigment-producing industries, metallurgical industries, and electroplating (Ghani, 2011; Tchounwou et al., 2012; Jaishankar et al., 2014). All these sources increase the concentration of chromium significantly in the soil and water system, which subsequently affects the vegetation and other life forms growing in that particular environment. Plants accumulate chromium not only in their cells but also the soil particles sticks to the outer surface of the plant, which are removed by washing and from where it can enter the food chain. Humans get exposed to chromium when they consume these plants growing in chromium contaminated soil.

Cr^{6+} is a highly toxic form of the chromium metal. It is a strong oxidizing agent and can easily penetrate the cell membrane. On the other hand, due to its weak membrane penetration power, Cr^{3+} is less toxic. Cr^{6+} is a human carcinogen, and its carcinogenic nature is attributed towards its mutagenic properties. Being a strong oxidizing agent, Cr^{6+} gets easily reduced and generates several ROS such as OH^- and superoxide ions leading to oxidative stress in cells. These ROS damages cellular DNA and proteins and impairs cell function (Stohs and Bagchi, 1995; O'Brien et al., 2003; De Mattia et al., 2004; Tchounwou et al., 2012; Jaishankar et al., 2014). The toxicity of chromium in humans is so severe that even a skin contact to chromium contaminated soil or other materials can cause skin ulcers which are very persistent.

9.4.2 EFFECT OF CADMIUM ON HUMANS

As discussed earlier, cadmium is a toxic metal and adversely affects human beings. Cadmium toxicity in humans affects major organs like the kidney, lungs, skeletal system, and brain (Sobha et al., 2007; Johri et al., 2010; Singh et al., 2011). Long-term exposure to cadmium leads to several chronic disorders of kidneys, lungs, bones, and brain. As cadmium is a severe pulmonary irritant inhaling cadmium polluted dust particles or working in areas having cadmium contaminated soil causes lung disease including cadmium pneumonitis. Apart from this, cadmium also leads to bone disorders like osteomalacia (Duruibe et al., 2007). Cadmium metabolism in the human body is very slow; therefore, it can accumulate for a very long time. More harmful effects of cadmium toxicity on bones include hypercalciuria and dysfunctional vitamin D metabolism (Johri et al., 2010). Cadmium toxicity in the human body results in the synthesis of ROS, which directly damages the DNA in the cell and inhibits normal cellular processes such as replication, transcription, and translation (Mitra, 1984; Stohs and Bagchi, 1995; Tchounwou et al., 2012). Heavy metal concentration has a very significant effect on the level of toxicity in the organism. It has been reported that cadmium at a concentration of 0.1 to 10 mM leads to DNA damage. Apart from these, cadmium induces several other effects; for example, it activates mechanism of protein degradation and affects cell signaling pathways (Durnam and Palmiter, 1981; Tchounwou et al., 2012).

9.4.3 EFFECT OF LEAD ON HUMANS

Lead is potentially one of the most toxic heavy metals to humans even at very low concentrations. It affects the major organs in the human body like kidney, liver, gastrointestinal tract. Humans can be exposed towards lead through ingesting contaminated food, soil, inhaling contaminated soil particles or dust and also through contaminated water (Ferner, 2001; Duruibe et al., 2007). Inside the body lead can cause serious damage to the human system. For example, lead toxicity can cause inhibition of haemoglobin synthesis. This can cause serious deficiency in the transport of oxygen in the blood, putting an extra amount of load on the cardio-vascular system and such condition also affects the nervous system both short term and long term (Ogwuebgu and Muhangha, 2005). Lead can also damage the gastrointestinal tract and cause serious ailments related to digestive system. In addition to this, Lead toxicity also affects children of a young age, causing developmental disorders, the most prominent being the development of brain. According to Morgan (2012), 10 μg/dl of lead in the blood can cause a serious reduction in the intelligence quotient of the children.

Lead inhibits the functioning of several important enzymes by incor-porating itself into different functional groups resulting in the change in configuration of enzymes and thus affecting their activity (Tchounwou et al., 2012). It can also act as a competitive binder for other metals acting as a cofactor such as calcium and disrupts major metabolic processes going in the cell (Flora et al., 2007). Affecting calcium function in the body lead can cause cell signaling disruption (Goldstein et al., 1993). Chronic exposure of lead to the people working in lead-contaminated areas results in several other effects, including damage to DNA and oxidative stress in the body.

9.4.4   EFFECT OF ARSENIC ON HUMAN HEALTH

Arsenic is a toxic and carcinogenic heavy metal found in nature. Arsenic, in its inorganic form of arsenite (As^{3+}) and arsenate (As^{5+}), is most harmful to humans. It can have both chronic and acute effects on the human being. Exposure of humans to arsenic can take place by natural sources such as by dissolving in groundwater or consumption of plants growing in arsenic contaminated soils. Arsenic is used to manufacture many kinds of herbi-cides, insecticides, and fungicides, which are one of the major sources of increasing the arsenic concentration in soil and contaminating it and also the vegetation and other life forms growing in it (Tchounwou et al., 2012). Arsenic affects most of the organs of the human body, but mostly it affects the kidney. It also affects ATP synthesis in the cell, thus damaging cellular respiration and other important processes (Duruibe et al., 2007). In addi-tion to this, arsenic toxicity is a major cause of skin diseases such as the development of lesions on the skin, which later on induces skin cancer (Lage et al., 2006; McCarty et al., 2007; Jomova et al., 2011). In humans, inorganic arsenic gets methylated and is biotransformed into monomethylarsonic acid and dimethylarsinic acid. Monomethylarsonic acid is not excreted through urine and gets accumulated inside the body leading to be a potential reason for arsenic-induced cancer (Singh et al., 2007; Jaishankar, 2014). Arsenic also inhibits enzymes responsible for important processes like oxidative phosphorylation, beta-oxidation of fatty acids (Belton et al., 1985). Furthermore, arsenic toxicity damages cellular DNA and also inhibits cellular division in mammalian cells (Hartmann and Speit, 1994; Banu et al., 2001).

9.5   METHODS FOR BIOREMEDIATION OF METAL CONTAMINATED SOILS

Physiochemical analysis of soil is a crucial step while designing a bioremediation protocol for any site. The efficiency of phytoremediation and microbial remediation depends on the physical and chemical state of native soil. Various parameters are evaluated to check the quality of the soil. Physical indicators include water holding capacity, density, and texture. The chemical analysis involves pH, conductivity, C, S, P levels, and biological properties include microbial biomass, including mainly nitrogen and carbon (Gil-Sotres et al., 2005; Nanda and Abraham, 2013). The enzymatic activities are also analyzed to check the status of nutrient cycling, and it includes an assay of urease, phosphatase, and β-galactosidase. The growth of plants and the activity of microbes are highly influenced by the pH of the soil. The presence of metal ions is also affected by the local conditions of the spoil. With the decrease in pH, solubility of metal ions was found to increase (Sanders and Adams, 1987).

9.5.1   PLANTS

Phytoremediation is the use of plants for the removal of pollutants from contaminated sites. Worldwide research has been carried out in this area involving a large number of plant species. Among all the plant species, plants belonging to the category of hyperaccumulators showed promising results. Hyperaccumulators are those species that have the ability to accumulate certain chemicals, heavy metals at very high concentrations without themselves being harmed. No impact of such elevated level of heavy metal on the growth of these plants. The hyperaccumulators can be distinguished from nonhyperaccumulators type based on mainly three properties, i.e., increased uptake of heavy metals, rapid translocation from root to shoot (Rasico and Navari-Izzo, 2011) and more detoxification rate of heavy metals via leaves of such plants. The plants belonging to the Brassicaceae family have been the most widely exploited species used for their phytoremediation potential for the removal of heavy metals from different sites (Sims, 2006; Perfumo et al., 2007; Ahmad, 2017). Species belonging to *Brassica*, *Arabidopsis*, *Bornmuellera*, *Alyssum* have been reported to be successful hyperaccumulators (Ahmad, 2017). In some plants, metal sequestrations have found to be carried out by the presence of specific phytochelators (Cobbett and Goldsbrough, 2002). Metallothioneins are the other classes of molecules responsible for the successful removal of metal ions from the contaminated sites. Organic fertilizers were considered as a better source of nutrients for plants due to their nontoxic behavior. There were several points that make them popular as compared to chemical fertilizers. But recent studies have reported their role in augmenting the natural microflora of the soil (Masto et al., 2006). The careful selection of biofertilizer would further help in increasing the efficiency of bioremediation. Juwarkar et al. (2008) reported the use of *Azotobacter* resulted in an increased rate of removal of heavy metal from the soil as compared to the initial studies where plants were used alone for the removal of contaminants.

Plants employed various mechanisms for the removal and immobilization of heavy metals in the soil. Phytotransformation and phytodegradation have been successfully reported by many plants in which certain enzymes carried out the conversion or breakdown of potential pollutant into nonreactive form (McCutcheon and Schnoor, 2003). Rhizofilteration is another approach where contaminants are absorbed or precipitated or absorbed by the action of root exudates and enzymes on the surface of roots (McCutcheon and Schnoor, 2003).

The increased solubility resulted in increased bioavailability and hence, enhanced uptake by the plants and microbes. The results of the enzymatic analysis can be used as indicators of not only the biogeochemical cycling of nutrients but also the index of microbial activity. The enzymes present in soil are mainly contributed by the microbial population (Ladd, 1978), followed by plant and animal origin (Tabatabai, 1994). The concentration and solubility of heavy metals affects the enzymatic activity of various microbes by influencing the availability of substrates, feedback inhibition, and chemical inhibition. On the other hand, some metals act as cofactors and in turn, enhanced the activity of certain enzymes (Nanda and Abraham, 2013). Hence the addition of biofertilizers can be used to increase the bioremediation potential of both the microbes and plants. The sludge from various sources like dairy sludge, effluents having high organic matter content, and metal content can act as a potential source of organic nitrogen, phosphorus, and carbon and also act as a source of microbial stock (Juvarkar et al., 2008).

9.5.1.1   PHYTOEXTRACTION

Phytoextraction is another technique based on the use of plants for the removal of trace elements from the soil. It is based on translocation of elements into the harvestable part of the plant followed by post-harvest treatment of the recovered biomass by employing methods like thermal treatment and composting for reducing the biomass volume and recovery of trace elements from the selected plant (Vangronsveld et al., 2009). But the main disadvantage of the above method is the longer time period required for the growth of plant and bioavailability of the element for translocation to the harvestable part of the plant. Vangronsveld et al. (2009) reported the use of phytoextraction potential of hyperaccumulators like *Thlaspi* sp. and *Alyssum* sp. for the removal of trace elements. In a study carried out by Kunito et al. (2001) in copper contaminated sites in rhizospheric and non rhizospheric area of Phragmites, significant difference in the bacterial communities was observed. Rhizospheric bacteria were observed to produce certain exopolymers that bind to trace elements and thus facilitates their translocation into plants. Copper resistant *Bacillus* sp. dominated the rhizosphere region of Phragmites, while the non-rhizospheric region was dominated by *Methylobacterium* sp. (Kunito et al., 1997). In another study carried out by Mengoni et al. (2004) using a cultivation-independent approach, Proteobacteria dominated the rhizospheric area of *A. bertolonii*, while alpha and gamma proteobacteria dominated the leaf community of the plant. The culture-dependent approach reported different communities of bacteria related to genera *Bacillus, Staphylococcus, Microbacterium* in *A. bertolonii* (Barzanti et al., 2007). Thus both culture-dependent and independent studies should be studied together to have a better understanding of the bacterial communities associated with these types of plants. This will further help in designing protocols for bioremediation.

9.5.1.2   PHYTOSTABILIZATION

It is a method being employed to reduce the bioavailability of heavy metals in the soil by erosion. It involves the use of plants to trap and hold heavy metals and not allowing their movement through the soil, thus reducing their cover zone and also preventing their leaching into groundwater (Salt et al., 1995; Chubuike and Obiora et al., 2014). It is an alternative method to phytoextraction (McGrath and Zhao, 2003). It has been observed that in phytostabilization, plants convert more toxic form of heavy metals to a less toxic form and immobilize them using their roots and thus preventing their movement in the soil and contaminating it further (Wu et al., 2010; Ali et al., 2013). Continuous studies are being performed to try and test new plant species which could prove to be a better phytostabilizer plant. One such exploration is being done with *Vossia cuspidate* and macrophtye. Galal et al. (2017) used this plant to against common heavy metals including chromium, copper, lead, and aluminum. They found accumulation of these metals in their root and shoot system and thus preventing their leaching into soil and thus prove to be a good phytostabilizer. This method can be used in combination with microorganisms which improves their efficiency. For example, Ahsan et al. (2017) make use of a combination of plant and bacteria for remediation of soil contaminated with lead and uranium. In this approach, they used different bacterial endophytes, including *Enterobacter* sp. HU38, *Microbacterium arborescens* HU33 and *Pantoea stewartii* ASI11 in combination with *Leptochloa fusca*. Results showed an enhance growth of plant in contaminated soil due to bacterial endophytes. Not only bacterial endophytes help in growth enhancement but also increased the metal uptake by plants and thus stabilizing the soil. Phytostabilization as a method for remediation of the heavy metal contaminated environment has proven its efficiency, but as it only limits the movement of heavy metals in the environment and does not remove, it can be a problem.

9.5.1.3   PHYTOVOLATILIZATION

It is a type of method involving plants that can uptake heavy metals from the contaminated soil, converting them into volatile form and then releasing them into the atmosphere (He et al., 2015). This method has been mostly employed for the treatment of mercury-contaminated environment in which using specific plant species, mercuric ion is converted to elemental mercury (US Environment Protection Agency, 2000). Making the method more efficient genetic engineering has been used to develop transgenic plant species to treat contaminated soil. These are *Arabidopsis thaliana*, *Nicotiana tabacum* (Rugh et al., 1998; Meagher et al., 1999; Chubuike and Obiora, 2014). The problem with this method is that the metal contaminant is only converted from one form to another and is not completely removed from the environment; ultimately, the metal remains in the environment, which can be get resuspended into soil by rainfall. Table 9.1 summarizes the different plants being used to bioremediate heavy metal contamination in soil.

TABLE 9.1   Different Plant Species Being Used to Treat Heavy Metal Contamination in Soil

Plant Species	Heavy Metal	References
Pteris vittata	As	Ma et al., 2001
Aeollanthus subacaulis	Cu	Chubuike and Obiora, 2014
Agrostis tenuis	Pb	Chubuike and Obiora, 2014
Solanum nigrum	Zn	Marques et al., 2008
Brassica juncea	Se	Laperche et al., 1997
Sedum alfredii	Zn, Cd	Yang et al., 2004

9.5.2   MICROBES

Microorganisms can be used as an effective agent for cleaning up the contaminated sites. A wide variety of microorganisms can be isolated from extreme regions of the environment. As they are continuously exposed to these extreme conditions, they developed certain unique features which can be exploited for various purposes. In the same way, microorganisms residing in polluting sites have more potential to degrade the novel xenobiotic compounds, which are otherwise difficult to degrade by other means. The novel microorganisms can be developed in laboratory with the use of genetic engineering as per the requirement of the particular site such microorganisms, which are popularly known as GEM (genetically engineered microbes) have more potential and better survival and success rate as compare to naturally isolated microbes. One such successful use of superbug *Pseudomonas putida* engineered by Chakrobarty showed a high success rate in removal of oil spills. Various techniques are available for introducing microbes to the rare of contamination. The best approach to add consortia rather than single microbe as the former had a better survival rate due to efficient competition they could give to native microflora without themselves getting inhibited due to limitation of nutrients and other environmental factors. As already discussed, bioaugmentation, i.e., addition of microbes from outside into the site, biostimulation which is the addition of nutrients to stimulate the activity of already occurring microbes and some of the famous approaches being used for *in situ* bioremediation. Biosorption is another successful technique employed in bioremediation where metal species are adsorbed or absorbed by the microbial cells. Depending on the type of microorganism and metal species, metal may be absorbed or precipitated by the action of membrane transport pumps, enzymatic actions, or they may be translocated to specific sites in that organism (Gadd, 1996). Li et al. (2013) reported the successful use of microbes in sequestration of heavy metals from highly stable states. The microorganisms along with plants have more potential for remediation as compared to that case when both approaches are being used alone. The rhizosphere is the region where majority of the microbes interact with the plants. Most of the nutrients and minerals are taken by the plant from the rhizosphere area. Hence the region has the high potential to be explored for the bioremediation. The roots of the plants secrete variety of exudates ranging from organic acid, sugars, vitamins, enzymes to siderophores. The rhizospheric microbes have synergistic associations to increase the bioavailability of certain nutrients by the microbial action which in turn get the nutrients from the plants (Barea et al., 2002; Yang et al., 2009). Varieties of metal chelators are secreted by the roots of higher plants. Such chelators facilities the release of metal ions bounded to the soil particles by forming complexes with them and hence increased their bioavailability. Dakora and Philip (2002) reported the use of phytochelators for removal of heavy metals from the soil. The plant growth-promoting bacteria (PGPB) popularly known as PGPR have been widely studied and known to promote growth of the associated plants. Several mechanisms are responsible for the action including competition for binding site in the root region, production of cell wall lysing enzymes and antibiotics (Glick et al., 1995; Glick et al., 2007).

Apart from promoting the growth of plants by the above mechanisms, they are also reported to be effective biocontrol agents. The attention is now diverted for exploring the application of these rhizobacteria in removal of contaminants from the soil. The degradation pathway of poly-chlorinated biphenyls and the genes responsible for them were well studied in bacteria isolated from the rhizophoric region of the plants. The only lacunae observed in this case were loss of activity of these bacteria under nutrient-deficient conditions (Brazil et al., 1995; Normander et al., 1999). Endophytic bacteria are other bacteria with a great potential to facilitate the removal of heavy metals by increasing the rate of phytoremediation in host plants. In a study carried out by Idris et al. (2004) in *Thlaspi goesingense*, a plant known for its nickel hyperaccumulation, endophytes showed higher resistance to elevated levels of heavy metals as compared to rhizobacteria. The major problem associated with the use of endophytes in bioremediation programme is that most of they are non-culturable. The more studies are required in this area to explore the actual mechanism, involved in the role of endophytes in promoting accumulation of metals in plants at such a high concentration (Weyen et al., 2009). Besides using bacteria for remediation studies, mychorrhizae was also reported for their potential to be used for removal of pollutants from the soil. They were known to provide resistance to the host plant from the toxic effect of pollutants by forming protective sheath around the plant roots and also affect the transport of pollutant by modifying the soil moisture content (Mehrag and Cairney, 2000). Table 9.2 summarizes the common contaminants and the microorganisms used against them.

9.5.3   BIOCHAR

Due to the limitations of methods like phytostabilization and phytovolatilization, new methods are being taken into consideration. One such method is the use of biochar. It is a carbon-rich organic compound produced by pyrolysis of biological material such as wood, manure, and feedstock. Biochar increases the soil pH due to which the mobility and bioavailability of heavy metals in the soil decreases, and also their uptake by the plants also decreases (Ferreiro et al., 2013; Ahmad et al., 2014). In addition to this biochar uses several other mechanisms to decrease mobility or prevent the uptake of heavy metals by plants including complex formation of heavy metals and different chemical groups present in biochar, exchange of heavy metals with the cations present in biochar, electrostatic attraction between positively charged heavy metals and negatively charged biochar components or vice versa (Uchimiya et al., 2011).

TABLE 9.2   Common Contaminants in Soil and Microorganisms Being Used to Treat Them

Contaminant	Microorganisms	References
Dyes	*Clostridium perfringes,* *Pseudomonas* spp., *Enterococcus faecalis*	Shah, 2014
Oil spills	*Alcanivorax borkumensis,* *Oleiphilus, Thalassolituus,* *Oleispira*	Brooijmans et al., 2009
Mercury	*Pseudomonas aeruginosa,* *Aspergillus niger, Bacillus* spp.	Rajendran et al., 2003
Iron	*Thiobacillus ferroxidans*	Sand et al., 1992
Sulfur	*Leptospirillum ferroxidans*	Sand et al., 1992
Cadmium	*Alcaligenes xylosoxidans,* *Alcaligenes eutrophus,* *Staphylococcus aureus*	Schmidt and Schlegel, 1994
Selenium	*Bacillus* spp.	Prakash et al., 2010
Nickel, Zinc	*Penicillium funiculosum*	Bosecker, 1993
Gold	*Chlorella vulgaris*	Gupta et al., 2000
Aromatic compounds like Pyrene, Phenanthrene, Carbazolr, Naphthalene, Benzopyrene, and Fluoranthene	*Arthrobacter* sp., *Burkholderia* sp., *Mycobacterium* sp., *Pseudomonas* sp., *Stenotrophomonas maltophilia*	Grifoll et al., 1994; Schneider et al., 1996; Juhasz et al., 2000; Seo et al., 2006

It has been well reported that Biochar decreases the bioavailability for plant uptake of certain heavy metals like cadmium, arsenic, and lead in soil (Namgay et al., 2010). Lu et al. (2017) conducted a research to study the effect of biochars produced from bamboo and rice straw on heavy metals like zinc, cadmium, copper, and lead. Biochars from both bamboo and rice straw were applied at different concentrations on sandy loam soil contaminated with above-mentioned heavy metals. It was found that there was a decrease in the mobility of the heavy metals with both the types of biochars. The decrease can be linked to the increase in soil pH by application of biochar. However, the effect of biochar on metal mobility varies with the type of biochar applied. Biochar applied for removal of metal of one type may not be that much efficient in removing the other type of metal. In a study, two types of biochars were used, one chicken manure and other from green waste. It was found that the former was effective in reducing cadmium and lead but no significant effect was observed in case of copper whereas the latter significantly reduced the concentration of all the heavy metals, i.e., cadmium, lead, and copper (Park et al., 2011; Ferreiro et al., 2013). Table 9.3 summarizes the different types of biochars used against different heavy metals.

TABLE 9.3   Types of Biochar Being Used for Remediation of Heavy Metal Contaminated Soils

Biochar	Heavy Metal	References
Rice straw	Cu, Pb, Cd, Zn	Lu et al., 2017
Bamboo	Cu, Pb, Cd, Zn	Lu et al., 2017
Eucalyptus	As, Cd, Cu, Zn, Pb	Namgay et al., 2010
Chicken manure	Cd, Cu, Pb	Park et al., 2011
Hardwood	Cd, Zn	Beesley et al., 2010
Green waste	Cd, Cu, Pb	Park et al., 2011
Dairy manure	Pb	Cao et al., 2011
Oakwood	Pb	Ahmad et al., 2012

9.6   CONCLUSION

Based on the above studies, we suggest the use of metagenomics along with culture-dependent studies for the selection of high performing strains associated with plant and soil communities. The selected strains can be further evaluated under lab conditions for the bioremediation potential. Such studies will further pave the way for designing advanced genetically engineered strains with high potential. They should be evaluated on all the spheres, including their colonization potential, survival potential in non-native soils, and ecotoxicological studies. The above parameters would help in designing promising bioremediation techniques by employing both plants and microbes with high potential and sustainable approach. Bioremediation methods based on either microbes or plants alone usually suffered from many disadvantages. The main problems observed during the above studies were poor survival rates of augmented microbes in actual sites of contamination due to varying environmental factors and competition from native microflora for limited resources. The phytoremediation studies mainly suffered due to poor bioavailability and translocation of contaminants in the plants. The above bottlenecks can be overcome by employing more studies involving more contaminated sites and exploring the role of plant-microbe interactions in the degradation of a variety of pollutants.

KEYWORDS

- **bioaugmentation**
- **biochar**
- **bioremediation**
- **biosorption**
- **biostimulation**
- **metal contamination**
- **phytoextraction**
- **phytoremediation**
- **phytotransformation**
- **phytovolatilization**

REFERENCES

Ahmad, I., Akhtar, M. J., Zahir, Z. A., & Jamil, A. Effect of cadmium on seed germination and seedling growth of four wheat (*Triticum aestivum* L.) cultivars. *Pak. J. Bot.*, **2012**, *44*, 1569–1574.

Ahmad, M., Lee, S. S., Yang, J. E., Ro, H. M., Lee, Y. H., & Ok, Y. S. Effects of soil dilution and amendments (mussel shell, cow bone, and biochar) on Pb availability and phytotoxicity in military shooting range soil. *Ecotoxicol. Environ. Saf.*, **2012**, *79*, 225–231.

Ahmad, M., Rajapaksha, A. U., Lim, J. E., Zhang, M., Bolan, N., Mohan, D., Vithanage, M., Lee, S. S., & Ok, Y. S. Biochar as a sorbent for contaminant management in soil and water: A review. *Chemosphere*, 2014, *99*, 19–33.

Ahmad, P. *Oilseed Crops: Yield and Adaptations Under Environmental Stress*. John Wiley and Sons Ltd, Hoboken, New Jersey, **2017**.

Ahsan, M. T., Najam-Ul-Haq, M., Idrees, M., Ullah, I., & Afzal, M. Bacterial endophytes enhance phytostabilization in soils contaminated with uranium and lead. *Int. J. Phytoremediation*, **2017**, *19*, 937–946.

Ali, H., Khan, E., & Sajad, M. A. Phytoremediation of heavy metals – concepts and applications. *Chemosphere*, **2013**, *91*, 869–881.

Assche, F. V., & Clijsters, H. Effects of metals on enzyme activity in plants. *Plant Cell Environ.*, **1990**, *13*, 195–206.

Banu, B. S., Danadevi, K., Jamil, K., Ahuja, Y. R., Rao, K. V., & Ishaq, M. *In vivo* genotoxic effect of arsenic trioxide in mice using comet assay. *Toxicology*, **2001**, *162*, 171–177.

Barea, J. M., Azcón, R., & Azcón-Aguilar, C. Mycorrhizosphere interactions to improve plant fitness and soil quality. *Antonie Leeuwenhoek*, **2002**, *81*, 343–351.

Barzanti, R., Ozino, F., Bazzicalupo, M., Gabbrielli, R., Galardi, F., Gonnelli, C., & Mengoni, A. Isolation and characterization of endophytic bacteria from the nickel hyper-accumulator plant *Alyssum bertolonii*. *Microbial. Ecol.*, **2007**, *53*, 306–316.

Beesley, L., Moreno-Jiménez, E., & Gomez-Eyles, J. L. Effects of biochar and greenwaste compost amendments on mobility, bioavailability and toxicity of inorganic and organic contaminants in a multi-element polluted soil. *Environ. Pollut.*, **2010**, *158*, 2282–2287.

Belton, J. C., Benson, N. C., Hanna, M. L., & Taylor, R. T. Growth inhibitory and cytotoxic effects of three arsenic compounds on cultured Chinese hamster ovary cells. *J. Environ. Sci. Health A.*, **1985**, *20*, 37–72.

Bosecker, K. Biosorption of heavy metals by filamentous fungi. *Biohydrometallurgical Technologies*, **1993**, *2*, 55–64.

Brazil, G. M., Kenefick, L., Callanan, M., Haro, A., De Lorenzo, V., Dowling, D. N., & O'Gara, F. Construction of a rhizosphere pseudomonad with potential to degrade polychlorinated biphenyls and detection of bph gene expression in the rhizosphere. *Appl. Environ. Microbiol.*, **1995**, *61*, 1946–1952.

Brooijmans, R. J., Pastink, M. I., & Siezen, R. J. Hydrocarbon-degrading bacteria: The oil-spill clean-up crew. *Microbial Biotechnology*, **2009**, *2*, 587–594.

Cao, X., Ma, L., Liang, Y., Gao, B., & Harris, W. Simultaneous immobilization of lead and atrazine in contaminated soils using dairy-manure biochar. *Environ. Sci. Technol.*, **2011**, *45*, 4884–4889.

Chibuike, G. U., & Obiora, S. C. Heavy metal polluted soils: Effect on plants and bioremediation methods. *Appl. Environ. Soil Sci.*, **2014**, 752708, http://dx.doi.org/10.1155/2014/752708 (Accessed on 7 October 2019).

Chodak, M., Gołębiewski, M., Morawska-Płoskonka, J., Kuduk, K., & Niklińska, M. Diversity of microorganisms from forest soils differently polluted with heavy metals. *Appl. Soil Ecol.*, **2013**, *64*, 7–14.

Cobbett, C., & Goldsbrough, P. Phytochelatins and metallothioneins: Roles in heavy metal detoxification and homeostasis. *Annu. Rev. Plant Biol.*, **2002**, *53*, 159–182.

Dakora, F. D., & Phillips, D. A. Root exudates as mediators of mineral acquisition in low-nutrient environments. *Plant Soil*, **2002**, *245*, 35–47.

Duan, N., Wang, X. L., Liu, X. D., Lin, C., & Hou, J. Effect of anaerobic fermentation residues on a chromium-contaminated soil-vegetable system. *Procedia Environ. Sci.*, **2010**, *2*, 1585–1597.

Durnam, D. M., & Palmiter, R. D. Transcriptional regulation of the mouse metallothionein-I gene by heavy metals. *J. Biol. Chem.*, **1981**, *256*, 5712–5716.

Duruibe, J. O., Ogwuegbu, M. O. C., & Egwurugwu, J. N. Heavy metal pollution and human biotoxic effects. *Int. J. Phys. Sci.*, **2007**, *2*, 112–118.

Ferner, D. J. Toxicity, heavy metals. *eMed. J.*, **2001**, *2*(5), 1.

Finnegan, P., & Chen, W. Arsenic toxicity: The effects on plant metabolism. *Front Physiol.*, **2012**, *3*, 182.

Flora, S. J. S., Saxena, G., Gautam, P., Kaur, P., & Gill, K. D. Response of lead-induced oxidative stress and alterations in biogenic amines in different rat brain regions to combined administration of DMSA and MiADMSA. *Chem. Biol. Interact*, **2007**, *170*, 209–220.

Fusconi, A., Gallo, C., & Camusso, W. Effects of cadmium on root apical meristems of *Pisum sativum* L.: Cell viability, cell proliferation and microtubule pattern as suitable markers for assessment of stress pollution. *Mutat. Res. Genet Toxicol. Environ. Mutagen.*, **2007**, *632*, 9–19.

Gadd, G. M. Influence of microorganisms on the environmental fate of radionuclides. *Endeavour*, **1996**, *20*, 150–156.

Galal, T. M., Gharib, F. A., Ghazi, S. M., & Mansour, K. H. Phytostabilization of heavy metals by the emergent macrophyte *Vossia cuspidata* (Roxb.) Griff.: A phytoremediation approach. *Int. J. Phytoremediation*, **2017**, *19*(11), 992–999, doi:10.1080/15226514.2017.1303816.

Gans, J., Wolinsky, M., & Dunbar, J. Computational improvements reveal great bacterial diversity and high metal toxicity in soil. *Science*, **2005**, *309*, 1387–1390.

Ghani, A. Effect of chromium toxicity on growth, chlorophyll and some mineral nutrients of *Brassica juncea* L. *Egyptian Acad. J. Biol. Sci.*, **2011**, *2*, 9–15.

Gil-Sotres, F., Trasar-Cepeda, C., Leirós, M. C., & Seoane, S. Different approaches to evaluating soil quality using biochemical properties. *Soil Biol. Biochem.*, **2005**, *37*, 877–887.

Glick, B. R., Karaturovíc, D. M., & Newell, P. C. A novel procedure for rapid isolation of plant growth promoting pseudomonads. *Can. J. Microbiol.*, **1995**, *41*, 533–536.

Glick, B. R., Todorovic, B., Czarny, J., Cheng, Z., Duan, J., & McConkey, B. Promotion of plant growth by bacterial ACC deaminase. *Crit. Rev. Plant Sci.*, **2007**, *26*, 227–242.

Goldstein, G. W. Evidence that lead acts as a calcium substitute in second messenger metabolism. *Neurotoxicology*, **1993**, *14*, 97–101.

Gosavi, K., & Sammut, J. Gifford, S., & Jankowski, J. Macroalgal biomonitors of trace metal contamination in acid sulfate soil aquaculture ponds. *Sci. Total Environ.*, **2004**, *324*, 25–39.

Grifoll, M., Selifonov, S. A., & Chapman, P. J. Evidence for a novel pathway in the degradation of fluorene by *Pseudomonas* sp. strain F274. *Appl. Environ. Microbiol.*, **1994**, *60*, 2438–2449.

Gupta, R., Ahuja, P., Khan, S., Saxena, R. K., & Mohapatra, H. Microbial biosorbents: Meeting challenges of heavy metal pollution in aqueous solutions. *Curr. Sci.*, 2000, *78*, 967–973.

Hartmann, A., & Speit, G. Comparative investigations of the genotoxic effects of metals in the single cell gel (SCG) assay and the sister chromatid exchange (SCE) test. *Environ. Mol. Mutagen*, **1994**, *23*, 299–305.

He, Z., Shen, J., Ni, Z., Tang, J., Song, S., Chen, J., & Zhao, L. Electrochemically created roughened lead plate for electrochemical reduction of aqueous CO_2. *Catal. Commun.*, **2015**, *72*, 38–42.

Idris, R., Trifonova, R., Puschenreiter, M., Wenzel, W. W., & Sessitsch, A. Bacterial communities associated with flowering plants of the Ni hyperaccumulator *Thlaspi goesingense*. *Appl. Environ. Microbiol.*, **2004**, *70*, 2667–2677.

Jain, R., Srivastava, S., & Madan, V. K. Influence of chromium on growth and cell division of sugarcane. *Indian J. Plant Physiol.*, **2000**, *5*, 228–231.

Jaishankar, M., Tseten, T., Anbalagan, N., Mathew, B. B., & Beeregowda, K. N. Toxicity, mechanism and health effects of some heavy metals. *Interdiscip. Toxicol.*, **2014**, *7*, 60–72.

Johri, N., Jacquillet, G., & Unwin, R. Heavy metal poisoning: the effects of cadmium on the kidney. *Biometals*, **2010**, *23*, 783–792.

Jomova, K., Jenisova, Z., Feszterova, M., Baros, S., Liska, J., Hudecova, D., Rhodes, C. J., & Valko, M. Arsenic: Toxicity, oxidative stress and human disease. *J. Appl. Toxicol.*, **2011**, *31*, 95–107.

Juhasz, A. L., Stanley, G. A., & Britz, M. L. Microbial degradation and detoxification of high molecular weight polycyclic aromatic hydrocarbons by *Stenotrophomonas maltophilia* strain VUN 10,003. *Lett. Appl. Microbiol.*, **2000**, *30*, 396–401.

Juwarkar, A. A., Yadav, S. K., Kumar, P., & Singh, S. K. Effect of biosludge and biofertilizer amendment on growth of *Jatropha curcas* in heavy metal contaminated soils. *Environmental Monit. Assess*, **2008**, *145*, 7–15.

Kavamura, V. N., & Esposito, E. Biotechnological strategies applied to the decontamination of soils polluted with heavy metals. *Biotechnol. Adv.*, **2010**, *28*, 61–69.

Kunito, T., Nagaoka, K., Tada, N., Saeki, K., Senoo, K., Oyaizu, H., & Matsumoto, S. Characterization of Cu-resistant bacterial communities in Cu-contaminated soils. *Soil Sci. Plant Nutr.*, **1997**, *43*, 709–717.

Kunito, T., Saeki, K., Nagaoka, K., Oyaizu, H., & Matsumoto, S. Characterization of copper-resistant bacterial community in rhizosphere of highly copper-contaminated soil. *Eur. J. Soil Biol.*, **2001**, *37*, 95–102.

Ladd, J. N. Origin and range of enzymes in soil. In: Burns, R. G., (ed.), *Soil Enzymes* (pp. 1–380). Academic Press, London, UK, **1978**.

Lage, C. R., Nayak, A., & Kim, C. H. Arsenic ecotoxicology and innate immunity. *Integr. Comp. Biol.*, **2006**, *46*, 1040–1054.

Laperche, V., Logan, T. J., Gaddam, P., & Traina, S. J. Effect of apatite amendments on plant uptake of lead from contaminated soil. *Environ. Sci. Technol.*, **1997**, *31*, 2745–2753.

Leyval, C., Turnau, K., & Haselwandter, K. Effect of heavy metal pollution on mycorrhizal colonization and function: Physiological, ecological and applied aspects. *Mycorrhiza*, **1997**, *7*, 139–153.

Li, M., Cheng, X., & Guo, H. Heavy metal removal by biomineralization of urease producing bacteria isolated from soil. *Int. Biodeterior. Biodegradation*, **2013**, *76*, 81–85.

Lu, K., Yang, X., Gielen, G., Bolan, N., Ok, Y. S., Niazi, N. K., Xu, S., Yuan, G., Chen, X., Zhang, X., & Liu, D. Effect of bamboo and rice straw biochars on the mobility and redistribution of heavy metals (Cd, Cu, Pb and Zn) in contaminated soil. *J. Environ. Manage*, **2017**, *186*, 285–292.

Ma, L. Q., Komar, K. M., Tu, C., Zhang, W., Cai, Y., & Kennelley, E. D. A fern that hyperaccumulates arsenic. *Nature*, **2001**, *409*, 579–579.

Marques, A. P., Oliveira, R. S., Rangel, A. O., & Castro, P. M. Application of manure and compost to contaminated soils and its effect on zinc accumulation by *Solanum nigrum* inoculated with arbuscular mycorrhizal fungi. *Environ. Pollut.*, **2008**, *151*, 608–620.

Martin, S., & Griswold, W. Human health effects of heavy metals. *Environ. Sci. Technol. Brief Cit.*, **2009**, *15*, 1–6.

Masto, R. E., Chhonkar, P. K., Singh, D., & Patra, A. K. Changes in soil biological and biochemical characteristics in a long-term field trial on a sub-tropical inceptisol. *Soil Biol. Biochem.*, **2006**, *38*, 1577–1582.

Mattia, G. D., Bravi, M. C., Laurenti, O., Luca, O. D., Palmeri, A., Sabatucci, A., Mendico, G., & Ghiselli, A. Impairment of cell and plasma redox state in subjects professionally exposed to chromium. *Am. J. Ind. Med.*, **2004**, *46*, 120–125.

McCarty, K. M., Chen, Y. C., Quamruzzaman, Q., Rahman, M., Mahiuddin, G., Hsueh, Y. M., Su, L., Smith, T., Ryan, L., & Christiani, D. C. Arsenic methylation, GSTT1,

GSTM1, GSTP1 polymorphisms, and skin lesions. *Environmental Health Perspectives*, **2007**, *115*, 341–345.

McCutcheon, S. C., & Schnoor, J. L. *Phytoremediation: Transformation and Control of Contaminants* (pp. 1–939). Wiley-Interscience Inc., Hoboken, NJ, **2003**.

McGrath, S. P., & Zhao, F. J. Phytoextraction of metals and metalloids from contaminated soils. *Curr. Opin. Biotechnol.*, **2003**, *14*, 277–282.

Meagher, R. B., Rugh, C. L., Kandasamy, M. K., Gragson, G., & Wang, N. J. Engineered phytoremediation of mercury pollution in soil and water using bacterial genes. In: Terry, N., & Banuelos, G., (eds.), *Phytoremediation of Contaminated Soil and Water* (pp. 202–233). CRC Press, Florida, **1999**.

Meharg, A. A., & Cairney, J. W. Ectomycorrhizas-extending the capabilities of rhizosphere remediation? *Soil Biol. Biochem.*, **2000**, *32*, 1475–1484.

Mengoni, A., Grassi, E., Barzanti, R., Biondi, E. G., Gonnelli, C., Kim, C. K., & Bazzi-calupo, M. Genetic diversity of bacterial communities of serpentine soil and of rhizo-sphere of the nickel-hyperaccumulator plant *Alyssum bertolonii*. *Microb. Ecol.*, **2004**, *48*, 209–217.

Mitra, R. S. Protein synthesis in *Escherichia coli* during recovery from exposure to low levels of Cd^{2+}. *Appl. Environ. Microbiol.*, **1984**, *47*, 1012–1016.

Mohanty, M., & Patra, H. K. Effect of ionic and chelate assisted hexavalent chromium on mung bean seedlings (*Vigna radiata* L. Wilczek. var k-851) during seedling growth. *J. Stress Physiol. Biochem*, **2013**, *9*, 232–241.

Morgan, R. Soil, heavy metals, and human health. In: Brevik, E. C., & Burgess, L. C., (eds.), *Soils and Human Health* (pp. 59–82). CRC Press, Florida, **2012**.

Nagajyoti, P. C., Lee, K. D., & Sreekanth, T. V. M. Heavy metals, occurrence and toxicity for plants: A review. *Environ. Chem. Lett.*, **2010**, *8*, 199–216.

Najeeb, U., Ahmad, W., Zia, M. H., Zaffar, M., & Zhou, W. Enhancing the lead phytostabilization in wetland plant *Juncus effusus* L. through somaclonal manipulation and EDTA enrichment. *Arab J. Chem.*, **2017** *10*(2), 10–17.

Namgay, T., Singh, B., & Singh, B. P. Influence of biochar application to soil on the availability of As, Cd, Cu, Pb, and Zn to maize (*Zea mays* L.). *Soil Res.*, **2010**, *48*, 638–647.

Nanda, S., & Abraham, J. Remediation of heavy metal contaminated soil. *Afr. J. Biotechnol.*, **2013**, *12*, 3099–3109.

Normander, B., Hendriksen, N. B., & Nybroe, O. Green fluorescent protein-marked *Pseudomonas fluorescens*: Localization, viability, and activity in the natural barley rhizosphere. *Appl. Environ. Microbiol.*, **1999**, *65*, 4646–4651.

O'Brien, T. J., Ceryak, S., & Patierno, S. R. Complexities of chromium carcinogenesis: Role of cellular response, repair and recovery mechanisms. *Mutat. Res. Fundam. Mol. Mech. Mutagen.*, **2003**, *533*, 3–36.

O'Loughlin, E. J., Traina, S. J., & Sims, G. K. Effects of sorption on the biodegradation of 2-methylpyridine in aqueous suspensions of reference clay minerals. *Environmental Toxicology and Chemistry*, **2000**, *19*, 2168–2174.

Ogwuegbu, M. O. C., & Muhanga, W. Investigation of lead concentration in the blood of people in the Copper belt Province of Zambia. *J. Environ.*, **2005**, *1*, 66–75.

Panda, S. K., & Choudhury, S. Chromium stress in plants. *Braz. J. Plant Physiol.*, **2005**, *17*, 95–102.

Park, J. H., Choppala, G. K., Bolan, N. S., Chung, J. W., & Chuasavathi, T. Biochar reduces the bioavailability and phytotoxicity of heavy metals. *Plant Soil*, **2011**, *348*(1/2), 439–451.

Paz-Ferreiro, J., Lu, H., Fu, S., Méndez, A., & Gascó, G. Use of phytoremediation and biochar to remediate heavy metal polluted soils: A review. *Solid Earth Discus*, **2013**, *5*, 2155–2179.

Perfumo, A., Banat, I. M., Marchant, R., & Vezzulli, L. Thermally enhanced approaches for bioremediation of hydrocarbon-contaminated soils. *Chemosphere*, **2007**, *66*, 179–184.

Peters, R. W. In: Nathanson, J. A., & Schneider, R. A., (eds.), *Basic Environmental Technology: Water Supply, Waste Management, and Pollution Control* (pp. 5, 6). Prentice-Hall/Pearson Education, Inc., Boston, MA, **2015**.

Pourrut, B., Shahid, M., Dumat, C., Winterton, P., & Pinelli, E. Lead uptake, toxicity, and detoxification in plants. In: *Reviews of Environmental Contamination and Toxicology* (Vol. 213, pp. 113–136). Springer, New York, **2011**.

Prakash, N. T., Sharma, N., Prakash, R., & Acharya, R. Removal of selenium from Se enriched natural soils by a consortium of *Bacillus* isolates. *Bull. Environ. Contam. Toxicol.*, **2010**, *85*, 214–218.

Rajendran, P., Muthukrishnan, J., & Gunasekaran, P. Microbes in heavy metal remediation. *Indian J. Exp. Biol.*, **2003**, *41*, 935–944.

Rascio, N., & Navari-Izzo, F. Heavy metal hyperaccumulating plants: How and why do they do it? And what makes them so interesting? *Plant Sci.*, **2011**, *180*, 169–181.

Rascio, N., Dalla, V. F., La Rocca, N., Barbato, R., Pagliano, C., Raviolo, M., Gonnelli, C., & Gabbrielli, R. Metal accumulation and damage in rice (cv. *Vialone nano*) seedlings exposed to cadmium. *Environ. Exp. Bot.*, **2008**, *62*, 267–278.

Rodríguez, M. C., Barsanti, L., Passarelli, V., Evangelista, V., Conforti, V., & Gualtieri, P. Effects of chromium on photosynthetic and photoreceptive apparatus of the alga *Chlamydomonas reinhardtii*. *Environ. Res.*, **2007**, *105*, 234–239.

Rout, G. R., Samantaray, S., & Das, P. Effects of chromium and nickel on germination and growth in tolerant and non-tolerant populations of *Echinochloa colona* (L.) Link. *Chemosphere*, **2000**, *40*, 855–859.

Rugh, C. L., Senecoff, J. F., Meagher, R. B., & Merkle, S. A. Development of transgenic yellow poplar for mercury phytoremediation. *Nat. Biotechnol.*, **1998**, *16*, 925–928.

Salt, D. E., Blaylock, M., Kumar, N. P., Dushenkov, V., Ensley, B. D., Chet, I., & Raskin, I. Phytoremediation: A novel strategy for the removal of toxic metals from the environment using plants. *Nat. Biotechnol.*, **1995**, *13*, 468–474.

Sand, W., Rohde, K., Sobotke, B., & Zenneck, C. Evaluation of *Leptospirillum ferrooxidans* for leaching. *Appl. Environ. Microbiol.*, **1992**, *58*, 85–92.

Sanders, J. R., & Adams, T. M. The effects of pH and soil type on concentration of zinc, copper and nickel extracted by calcium chloride from sewage sludge-treated soils. *Environ. Pollut.*, **1987**, *43*, 219–228.

Schmidt, T., & Schlegel, H. G. Combined nickel-cobalt-cadmium resistance encoded by the ncc locus of *Alcaligenes xylosoxidans* 31A. *J. Bacteriol.*, **1994**, *176*, 7045–7054.

Schneider, J., Grosser, R., Jayasimhulu, K., Xue, W., & Warshawsky, D. Degradation of pyrene, benz[*a*]anthracene, and benzo[*a*]pyrene by *Mycobacterium* sp. strain RJGII-135, isolated from a former coal gasification site. *Appl. Environ. Microbiol.*, **1996**, *62*, 13–19.

Seo, J. S., Keum, Y. S., Hu, Y., Lee, S. E., & Li, Q. X. Phenanthrene degradation in *Arthrobacter* sp. P1-1: initial 1,2-, 3,4- and 9,10-dioxygenation, and meta- and ortho-cleavages of naphthalene-1, 2-diol after its formation from naphthalene-1,2-dicarboxylic acid and hydroxyl naphthoic acids. *Chemosphere*, **2006**, *65*, 2388–2394.

Shah, M. P. Microbiological removal of phenol by an application of *Pseudomonas* spp. ETL: An innovative biotechnological approach providing answers to the problems of FETP. *J. Appl. Environ. Microbiol.*, **2014**, *2*, 6–11.

Shah, M. P., Patel, K. A., Nair, S. S., & Darji, A. M. An innovative approach to biodegradation of textile dye (Remazol Black B) by *Bacillus* spp. *Int. J. Environ. Bioremediat. Biodegrad.*, **2013c**, *1*, 43–48.

Shah, M. P., Patel, K. A., Nair, S. S., & Darji, A. M. Decolorization of remazol black-B by three bacterial isolates. *Int. J. Environ. Bioremediat. Biodegrad.*, **2014**, *2*, 44–49.

Shah, M. P., Patel, K. A., Nair, S. S., & Darji, A. M. Microbial decolorization of methyl orange dye by *Pseudomonas* spp. ETL-M. *Int. J. Environ. Bioremediat. Biodegrad.*, **2013a**, *1*, 54–59.

Shah, M. P., Patel, K. A., Nair, S. S., & Darji, A. M. Microbial degradation and decolorization of reactive orange dye by strain of *Pseudomonas* spp. *Int. J. Environ. Bioremediat. Biodegrad.*, **2013b**, *1*, 1–5.

Shri, M., Kumar, S., Chakrabarty, D., Trivedi, P. K., Mallick, S., Misra, P., Shukla, D., Mishra, S., Srivastava, S., Tripathi, R. D., & Tuli, R. Effect of arsenic on growth, oxidative stress, and antioxidant system in rice seedlings. *Ecotoxicol. Environ. Saf.*, **2009**, *72*, 1102–1110.

Siedlecka, A., Samuelsson, G., Gardenstrom, P., Kleczkowski, L. A., & Krupa, Z. The activatory model of plant response to moderate cadmium stress-relationship between carbonic anhydrase and Rubisco. *Photosynthesis: Mechanisms and Effects*, **1998**, *4*, 2677–2680.

Sims, G. K. Nitrogen starvation promotes biodegradation of N-heterocyclic compounds in soil. *Soil Biol. Biochem.*, **2006**, *38*, 2478–2480.

Singh, N., Kumar, D., & Sahu, A. P. Arsenic in the environment: Effects on human health and possible prevention. *J. Environ. Biol.*, **2007**, *28*, 359.

Singh, R., Gautam, N., Mishra, A., & Gupta, R. Heavy metals and living systems: An overview. *Indian J. Pharmacol.*, **2011**, *43*, 246.

Singh, U., Rani, B., Chauhan, A. K., Maheshwari, R., & Vyas, M. K. Role of environmental biotechnology in decontaminating polluted water. *Int. J. Life Sci. Biotechnol. Pharma. Res.*, **2012**, *1*, 32–46.

Sobha, K., Poornima, A., Harini, P., & Veeraiah, K. A study on biochemical changes in the freshwater fish, *Catla catla* (Hamilton) exposed to the heavy metal toxicant cadmium chloride. *Kathmandu University Journal of Science, Engineering and Technology*, **2007**, *3*, 1–11.

Stohs, S. J., & Bagchi, D. Oxidative mechanisms in the toxicity of metal ions. *Free Radic. Biol. Med.*, **1995**, *18*, 321–336.

Tabatabai, M. A. Soil enzymes. In: Weaver, R. W., Angle, J. S., & Bottomley, P. S., (eds.), *Methods of Soil Analysis. Part 2: Microbiological and Biochemical Properties* (pp. 775–833). SSSA Book Series No. 5, Soil Science Society of America, Madison, WI, **1994**.

Tchounwou, P. B., Yedjou, C. G., Patlolla, A. K., & Sutton, D. J. Heavy metal toxicity and the environment. In: Luch, A., (ed.), *Molecular, Clinical and Environmental Toxicology* (pp. 133–164). Springer, Basel, **2012**.

Thakur, I. S. *Environmental Biotechnology: Basic Concepts and Applications*. IK International, New Delhi, **2011**.

Tran, T. A., & Popova, L. P. Functions and toxicity of cadmium in plants: Recent advances and future prospects. *Turk. J. Bot.*, **2013**, *37*, 1–13.

Uchimiya, M., Wartelle, L. H., Klasson, K. T., Fortier, C. A., & Lima, I. M. Influence of pyrolysis temperature on biochar property and function as a heavy metal sorbent in soil. *J. Agric. Food Chem.*, **2011**, *59*, 2501–2510.

Vangronsveld, J., Herzig, R., Weyens, N., Boulet, J., Adriaensen, K., Ruttens, A., Thewys, T., Vassilev, A., Meers, E., Nehnevajova, E., & Van Der Lelie, D. Phytoremediation of contaminated soils and groundwater: Lessons from the field. *Environ. Sci. Pollut. Res.*, **2009**, *16*, 765–794.

Wang, F, Yao, J., Si, Y., Chen, H., Russel, M., Chen, K., Qian, Y., Zaray, G., & Bramanti, E. Short-time effect of heavy metals upon microbial community activity. *Journal of Hazardous Materials*, **2010**, *173*, 510–516.

Wang, H., Kimberley, M. O., & Schlegelmilch, M. Biosolids derived nitrogen mineralization and transformation in forest soils. *J. Environ. Qual.*, **2003**, *32*, 1851–1856.

Weyens, N., Van Der Lelie, D., Taghavi, S., & Vangronsveld, J. Phytoremediation: Plant-endophyte partnerships take the challenge. *Curr. Opin. Biotechnol.*, **2009**, *20*, 248–254.

Wu, G., Kang, H., Zhang, X., Shao, H., Chu, L., & Ruan, C. A critical review on the bio-removal of hazardous heavy metals from contaminated soils: Issues, progress, eco-environmental concerns, and opportunities. *J. Hazard Mater.*, **2010**, *174*, 1–8.

Yang, J., Kloepper, J. W., & Ryu, C. M. Rhizosphere bacteria help plants tolerate abiotic stress. *Trends Plant Sci.*, **2009**, *14*, 1–4.

Yang, X. E., Long, X. X., Ye, H. B., He, Z. L., Calvert, D. V., & Stoffella, P. J. Cadmium tolerance and hyperaccumulation in a new Zn-hyperaccumulating plant species (*Sedum alfredii* Hance). *Plant and Soil*, **2004**, *259*(1/2), 181–189.

Yerima, M. B., Umar, A. F., Shinkafi, S. A., & Ibrahim, M. L. Bioremediation of hydrocarbon pollution: a sustainable means of biodiversity conservation. *J. Sustain. Dev. Environ. Prot.*, **2012**, *2*, 43–50.

Zeid, I. M. Responses of *Phaseolus vulgaris* chromium and cobalt treatments. *Biologia Plantarum*, **2001**, *44*, 111–115.

CHAPTER 10

RADIATION-RESISTANT THERMOPHILES: FROM HIGH TEMPERATURE AND RADIATION TO ENGINEERED BIOREMEDIATION

PREETI RANAWAT[1] and SEEMA RAWAT[2]

[1]Department of Botany and Microbiology, Hemvati Nandan Bahuguna Garhwal University, Srinagar (Garhwal)–246174, Uttarakhand, India

[2]School of Life Sciences, Central University of Gujarat, Gandhinagar, Gujarat–382030, India

10.1 INTRODUCTION

Microorganisms are ubiquitous and competent enough to inhabit hostile niches on Earth-like subsurface, deep oceans, extreme acid, and extreme alkaline conditions, frozen glaciers, seawater, and hydrothermal systems (Rampellotto, 2010). The existence of microbial life in inhospitable environments, viz., geothermal regions, hydrothermal vents, acidic, and hot geysers have unveiled the presence of special category of microorganisms known as thermophiles or hyperthermophiles. The habitat of thermo (or hyperthermophilic) microorganisms is versatile and they have been reported from freshly fallen snow to pasteurized milk to geothermal springs (Ranawat and Rawat, 2017a). The famous abode of thermophiles include hot springs, viz., Yellowstone National Park, USA (Reysenbach et al., 1994, 2000; Slobodkin et al., 1997; Schaffer et al., 2004), in India from Bakreshwar hot spring, West Bengal (Ghosh et al., 2003), Tulasi Shyam hot spring, Gujarat (Gehlani et al., 2015), Unkeshwar hot spring of Maharashtra (Mehetre et al., 2015), Soldhar hot spring of Uttrakhand (Sharma et al., 2015), Deulajhari hot spring, Odhisha (Singh and Subudhi,

2016), Grensdalur hot spring, Iceland, Garga hot spring, Russia (Nazina et al., 2004), Nalychevskie, Oksinskie, Apapelskie, and Dachnye hot springs in Kamchatka Peninsula, Russia (Belkova et al., 2007), El Biban hot spring in Northeast of Algeria (Kecha et al., 2007), Rotoura hot spring, New Zealand (Niederberger et al., 2008), Tengchong hot spring, China (Hou et al., 2013). The continental and submarine volcanic areas, viz., solfatara fields, geothermal power plants, hydrothermal vents, and geothermally heated sea sediments also provide suitable conditions for the growth of heat-loving microorganisms and mainly comprised of a population of anaerobic thermophiles and hyperthermophiles (Erauso et al., 1993; Uemori et al., 1993). The variations in environmental parameters, viz., temperature, pH, oxygen, nutrients, and light intensity in the habitat of thermophiles leads to stress conditions for the inhabitants, which results in slow growth or cell death. In stress conditions, the thermophilic life forms exhibit a variety of physiological changes, which include the production of DNA binding proteins, activation of reactive oxygen species (ROS) detoxification system, accumulation of compatible solute, expression of heat shock proteins and alterations in morphology (Ranawat and Rawat, 2017a). Thermophiles are the center of curiosity amongst the scientific community from the viewpoint of the mechanisms, which provide the ability to cope up simultaneously with variety of stresses, viz., desiccation, radiation, and pressure along with elevated temperature (DiRuggiero et al., 1997; Beblo et al., 2011; Cavicchioli et al., 2011). The radiation stress at high temperatures is the most interesting phenomenon to explore as it reveals the adaptation mechanisms exhibited by thermophiles to radiations along with adaptations in the temperature range.

The exposure to ionizing radiations (IR) like α, β, γ adversely affects DNA, lipids, and proteins, along with the production of oxidative stress (Webb and DiRuggiero, 2012). IR can either cause direct damage or indirect damage. In 'indirect damage,' the ROS formed by the radiolysis of water and generate hydroxyl radicals (OH$^-$), superoxide (O$_2^-$), and hydrogen peroxide (H$_2$O$_2$). The water molecules associated with DNA undergo radiolysis become detrimental for DNA molecules and as a result, generate oxidized DNA bases and sugar moieties, single-strand breaks (SSBs), abasic sites, and cross-links in proteins (Imlay, 2006). The increased intensity of IR leads to damage in the linear density of DNA base and induces SSBs on both strands, which in turn results in double-strand breaks (DSBs). The proteins are the main target of ROS, which inactivates and denature proteins by introducing amino acid radical chain reactions and cross-linking of proteins (Stadtman and Levine, 2003; Imlay, 2006). O_2^- cannot react with DNA as well as with the majority of proteins and unable to cross membranes, but it can cause inactivation of enzymes with exposed 2Fe-2S or 4Fe-4S clusters. Fe^{2+} reacts with H_2O_2 and catalyzes the Fenton reaction (Imlay, 2006). It is now widely accepted that radiations specifically target proteins, and in order to survive damage from IR protection against oxidation is required (Du and Gebicki, 2004; Daly et al., 2007). IR also induces 'direct damage' when macromolecules absorb X-ray and γ-ray photon (Von Sontaag, 1987). A dose of IR typically causes 40 times more SSBs than DSBs (Von Sontaag, 1987; Daly et al., 1994). The radiation-resistant thermophiles counteract the damaging effects of IR by preventing the formation of ROS via superoxide dismutase and peroxidase. Resistant bacteria also accumulate Mn^{2+} ions that protect Fe-S cluster containing proteins from O_2^-. Thus, the release of Fe^{2+} from Fe-S cluster containing proteins can be prevented, minimizing the effects of Fenton chemistry and help enzymes to function efficiently (Ghosal et al., 2005; Imlay, 2006; Daly, 2009). The effects of IR on the cell are summarized in Figure 10.1. Thermophiles deal with the damaging effects of IR by using a detoxification system that scavenges ROS, mechanisms that repair DNA damage and accumulation of Mn^{2+} and trehalose (Makarova et al., 2007; Liedert et al., 2012; Webb and Di Ruggiero, 2012).

Radiation resistant thermophiles possess genes for detoxification and removal of toxic elements, which extends their application in the field of metal bioremediation also (Urmania, 2005). In this process, the radiation-resistant thermophiles not only perform degradation of the toxic metal waste but also oxidize a number of organic and inorganic substances along with metal ions, which results in the production of non-toxic metal ions from toxic ones (Sar et al., 2013). Therefore, these microorganisms are biotechnological assets that can be applied for bioremediation of heavy metals and radionuclides expelled out from nuclear power plants. The thermophilic *Clostridium thermosuccinogenes* can survive 6mM concentration of U(VI) (Wright et al., 2012) while mesophilic, *Serratia marcescens* has a minimum inhibitory concentration (MIC) of 4 mM (Kumar et al., 2011). Biotechnologically engineered thermophilic radiation-resistant strains like *Deinococcccus geothermalis* can reduce Fe(III)-nitrilotriacetic acid, U(VI) and Cr(VI). This suggests that thermophiles are better and efficient candidates than mesophilic microorganisms for bioremediation of radionuclide wastes (Brim et al., 2003). The present chapter deals with the mechanisms governing radiation-resistant in thermophiles and their

potential biotechnological applications ranging from exploitation of engineered strains for treatment of radionuclides contaminated environments
to astrobiology.

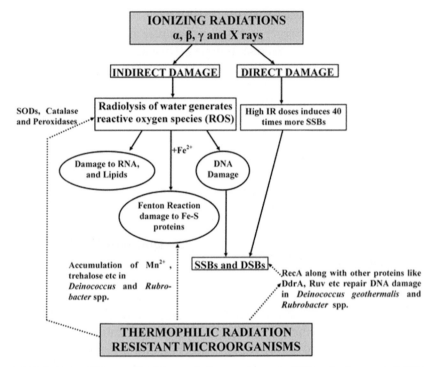

FIGURE 10.1 Effects of ionizing radiations on bacteria (SSBs: single-stranded DNA
breaks; DSBs: double-strand DNA breaks; SODs: superoxide dismutases; bold arrows
represent effects of ionizing radiations, and dotted arrows represent responses of thermophiles to radiation stress).

Source: Modified from Ranawat and Rawat, 2017b.

10.2   CO-EXISTENCE OF DESICCATION AND RADIATION RESISTANCE IN THERMOPHILES

The surviving conditions of early Earth resemble that of the present-day
habitat of thermophiles, which supports the hypothesis that thermophiles
were the first life forms on Earth (Di Giulio, 2000; Nisbet and Sleep, 2001).
However, the hypothesis of lithopanspermia, suggests that the organisms
which were carried by meteorites and survived the harsh environment of

space, originated life on Earth (Nicholson et al., 2000; Horneck et al., 2001). It has been observed that there is striking similarity in consequences of heat and water loss in the cell, which include DNA double-strand breaks and protein denaturation (Prestrelski et al., 1993; Mattimore and Battista, 1996). This suggests that microorganisms are able to survive in elevated temperature, and dehydration develops resistance for desiccation, and it leads to resistance for radiation as well. It has been speculated that the mechanisms that help *Pyrococcus furiosus* and *Thermococcus gammatolerans* to overcome DNA damage caused by high temperature are possibly involved in repair of damage caused by desiccation and radiation (Grogan, 2000; Williams et al., 2007; Beblo et al., 2009; Zivanovic et al., 2009).

In moderate thermophilic microorganism, *Halobacterium salinarum*, fast recovery from irradiation and desiccation has been reported by regeneration of intact chromosome from scattered fragments due to DSBs (Kottemann et al., 2005; Soppa, 2013). The presence of overlapping genome fragment is necessary for this repair mechanism, and thus, it works only in oligoploid, such as *D. radiodurans*, or polyploid species, such as Haloarchaea (Soppa, 2014). In *Deinococcus* spp. both stresses, i.e., desiccation, and radiation produce a similar type of DNA damage. It has been observed that radiation-sensitive mutants are desiccation sensitive also. Mattimore and Battista (1996), reported that the combination of UV and desiccation resistance in these microorganisms assist in their dispersal to hospitable environments. Proteins play an important role in the repair of damaged DNA, especially double-strand DNA breaks (DSBs). This protein repair system has been reported in *Halobacterium salinarum* NRC-1 where, Ral (tucHOG0456) helps in double-strand break repair and increases tolerance against radiations (Capes et al., 2012).

The other mechanisms that enhance radiation resistance in microorganisms include the usage of advanced antioxidant chemical and enzymatic defense systems which mitigate DNA and protein oxidation, protection of DNA from radical damage by using enhanced DNA repair enzymes and nucleoid condensation (Pavlopoulou et al., 2016). The plausible reason for radiation resistance in thermophiles is that these microorganisms continuously come across radioactive substances in their habitat, which may lead to radiation resistance. Elderfield and Schultz (1996) reported that some hydrothermal chimneys which exist in heavy metal-rich environments are naturally exposed to radioactivity doses, which are expected to be a hundred times higher than those observed in Earth's atmosphere. The radiotolerance in mesophilic microorganisms like *E. coli* can be developed comparable to that of *D. radiodurans* by genetic alterations due to exposure of cells to repetitive irradiation cycles (Harris et al., 2009; Byrne et al., 2014). The concept of genetic alterations possibly holds true for thermophiles also as in their natural habitats, they often encounter radioactive elements like Paralana hot springs (PHS), rich in uranium. The phylogenetically unrelatedness of radiation-resistant strains could be possibly due to the acquisition of radiotolerance during the evolution process (Cox and Battista, 2005).

10.3 MECHANISMS OF RADIATION RESISTANCE IN THERMOPHILES

Thermophiles deal with high temperature as well as radiations simultaneously which led to conclude that these microbes may have been present, and at times predominant throughout Earth's geologic history or represent lineages descendent from first living microorganisms on the planet. The mechanisms by which radiation-resistant thermophiles deal with stresses will give new insights into evolution and astrobiology (Ranawat and Rawat, 2017a). The microbes are subjected to different stresses like an extreme vacuum, desiccation, solar, and cosmic radiation, microgravity, and both extreme hot and cold temperatures in outer space environments (Nicholson et al., 2000). The spores of *Bacillus subtilis*, vegetative cells of *Deinococcus* and some halophilic archaea like *Halobacterium* spp. are the model organisms for study due to their high resistance to extreme conditions (Hecker and Völker, 1998; De Vera et al., 2003; Rampelotto et al., 2007, 2009; Nicholson, 2009; Rosa et al., 2009). Radiation resistant thermophiles include archaea like *Archaeoglobus fulgidus*, *Methanocaldococcus jannaschii*, *Pyrococcus furiosus*, *Thermococcus gammatolerans* and *Thermococcus radiotolerans*, actinobacteria like *Rubrobacter radiotolerans*, *Rubrobacter taiwanensis*, *Rubrobacter xylanophilus* and bacteria like *Deinococcus geothermalis*, *Deinococcus murrayi* and *Truepera radiovictrix* (Carreto et al., 1996; Di Ruggiero et al., 1997; Ferreira et al., 1997; Jolivet et al., 2003; Chen et al., 2004; Jolivet et al., 2004; Albuquerque et al., 2005; Beblo et al., 2011). It is very interesting to explore the science behind the processes that define radiation resistance and recovery from damage caused by IR in radiation-resistant thermophilic microorganisms. Thus, the upcoming sections present an overview of the DNA repair system, ROS detoxification mechanism, and other mechanisms that enable radiation resistance in thermophiles.

10.3.1 DNA REPAIR SYSTEM

IR exposure causes extensive damage to cellular macromolecules, especially DNA. The direct or indirect damage induces single-stranded and double-stranded DNA breaks. Hence, different radiation-resistant microorganisms respond in different ways to cope up with DNA damage and initiate an event of repair mechanisms, which are summarized in Figure 10.2.

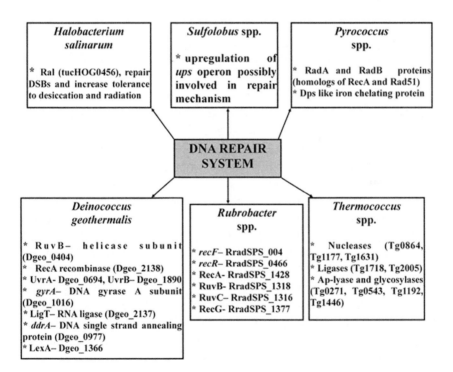

FIGURE 10.2   DNA repair system in radiation-resistant thermophiles.

Most of the members of the genus *Deinococcus* are resistant to radiation like *D. radiodurans* can resist a high dose of IR [≥12,000 Gy (gray; absorbed radiation dose)]. *D. geothermalis* is the thermophilic radiation-resistant bacteria which belonged to the genus *Deinococcus* and recovered from geothermal spring (Ferreira et al., 1997). The other radiation-resistant species of genus *Deinococcus* which can tolerate an average dose of 10kGy are *D. apachensis, D. grandis, D. hopiensis, D. indicus, D. maricopensis, D. murryai, D. pimensis, D. piscis, D. papagonensis, D. radiodurans, D. radiophilus, D. radiopugnans, D. sonorensis,* and *D. yunweiensis* (Ferreira et al., 1997; Pavlopoulou et al., 2016). Genus *Truepera* represents a distinct lineage of Phylum "*Deinococcus/Thermus*" and includes *Truepera radiovictrix,* which is also a radiation-resistant thermophile and 60% cells can survive 5 kGy. It also shares the trait of radiation resistance with genus *Deinococcus* (Ferreira et al., 1999).

Ferreira et al. (1997) reported that when *D. geothermalis* (DSM 11300) and *D. radiodurans* (ATCC BAA-816) were exposed to ⁶⁰Co and ultraviolet (UV) radiations (254 nm) at 50°C and 32°C, respectively both were observed to be equally resistant to radiations but since *D. geothermalis* is a thermophilic radiation-resistant bacteria its recovery was 1000 times greater at high temperature (50°C) than at mesophilic temperature (32°C). It was observed that *D. geothermalis* and *D. radiodurans* contain a conserved set of genes that confer extreme radiation resistance to these microorganisms. DR1289, a protein of RecQ helicase family present in *D. radiodurans,* contains three Helicase and RNAse D C-terminal (HRDC) domains while there is a single HRDC domain in other bacterial RecQ protein (Huang et al., 2007). In *D. geothermalis* no ortholog of *D. radiodurans* RecQ is reported, but it possesses another protein, Dgeo_1226, which has resemblance with corresponding domains of DR1289 and contains a helicase superfamily II C-terminal domain and a second HDRC domain. Thus, it is speculated that both proteins (DR1289 and D_geo1226) are the part of predicted resistance regulon (Huang et al., 2007). It was assumed that *D. radiodurans* possess non-homologous end joining (NHEJ) but later, studies reported that instead of NHEJ, *D. radiodurans* has DRB0100 which encode a ATP-dependent ligase (a phosphatase of H$_2$ macro superfamily), a polynucleotide kinase and a HD family phosphatase (Makarova et al., 2001, 2007; Levin-Zaidman et al., 2003; Liu et al., 2003; Englander et al., 2004; Narumi et al., 2004; Tanaka et al., 2004; Bowater and Doherty, 2006). In *D. geothermalis*, no orthologs of DRB0100 are reported but in *D. radiodurans* the induction of DRB0100 is induced by IR exposure. But it was also observed that DRB0100 does not exhibit DNA or RNA ligase function *in vitro* (Blasius et al., 2007).

In *D. radiodurans,* the exposure of cells to IR for 3 hours leads to increase in transcriptional induction by 2 folds (Liu et al., 2003). However,

45% genes induced in *D. radiodurans* following an exposure with IR have no orthologs in *D. geothermalis*. The plausible reason for this could be either all genes are not expressed in *D. radiodurans* or *D. geothermalis* possess a different set of radiation resistance factors (Makarova et al., 2007). *D. radiodurans* own a unique set of genes which code for DNA helicase RecQ (DR1289) and a transcriptional regulator (DR0171) but DNA single-strand annealing (SSA) protein, *pprA*-involved in DNA damage resistance mechanism and other 4 genes (*ddrC*-DR003/Dgeo_0047; *ddrB*-DR0070/Dgeo_0295; *ddrB*-DR0326/Dgeo_2186; *ddrA*-DR0423/Dgeo_0977) are exclusively reported from *Deinococcus* lineages which suggests that novel radiation resistance mechanisms are present in *Deinococcus* spp. (Liu et al., 2003; Tanaka et al., 2004). RecA is the crucial protein necessary for homologous DNA recombination repair following an exposure to IR (Cox and Battista, 2005). RecA of *D. radiodurans* first binds with DNA duplex and then with homologous single-stranded DNA while in other bacteria RecA first binds with single-stranded DNA and is a key protein involved in reassembly of damaged chromosome (Kim et al., 2002). Wanarska et al. (2011) reported that Rec A protein of *Deinococcus geothermalis* (*Dge*RecA) and of *Deinoccoccus murrayi* (*Dmu*RecA) are slightly thermostable which bind to ssDNA more readily as compared to dsDNA in the presence of Mg^{2+} ions and hydrolyze ATP and dATP in the presence of ssDNA (Kim et al., 2002).

RecA of *Deinococcus radiodurans* (*Dra*RecA) can enhance DNA strand exchange through an inverse pathway which is not present in RecA of *D. geothermalis* (*Dge*RecA) and RecA of *D. murryai* (*Dmu*RecA) (Kim et al., 2002). This suggests a unique property of *Dra*RecA which is not observed in other RecA protein (Wanarska et al., 2011). Makarova et al. (2007) reported that *D. radiodurans* and *D. geothermalis* encode *recJ* (a putative 5'-3' exonuclease) instead of *recBCE*. It has been speculated that *recJ* provides nuclease activity, which is not present in Klenow polymerase. In *D. radiodurans*, the regulation of gene expression while the cells are recovering from an exposure to IR has been deeply investigated. The irradiation leads to induction of RecA, which is regulated by IrrE/PprI protein. This protein is mainly comprised of two domains, a Xre-like HTH domain and a Zn-dependent protease (Earl et al., 2002; Hua et al., 2003). Gao et al. (2006) reported that gene *irrE* is present in upstream of folate biosynthesis operon and regulated individually in *D. geothermalis* and *D. radiodurans*. *D. geothermalis* has one *lexA* gene (DG1366), whereas there are two *lexA* paralogs (DRA0344, DRA0074) in *D. radiodurans,* but *lexA* genes are not induced after irradiation and therefore do not play any role in the induction of RecA in *D. radiodurans* (Naraumi et al., 2001). DdrO, a Xre family protein, is exclusively *Deinococcus* gene for predicted regulator and preceded by RDRM site. The arrangement of DdrO is identical to various stress response regulators like *lexA* genes of many species (Little et al., 1981). The DdrO putative regulator in *D. radiodurans* is DR2574, expressed by microarray experiment at lower doses (3000 Gy) while Dgeo_0336 is an ortholog of DdrO in *D. geothermalis* (Tanaka et al., 2004).

Radiation-desiccation response (RDR) regulon is present in both *D. geothermalis* and *D. radiodurans*. There are 20 operons in RDR regulon, which comprised of 25 genes in *D. geothermalis* and 29 genes in *D. radiodurans*. The dominating genes present in RDR regulon are DNA repair genes which also include recombination repair proteins RecA and RecQ, MutS, and MutL proteins (involved in mismatch repair (MMR) located in one operon in *D. geothermalis*) and UvrB and UvrC proteins (involved in nucleotide excision repair) (Kuzminov, 1999; Kunkel and Erie, 2005). Touati et al. (1996) reported that a transketolase gene is also present in RDR regulon, which is induced in stress like exposure to cold shock and mutagenic agents, which induces SOS response. In *D. radiodurans,* the DNA excision repair is reported to be facilitated by pentose-phosphate pathway (Zhang et al., 2003). The upstream regions of various genes, viz., DR0326, *ddrD*; DR0423, *ddrA*; DRA0346, *pprA*; DR0070, *ddrB* exhibited the presence of a strong palindromic motif known as radiation/ desiccation response motif (RDRM). Liu et al. (2003) and Tanaka et al. (2004) reported that RDR regulon is comprised of two groups: (i) orthologous genes present in *D. geothermalis* and *D. radiodurans* which contain RDRM; (ii) a exclusive set of genes present in *D. radiodurans* which contain RDRM and this gene set is upregulated when there is recovery of cells following an irradiation

Deinococcus spp. and *Thermus* spp. are members of group *Deinococcus-Thermus* which besides sharing various similarities like pigmentation and chemoorganotrophism varies in their responses to IR as *Deinococcus* genus has both radiation-resistant mesophile like *D. radiodurans* and thermophile like *D. geothermalis* while *Thermus* spp. are thermophilic and radiation-sensitive (Weisburg et al., 1989). *D. radiodurans* has two LexA, a SOS- response transcriptional repressor while *T. thermophilus* is devoid of that and also does not contain endonuclease VII (XseAB). However, it possesses photolyase (PhrB) and endonuclease

IV DNA polymerase III, indicating a very few similarities between the DNA repair systems of *D. radiodurans* and *T. thermophilus* (Omelchenko et al., 2005). *D. radiodurans* has an extensive repertoire of enzymes or proteins involved in DNA repair, which are necessary for IR resistance. It include DNA ligase, RNA ligase, double-strand break repair complex, a protein of HD family phosphatase, polynucleotide kinase, phosphatase of H_2 macro superfamily, double-stranded DNA binding protein, PprA, and DNA single-strand annealing protein, DdrA among all these proteins *Thermus* has only an ortholog of DdrA (Makarova et al., 2001; Liu et al., 2003; Harris et al., 2004; Martins and Shuman, 2004; Narumi et al., 2004). The megaplasmid of *T. thermophilus*, encodes a putative thermophilic-specific repair system involved in uncharacterized DNA repair. It was postulated that this complex is functionally analogous to the bacterial-eukaryotic system of translesion and mutagenic repair whose components are DNA polymerases of UmuC-DinB-Rad20-Rev1 superfamily, not present in thermophiles (Makarova et al., 2002). Henne et al. (2004) reported that megaplasmid of *T. thermophilus* HB27 encodes UV endonuclease (TTP0052), which repair DNA damages caused by UV. *T. thermophilus* HB27 encodes two-uracil-DNA glycosylases that remove uracil residues from U:G mispairs formed in DNA (Ohta et al., 2006).

Rubrobacter spp. is radiation-resistant thermophilic actinobacteria which can tolerate 8–12 kGy doses of radiations except *R. bracarensis* which is radiation-sensitive (Jurado et al., 2012). The complete genome of *R. radiotolerans* strain RSPS-4 was investigated for the genes which play a crucial role in DNA repair pathways and encode proteins required for homologous recombination, SSA, extended synthesis-dependent strand annealing or NHEJ (Egas et al., 2014). RSPS-4 contains all genes for RecFOR pathway, *recF* (RradSPS_0004), *recR* (RradSPS_0466), *recO* (RradSPS_1511), and *recJ* (RradSPS_0780). RecA, protein responsible for strand invasion and exchange, was encoded as a single copy (RradSPS_1428). Genes encoding the branch migration and resolution of Holliday junction proteins RuvA (RradSPS_1317), RuvB (RradSPS_1318), RuvC (RradSPS_1316), and RecG (RradSPS_1377) were detected, as were the homologs of the genes encoding the SbcD (*mre*11) (RradSPS_2355), and SbcC (Rad50) (RradSPS_2356) proteins. Egas et al. (2014) reported that *R. radiotolerans* strain RSPS-4 lacks *lexA,* which suggests the presence of alternative DNA repair mechanisms. The presence of gene copies of *mut*L (RradSPS_0036 and _0159) and *mut*S (RradSPS_0158) predict that MMR is probably active in RSPS-4 (Nowosielska and Marinus, 2008).

Thermophilic archaea like *Pyrococcus furiosus* and *Thermococcus gammatolerans* can resist 6kGy and 3kGy dose of radiations (Webb and Di Ruggiero, 2012). *T. gammatolerans* posses nucleases Tg0864, Tg1177, Tg1631, ligases (Tg1718, Tg2005), endonucleases, AP-lyases, and glycosylases Tg0271, Tg0543, Tg1192, Tg1446, Tg1637, Tg1814 which are involved in base excision repair and Tg0130, Tg0280, Tg1742, Tg1743, Tg1744 and Tg2074 are presumed to be involved in double-strand break repair (Zivanovic et al., 2009) while in *Pyrococcus furiosus* RadA and RadB proteins (homologs of RecA and Rad51 found in bacteria and eukaryotes) are involved in homologous recombination and also produce a putative Dps-like iron chelating protein (Komori et al., 2000; Gérard et al., 2001). In *Sulfolobus* spp., UV stress causes upregulation of UV-inducible pili operon of *Sulfolobus* (*ups* operon), which leads to the formation of type IV pili (T4P) involved in UV induced pili assembly, cellular aggregation and subsequent exchange of DNA between cells. The gene deletion analysis explained that UpsX (a membrane-localized protein encoded by *ups* operon) is involved in DNA transfer (via unknown mechanism) which is assumed to be the repair mechanism of the cell for UV induced DNA damage (Fröls et al., 2007, 2008; Götz et al., 2007). Ajon et al. (2011) reported that *ups* cluster encode UpsX, UpsE, a secretion ATPase; UpsF, a membrane protein; and UpsA and B, two-class III signal peptide-containing pilin subunits which are essentially required in DNA transfer and repair mechanism during UV stress.

10.3.2   ROS DETOXIFICATION SYSTEM

The exposure to IR leads to the formation of ROS, viz., superoxide (O_2^-), hydrogen peroxide (H_2O_2), or hydroxyl radical (OH^-). These ROS cause extreme damage to the cell as they have a high affinity for nucleic acids and proteins (Mostertz et al., 2004). The cells evade from ROS by a class of metalloenzymes, Superoxide dismutases (SODs), which detoxify oxygen radicals by breaking them to oxygen and hydrogen peroxide then catalase or peroxidase act on H_2O_2 and reduce it to oxygen and water (Fridovich, 1995). Fridovich (1995) and Youn et al. (1996) reported that on the basis of metal co-factors, SODs are classified into four types: manganese co-factored (Mn-SOD or SodA), iron co-factored (Fe-SOD or SodB), copper-zinc co-factored (Cu/Zn-SOD or SodC) and nickel co-factored (Ni-SOD or SodN). Another class of SODs from Fe/Mn SOD family is cambialistic SODs that can use either manganese or iron cata-lytically (Gabbianelli et al., 1995). These SODs have been reported from thermophilic photosynthetic bacterium *Chloroflexus aurantiacus,* which protects the bacterium from hypoxic environment subjected to high UV radiation fluxes (Lancaster et al., 2004). *Streptococcus thermophilus* and *Thermus filiformis* are other thermophiles having cambilalistic SODs while no such SODs have been reported in *Deinococcus* spp. (Mandelli et al., 2013; Krauss et al., 2015). *D. radiodurans* encodes Mn dependent SOD DR1279 and Cu/Zn dependent SODs, viz., DR1546, DRA0202 and DR0644 (Makarova et al., 2001) while *D. geothermalis* contain Mn depen-dent SOD Dgeo_830 and Cu/Zn dependent SOD Dgeo_0284 an ortholog of DR0644 (Makarova et al., 2007). The ROS detoxification mechanisms adopted by radiation-resistant thermophiles are described in Figure 10.3.

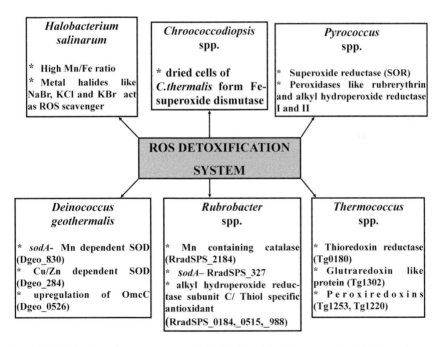

FIGURE 10.3 Reactive oxygen species (ROS) detoxification system in radiation resistant thermophiles.

In *D. geothermalis*, sugars like xylose are metabolized by a cluster of genes on megaplasmid DG574 (Ferreira et al., 1997; Makarova et al.,

2007). The plausible resistance-related functions that can be proposed for expanded families of Deinococci include hydrolases-degrade oxidized lipids; Yfit/DinB proteins-involved in cell damage related pathways (Makarova et al., 2001); subtilisn like proteases which degrade oxidized proteins (Makarova et al., 2000; Daly et al., 2007) nudix related hydrolases-Apn, a diadenosine polyphosphotases, form adenosine which is crucial for protection of cell from radiation and oxidative stress (Makarova et al., 2000; Daly et al., 2007). In comparison to *D. geothermalis*, *D. radiodurans* possess extra set of genes which code for Cu-Zn SOD, a peroxidase, HslJ-like heat shock proteins and other proteins that confer antibiotic resistance (Makarova et al., 2007). *D. radiodurans* lacks *nadABCD* genes while *D. geothermalis*, possess orthologs of *nadABCD* genes, essential for biosynthesis of nicotinamide adenine dinucleotide (NAD) (Venkateswaran et al., 2000; Makarova et al., 2001). In *Deinococcus geothermalis*, during oxidative stress the NADPH homeostasis is modulated by FprA (Dgeo_1014), a NADP-ferredoxin reductase. This similar defense barrier is known to be present in *E. coli* during oxidative stress (Krapp et al., 2002). Zhang et al. (2000) reported that the formation of ROS in *D. geothermalis* leads to retrieval of NADPH from carbon substrate by rechanneling central carbon metabolism. It can be done by channeling glucose to pentose phosphate pathway so that it forms NADPH instead of NADH. The NADPH is then used to accelerate the ROS scavenging activity of superoxide dismutase (SodA), KatA, and thioredoxin which ultimately lead to detoxification of ROS. *D. geothermalis* also produces Mn dependent superoxide dismutase, catalase, several protein repair enzymes and chaperones and also convert carbon substrates to succinate for scavenging of ROS. Thus, *D. geothermalis* deals with ROS by using manganese dependent enzymes. However, in the absence of manganese other ROS neutralizing metabolites, produced by carbon substrates through central carbon metabolism are used to neutralize ROS (Liedert et al., 2012).

Mn(II) ions play a crucial role in detoxification of ROS by scavenging O_2^-, protecting proteins from oxidative stress and acting as cofactors of ROS scavenging enzymes like catalases and SODs (Jakubovich and Jenkinson, 2001; Daly et al., 2007). Fe-S clusters are the most prolific and versatile enzyme cofactors, but these clusters are also a prime target of ROS (Imlay, 2008). Nachin et al. (2003) reported that there is induction of Suf Fe-S assembly protein system in nongrowing cells of *D. geothermalis*. In oxidative stress, the proteins of this family enhance ATP-coupled assembly and repair of proteins with exposed [Fe-S]x clusters.

Liedert et al. (2012) reported that in the presence of manganese a new round of cell division is initiated in nongrowing cells of *D. geothermalis*. *D. geothermalis*, under oxidative stress expresses one class II fumarase and peptidylprolyl isomerase (Dgeo_0070), which is a protein repair enzyme and helps in reversing covalent damage to proline residues (Visick and Clarke, 1995). Thus, in nongrowing cells with limited synthesis capacity, maintenance, and repair of damaged proteins is an important metabolic activity which helps in the revival of damaged cells. In *Deinococcus* spp., the radiation resistance is primarily determined by protecting proteins from oxidative damage (Daly et al., 2007). Lesniak et al. (2003) reported that in *D. geothermalis*, the protection against oxidative stress is provided by Dgeo_0526, which is a homolog of osmotically inducible protein (OsmC). Moreover, it has been observed that following an oxidative stress the genes which encode for pyridoxine biosynthesis protein (PdxS) are upregulated. PdxS is important in synthesis of vitamin B6 which is an effective antioxidant and a quencher of singlet oxygen (Bilski et al., 2001; Liedert et al., 2012). *Deinococcus* and *Thermus* and possess different ROS detoxification mechanisms (Omelchenko et al., 2005). In *T. thermophilus*, there is one SodA and one Mn dependent catalase while *D. radiodurans* has Mn dependent SOD DR1279 and Cu/Zn dependent SODs, viz., DR 1546, DRA0202 and DR0644 (Makarova et al., 2001). Peptide methionine sulfoxide reductases (PMSRs) are present in *D. radiodurans* but absent in *T. thermophilus* which suggests that in *T. thermophilus* PMSRs are exchanged with analogous enzymes which help the organism to survive at high temperatures. *T. thermophilus* also lacks the proteins of Dps/Ferritin family and desiccation resistance proteins (Omelchenko et al., 2005).

In *Rubrobacter* spp., the damage induced by ROS is prevented by manganese-containing catalase (RradSPS_2184) which is encoded by a single gene and SOD is encoded by *sodA* (RradSPS_327) (Makarova et al., 2001; Terato et al., 2011; Basu and Apte, 2012). Peroxiredoxins are encoded by six copies of alkyl hydroperoxide reductase subunit C/Thiol specific antioxidant (AhpC/TSA) (RradSPS_0148, _0515, _988, _1124, _2530, _2650) (Basu and Apte, 2012). In strain RSPS-4, the genes encoding for catalase and SODs are predicted to be encoding manganese-containing enzymes and two ABC-type Mn^{2+}/Zn^{2+} transport systems (Yuan et al., 2012). The archaeal members *Thermococcus* spp. and *Pyrococcus* spp. are anaerobic hyperthermophiles which possess low Mn/Fe ratios as compared to their thermophilic radiation-resistant counterparts (Daly et al., 2007). This is in contrast with the model of Mn^{2+}-dependent ROS scavenging for aerobic bacteria and archaea (Daly et al., 2007; Robinson et al., 2011; Slade and Radman, 2011).

However, few proteins in anaerobic microorganisms need iron, such as dehydrogenases and ferredoxin, an electron carrier that *P. furiosus* uses instead of NAD (Jenney and Adams, 2008; Lancaster et al., 2011; Schut et al., 2012). *P. furiosus* contains peroxidases including rubrerythrin, alkyl hydroperoxide reductase I and II and a nonheme iron-containing enzyme called superoxide reductase (SOR) in place of SOD which catalyzes the reduction of O_2^- into H_2O_2 (Jenney et al., 1999; Strand et al., 2010). The formation of O_2^- is faster in the absence of O_2 in comparison to its presence, which takes place in the one-step process as a free electron (e^-) reacts with O_2 (Lin et al., 2005). Webb and Di Ruggiero (2012) reported that in anaerobic conditions, the ultrafilterates of *P. furiosus* and *T. gammatolerans* displayed increased protection. *T. gammatolerans* cope up with oxidative stress with the help of cascade of proteins or enzymes which include thioredoxin reductase (Tg0180), a glutaredoxin-like protein (Tg1302) and two peroxiredoxins (Tg1253, Tg1220) while DNA damage can be recovered by constitutively expressed nucleases Tg0864, Tg1177, Tg1631, ligases (Tg1718, Tg2005), endonucleases, AP-lyases, and glycosylases Tg0271, Tg0543, Tg1192, Tg1446, Tg1637, Tg1814 which are putatively involved in base excision repair and Tg0130, Tg0280, Tg1742, Tg1743, Tg1744 and Tg2074 are presumed to be involved in DSB repair (Zivanovic et al., 2009). In *P. furiosus*, following an irradiation and formation of H_2O_2 there is no increase in the expression of genes responsible for SOR pathway and these genes are normally expressed in anaerobic environments also which prove that SOR pathway can function efficiently all time (Williams et al., 2007; Strand et al., 2010; Schut et al., 2012). In anaerobic conditions, the low level of ROS formed during irradiation combine with constitutively expressed detoxification system. This system is efficiently used by hyperthermophiles like *P. furiosus* and *T. gammatolerans* to resist radiations. The dependence on this system is quite high as it circumvents the process of accumulation of Mn-antioxidant complexes in these hyperthermophilic archaea (Webb and Di Ruggiero, 2012).

Halobacterium salinarum NRC-1 (a haloarchaea) combat ROS with the help of metal halides, viz., NaBr, KCl, and KBr. These metal halides besides scavenging ROS also provide enhanced protection against carbonylation of protein residues and modification of nucleotides. *H. salinarum*

NRC-1 possess a high Mn/Fe ratio identical to that of *Deinococcus radio-durans* and other radiation-resistant microorganisms. Thus, radiation resistance in microorganisms is a combined function of mechanisms which provide cytoprotection, detoxification, and maintenance as well as repair of biomolecules (Kish et al., 2009).

10.3.3 OTHER MECHANISMS GOVERNING RADIATION RESISTANCE

The other mechanisms employed by microorganisms to survive in stress conditions like osmotic shock and radiations include accumulation of compatible solutes. Trehalose and mannoglycerate are accumulated by *Rubrobacter xylanophilus* and *R. radiotolerans* RSPS-4 to cope up with radiations (Empadinhas et al., 2007; Nobre et al., 2008). It has been reported that there is a constitutive accumulation of compatible solutes in response to various stress conditions which effect the survival of cells (Empadinhas et al., 2007).

   Chroococcidiopsis thermalis has the ability to tolerate 5kGy dose of IR as well as it is resistant to desiccation (Billi et al., 2001). This cyanobacterium can survive in a range of habitats like water deficit conditions, microscopic fissures of weathering rocks and form biofilms at the stone-soil interface under pebbles in desert pavements (Friedmann, 1993). For survival in dehydrated and high ionizing radiation-exposed environments, it accumulates trehalose, sucrose, and replaces water of cellular components by non-reducing sugars. It also forms exocellular polysaccharides which play central role in desiccation tolerance of cells by regulating loss and uptake of water (Potts, 1999; Hoiczyk and Hansel, 2000). Dried cells in dehydrating conditions exhibited deposition of acid-, sulfate-, and beta-linked polysaccharides in cell envelope (Caiola et al., 1996a, b). Fe-superoxide dismutase reduces risk of OH^- formation in *C. thermalis*. It has been speculated that *C. thermalis* resist gamma radiation due to its ability to cope up in a desiccated environment (Potts, 1999).

10.4 APPLICATIONS

Thermophilic radiation-resistant microorganisms have a wide range of applications ranging from bioremediation, astrobiology, and radiation therapies to basic and applied biology studies. Figure 10.4 presents an overview of different applications of radiation-resistant thermophiles.

US Department of Energy (DOE) originated the idea of engineered bioremediation where engineered strains of *D. radiodurans* were used for cleaning up radioactive environmental waste left from nuclear weapons (Macilwain, 1996). The engineering of *D. radiodurans* was significant to treat radionuclide wastes but the temperature restricted bioremediation process to 39°C as the contaminated waste was thermally insulated and the engineering of a radiation-resistant thermophile was essential to decay radionuclides like ^{137}Cs and ^{90}Sr (Lange et al., 1998; Brim et al., 2000, 2003; Daly, 2000). Brim et al. (2003) reported efficient expression of gene systems developed for *D. radiodurans* in thermophile *D. geothermalis* thus, gene technology developed for *D. radiodurans* was readily transferrable to *D. geothermalis*. This could be applied in cleaning up radioactive environments, which usually works at high temperatures. The wild and engineered *D. geothermalis* has the ability to reduce number of metals, viz., U(VI), Cr(VI), Hg(II), Tc(VII), Fe(III) and Mn(III, IV) so US DOE predicted that it can be used for designing of efficient bioremediation systems which include cleaning up of radioactive wastes sites (Brim et al., 2000, 2003, 2006; Daly, 2000; Fredrickson et al., 2000). It is proposed that *D. geothermalis* strain T27 can be used for cleanup of xenobiotic compounds from contaminated environments as this strain has the ability to survive at elevated concentrations of diethylpthalate, ethyl acetate and toluene (Kongpol et al., 2008). Other radiation-resistant thermophiles that have made their way to environmental cleaning technology are thermophilic bacteria like *Pyrobaculum islandicum*, *Tepidibacter thalassicus,* and *Thermoterrabacterium ferrireducens* and archaea like *Thermococcus pacificus* and *Thermoproteus uzoniensis* can reduce ^{99}Tc (VII) to insoluble Tc (IV) precipitates and thus immobilizing Tc, a β-emitting fission product of ^{235}U which can be hazardous as it gets accumulated in food chain (Chernyh et al., 2007; Sar et al., 2013). Thus, thermophilic radiation-resistant microbes, either in their wild type or recombinant form, are the best candidates for environmental cleanup of radioactive wastes contaminated sites (Chernyh et al., 2007; Sar et al., 2013).

The radiation-resistant thermophiles are finding an important place in astrobiological studies. The recovery of *Deinococcus phoenicis* from phoenix spacecrafts indicated that microbes are ubiquitous (Stepanov et al., 2014). Life in outer space encounters multiple stresses, viz., high IR and temperature variation from hot to cold or *vice versa*. The radiation-resistant thermophiles, being a robust organism can survive high temperature, IR, desiccation, and even reported to be brought back to life even after 250 million years indicated that life could be extraterrestrial in origin (Vreeland et al., 2000; Rampelotto, 2010). Thus, exploration of life that thrives at high temperatures may lead to exciting discoveries and would also pave the path for astrobiological studies to relate high-temperature radiation-resistant survivors as the candidates that originated life on Earth.

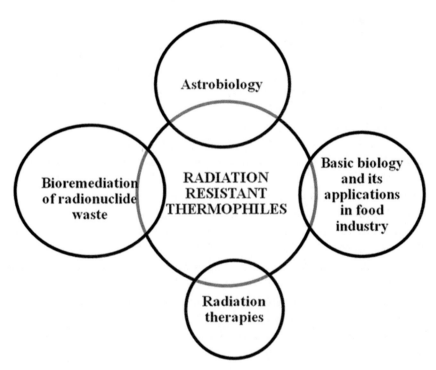

FIGURE 10.4   Applications of radiation-resistant thermophiles.
Source: Modified from Ranawat and Rawat, 2017b.

Compounds like amino acids, betain, sugar, and heteroside derivatives synthesized by extremophilic microorganisms help them to survive in high radiations (Letzen and Schwaz, 2006; Graf et al., 2008). These compounds are "extremolytes" synthesized by radiation-resistant microbes and can be used for pharmaceutical as well as biotechnological industries (Singh and Gabani, 2011). It is also hypothesized that extemolytes from ultra-violet radiation (UV-R)-resistant microorganisms can be employed for the development of anticancer drugs to prevent skin damage from UV-R as these molecules are inert in nature and also resist effects of radiations. Thus, metabolites from UV-R-resistant microorganisms could be a possible source of efficient human therapeutics which include anticancer drugs, anti-oxidants, and cell cycle blocking agents (Singh and Gabani, 2011). Pavlopoulou et al. (2016) reported that proteins of two families play a crucial role in enhancing radiation resistance. These include DNA histone-like DNA binding protein HU, which can condense DNA and form nucleoid-like structure and the other one is group of DNA repair proteins. The new insights into the radiation resistance biology can be provided by coupling analytical structural analysis with next-generation sequencing so that new parameters which confer radiation resistance in thermophilic microorganisms could be unveiled. These studies will help to develop ways and strategies to overcome tolerance to IR and improve radiation therapies.

Radiation resistant thermophiles like *D. geothermalis*, *D. radiodurans* and *Sulfolobus solfatricus* are studied for transfer and expression of genes in mesophiles like *E. coli* (Brim et al., 2003). *D. radiodurans* was engineered for bioremediation by cloning the *mer* operon of *E. coli* on plasmid. This plasmid cloned and expressed in *D. radiodurans* was successfully functional in *D. geothermalis,* which was able to grow at 50°C and could reduce mercury at high temperatures (Brim et al., 2003). The *aspATS* gene of thermophilic *Sulfolobus solfatricus,* which code for enzyme aspartate aminotransferase gene, was expressed in *E. coli* (Arnone et al., 1992). This suggests the complexity of mechanisms governing thermophilicity and radiation resistance and also provides evidence which suggests the exchange of genes between mesophilic and thermophilic microorganisms. Moreover, the study of radiation-resistant thermophiles can be effectively employed in the study of molecular switches like operons and regulon. There are platforms like "Radio P1" – a database of radiation resistance prokaryotes that provide a vast knowledge of radiation-resistant microorganisms, including thermophiles (Ben-Hamda et al., 2015). It presents an overview of predicted genes that are present during oxidative stress, DNA repair mechanisms, and potential uses of radiation-tolerant prokaryotes in biotechnology. These databases are important for the studies dedicated to evolutionary biology, therapeutics, and biotechnological applications of radiation-resistant thermophiles. This can be achieved by linking databases with the Kyoto Encyclopedia of Genes and Genomes (KEGG) and COGs (Cluster of Orthologs Genes), which will further analyze prokaryotes and categorize functional units from a new perspective (Ben-Hamda et al., 2015).

10.5 CONCLUSION

Thermophiles are the most interesting and most explored extremophiles. These microorganisms have drawn enormous attention from the scientific community due to their potential to produce extremozyes. However, the applications of thermophiles are not only restricted to the production of thermostable enzymes, but these microorganisms also have applications in bioremediation of heavy metals and radionuclides, textile effluents, antibiotic production, and so on. The discovery of thermophilic radiation-resistant members of genera *Deinococcus* and *Rubrobacter* has attracted the attention of the scientific world to understand the mechanisms which assist these microorganisms to deal with multiple stresses simultaneously. Radiation-resistant thermophiles, viz., *D. geothermalis*, *Rubrobacter* spp., are used as model microorganisms in various research laboratories to study the effects of cytotoxic compounds formed due to an exposure to radiations and cause damage to DNA and proteins. The understanding of mechanisms which govern radiation resistance in thermophiles has increased the application of these microorganisms and now they can be employed for bioremediation of environments which are thermally insulated and contaminated with radionuclide wastes. In the future, with the help of recombinant DNA technology, mesophilic microorganisms can be engineered by the transfer of radiation-resistant machinery of thermophiles, and thus, mesophiles can also be employed for cleanup of sites with radioactive wastes. The radiation resistance and high temperature in these microbes would provide an astrobiological link of thermophiles and pave the way for their possible use in radiation therapies to treat skin cancer. System biology-based approaches like genomics, proteomics, metagenomics, and metatranscriptomics are necessary to further develop the knowledge on the stress biology of thermophilic radiation-resistant microorganisms. This will open up horizons for exploring thermophiles and provide insights into radiation biology and radiation resistance in thermophiles.

KEYWORDS

- *Deinococcus geothermalis*
- *Deinococcus radiodurans*
- **desiccation resistance**
- **ionizing radiations**
- **metatranscriptomics**
- **proteomics**
- **reactive oxygen species**
- *Rubrobacter xylanophilus*
- *Sulfolobus solfatricus*
- *Taq* **polymerase**
- **thermophiles**
- *Thermus thermophilus*

REFERENCES

Ajon, M., Fröls, S., Van Wolferen, M., Stoecker, K., Teichmann, D., Driessen, A. J., Grogan, D. W., Albers, S. V., & Schleper, C. UV-inducible DNA exchange in hyperthermophilic archaea mediated by type IV pili. *Mol. Microbiol.*, **2011**, *82*, 807–817.

Albuquerque, L., Simões, C., Fernanda, N. M., Pino, N. M., Battista, J. R., Silva, M. T., Rainey, F., & Da Costa, M. S. *Truepera radiovictrix* gen. nov., sp. nov., a new radiation resistant species and proposal of *Trueperaceae* fam. nov. *FEMS Microbiology Letters*, **2005**, *247*, 161–169.

Arnone, M. I., Birolo, L., Cubellis, M. V., Nitti, G., Marino, G., & Sannia, G. Expression of hyperthermophilic aspartate aminotransferase in *Escherichia coli*. *Biochim. Biophys. Acta.*, **1992**, *1160*, 206–212.

Barrineau, P., Gilbert, P., Jackson, W. J., Jones, C. S., Summers, A. O., & Wisdom, S. The structure of *mer* operon. *Basic Life Sci.*, **1985**, *30*, 707–718.

Basu, B., & Apte, S. K. Gamma radiation-induced proteome of *Deinococcus radiodurans* primarily targets DNA repair and oxidative stress alleviation. *Mol. Cell Proteomics*, **2012**, *11*(1), M111.011734, doi: 10.1074/mcp.M111.011734.

Beblo, K., Douki, T., Schmalz, G., Rachel, R., Wirth, R., Huber, H., Reitz, G., & Rettberg, P. Survival of thermophilic and hyperthermophilic microorganisms after exposure to UV-C, ionizing radiation and desiccation. *Arch. Microbiol.*, **2011**, *193*, 797–809.

Beblo, K., Rabbow, E., Rachel, R., Huber, H., & Rettberg, P. Tolerance of thermophilic and hyperthermophilic microorganisms to desiccation. *Extremophiles*, **2009**, *13*, 521–531.

Belkova, N. L., Tazaki, K., Zakharova, J. R., & Parfenova, V. V. Activity of bacteria in water of hot springs from Southern and Central Kamchatskaya geothermal provinces, Kamchatka Peninsula, Russia. *Microbiol. Res.*, **2007**, *162*, 99–107.

Ben-Hamda, C., Benkahla, A., Ben, M. S., Ouled-Haddar, H., Montero-Calasanz, M. C., Gtari, M., Cherif, A., Hofner, B., Ghedira, K., & Sghaier, H. The RadioP1: An integrative web resource for radioresistant prokaryotes. In: Nenoi, M., (ed.), *Evolution of Ionizing Radiation Research* (pp. 89–105). InTech, Croatia, **2015**.

Billi, D., Friedmann, E. I., Helm, R. F., & Potts, M. Gene transfer to the desiccation tolerant cyanobacterium *Chroococcidiopsis*. *J. Bacteriol.*, **2001**, 2298–2305.

Bilski, P., Li, M. Y., Ehrenshaft, M., Daub, M. E., & Chignell, C. F. Vitamin B6 (pyridoxine) and its derivatives are efficient singlet oxygen quenchers and potential fungal antioxidants. *Photochem. Photobiol.*, **2000**, *71*, 129–134.

Blasius, M., Buob, R., Shevelev, I. V., & Hubscher, U. Enzymes involved in DNA ligation and end-healing in the radioresistant bacterium *Deinococcus radiodurans*. *BMC Mol. Biol.*, **2007**, *8*, 69, doi: 10.**1186**/1471–2199–8–69.

Bowater, R., & Doherty, A. J. Making ends meet: Repairing breaks in bacterial DNA by non-homologous end-joining. *PLoS Genet*, **2006**, *2*, e8, https://doi.org/10.1371/journal.pgen.0020008 (Accessed on 7 October 2019).

Brim, H., McFarlan, S. C., Fredrickson, J. K., Minton, K. W., Zhai, M., Wackett, L. P., & Daly, M. J. Engineering *Deinococcus radiodurans* for metal remediation in radioactive mixed waste environments. *Nat. Biotechnol.*, **2000**, *18*, 85–90.

Brim, H., Osborne, J. P., Kostandarithes, H. M., Fredrickson, J. K., Wackett, L. P., & Daly, M. J. *Deinococcus radiodurans* engineered for complete toluene degradation facilitates Cr(VI) reduction. *Microbiology*, **2006**, *152*, 2469–2477.

Brim, H., Venkateswaran, A., Kostandarithes, H. M., Fredrickson, J. K., & Daly, M. J. Engineering *Deinococcus geothermalis* for bioremediation of high-temperature radioactive waste environments. *Appl. Environ. Microbiol.*, **2003**, *69*, 4575–4582.

Byrne, R. T., Klingele, A. J., Cabot, E. L., Schackwitz, W. S., Martin, J. A., Martin, J., Wang, Z., Wood, E. A., Pennacchio, C., Pennacchio, L. A., Perna, N. T., Battista, J. R., & Cox, M. M. Evolution of extreme resistance to ionizing radiation via genetic adaptation of DNA repair. *Elife*, **2014**, *3*, e**0132**2, doi: 10.7554/eLife.01322.

Caiola, M. G., Billi, D., & Imre, F. E. Effect of desiccation on envelopes of the cyanobacterium *Chroococcidiopsis* sp. (Chroococcales). *Eur. J. Phycol.*, **1996a**, *31*, 97–105.

Caiola, M. G., Ocampo-Friedmann, R., & Friedmann, E. I. Cytology of long-term desiccation in the desert cyanobacterium *Chroococcidiopsis* sp. (Chroococcales). *Phycologia*, **1996b**, *32*, 315–322.

Capes, M. D., DasSarma, P., & DasSarma, S. The core and unique proteins of haloarchaea. *BMC Genomics*, **2012**, *13*, 39, https://doi.org/10.1186/1471–2164–13–39 (Accessed on 7 October 2019).

Carreto, L., Moore, E., Nobre, M. F., Waite, R., Riley, P. W., Sharp, R. J., & Da Costa, M. S. *Rubrobacter xylanophilus* sp. nov., a new thermophilic species isolated from a thermally polluted effluent. *Int. J. Syst. Bacteriol.*, **1996**, *46*, 460–465.

Cavicchioli, R., Amils, R., Wagner, D., & McGenity, T. Life and applications of extremophiles. *Environmental Microbiology*, **2011**, *13*(8), 1903–1907.

Chen, M. Y., Wu, S. H., Lin, G. H., Lu, C. P., Lin, Y. T., Chang, W. C., & Tsay, S. S. *Rubrobacter taiwanensis* sp. nov., a novel thermophilic, radiation-resistant species isolated from hot springs. *Int. J. Syst. Evol. Microbiol.*, **2004**, *54*, 1849–1855.

Chernyh, N. A., Gavrilov, S. N., Sorokin, V. V., German, K. E., Sergeant, C., Simonoff, M., Robb, F., & Slobodkin, A. L. Characterization of technetium (VII) reduction by cell suspensions of thermophilic bacteria and archaea *Appl. Microbiol. Biotechnol.*, **2007**, *76*, 467–472.

Cox, M. M., & Battista, J. R. *Deinococcus radiodurans* – the consummate survivor. *Nat. Rev. Microbiol.*, **2005**, *3*, 882–892.

Daly, M. J. A new perspective on radiation resistance based on *Deinococcus radiodurans*. *Nature Reviews*, **2009**, *7*, 237–245.

Daly, M. J. Engineering radiation resistant bacteria for environmental biotechnology. *Curr. Opin. Biotechnol.*, **2000**, *11*, 280–285.

Daly, M. J., Gaidamakova, E. K., Matrosova, V. Y., Vasilenko, A., & Zhai, M. Protein oxidation implicated as the primary determinant of bacterial radioresistance. *PLoS Biol.*, **2007**, *5*, 769–779.

Daly, M. J., Ouyang, L., Fuchs, P., & Minton, K. W. *In vivo* damage and *recA*-dependent repair of plasmid and chromosomal DNA in the radiation-resistant bacterium *Deinococcus radiodurans*. *J. Bacteriol.*, **1994**, *176*, 3508–3517.

De Vera, J. P., Horneck, G., Rettberg, P., & Ott, S. The potential of the lichen symbiosis to cope with extreme conditions of outer space–I. Influence of UV radiation and space vacuum on the vitality of lichen symbiosis and germination capacity. *Intern J. Astrobiol.*, **2003**, *1*, 285–293.

Di Giulio, M. The universal ancestor lives in a thermophilic or hyperthermophilic environment. *J. Theor. Biol.*, **2000**, *203*, 203–213.

Di Ruggiero, J., Santangelo, N., Nackerdien, Z., Ravel, J., & Robb, F. T. Repair of extensive ionizing-radiation DNA damage at 95C in the hyperthermophilic archaeon *Pyrococcus furiosus*. *J. Bacteriol.*, **1997**, *179*(14), 4643–4645.

Du, J., & Gebicki, J. M. Proteins are major initial cell targets of hydroxyl free radicals. *International Journal of Biochemistry and Cell Biology*, **2004**, *36*(11), 2334–2343.

Earl, A. M., Mohundro, M. M., Mian, I. S., & Battista, J. R. The IrrE protein of *Deinococcus radiodurans* R1 is a novel regulator of recA expression. *J. Bacteriol.*, **2002**, *184*, 6216–6224.

Egas, C., Barroso, C., Froufe, H. J. C., Pacheo, J., Albuquerque, L., & Da Costa, M. S. Complete genome sequence of the radiation-resistant bacterium *Rubrobacter radiotolerans* RSPS-4. *Standard in Genomic Sciences*, **2014**, *9*, 1062–1075.

Elderfield, H., & Schultz, A. Mid-ocean ridge hydrothermal fluxes and chemical composition of the ocean. *Annual Review of Earth and Planetary Sciences*, **1996**, *24*, 191–224.

Empadinhas, N., Mendes, V., Simões, C., Santos, M. S., Mingote, A., Lamosa, P., Santos, H., & Costa, M. S. Organic solutes in *Rubrobacter xylanophilus*: The first example of di-myo-inositol-phosphate in a thermophile. *Extremophiles*, **2007**, *11*, 667–673.

Englander, J., Klein, E., Brumfeld, V., Sharma, A. K., Doherty, A. J., & Minsky, A. DNA toroids: Framework for DNA repair in *Deinococcus radiodurans* and in germinating bacterial spores. *J. Bacteriol.*, **2004**, *186*, 5973–5977.

Erauso, G., Reysenbach, A. L., Godfroy, A., Meunier, J. R., Crump, B., Partensky, F., Baross, J. A., Marteinsson, V., Barbier, G., Pace, N. R., & Prieur, D. *Pyrococcus abyssi* sp. nov., a new hyperthermophilic archaeon isolated from a deep-sea hydrothermal vent. *Arch. Microbiol.*, **1993**, *160*, 338–349.

Ferreira, A. C., Nobre, M. F., Moore, E., Rainey, F. A., Battista, J. R., & Da Costa, M. S. Characterization and radiation resistance of new isolates of *Rubrobacter radiotolerans* and *Rubrobacter xylanophilus*. *Extremophiles*, **1999**, *3*, 235–238.

Ferreira, A. C., Nobre, M. F., Rainey, F. A., Silva, M. T., Wait, R., Burghardt, J., Chung, A. P., & Da Costa, M. S. *Deinococcus geothermalis* sp. nov. and *Deinococcus murrayi* sp. nov., two extremely radiation resistant and slightly thermophilic species from hot springs. *Int. J. Syst. Bacteriol.*, **1997**, *47*, 939–947.

Fisher, D. I., Cartwright, J. L., & McLennan, A. G. Characterization of the Mn^{2+}-stimulated (di)adenosine polyphosphate hydrolase encoded by the *Deinococcus radiodurans* DR2356 *nudix* gene. *Arch. Microbiol.*, **2006**, *186*, 415–424.

Fredrickson, J. K., Kostandarithes, H. M., Li, S. W., Plymale, A. E., & Daly, M. J. Reduction of Fe(III), Cr(VI), U(VI), and Tc(VII) by *Deinococcus radiodurans* R1. *Appl. Environ. Microbiol.*, **2000**, *66*, 2006–2011.

Fridovich, I. Superoxide radical and superoxide dismutases. *Annu. Rev. Biochem.*, **1995**, *64*, 97–112.

Friedmann, E. I. Extreme environments, limits of adaptation and extinction. In: Guerrero, R., & Pedros-Alio, (eds.), *Trends in Microbial Ecology* (pp. 8–12). Spanish Society for Microbiology, Barcelona, Spain, **1993**.

Fröls, S., Ajon, M., Wagner, M., Teichmann, D., Zolghadr, B., Folea, M., Boekema, E. J., Driessen, A. J., Schleper, C., & Albers, S. V. UV-inducible cellular aggregation of the hyperthermophilic archaeon *Sulfolobus solfataricus* is mediated by pili formation. *Mol. Microbiol.*, **2008**, *70*, 938–952.

Fröls, S., Gordon, P. M. K., Panlilio, M. A., Duggin, I. G., Bell, S. D., Sensen, C. W., & Schleper, C. Response of the hyperthermophilic archaeon *Sulfolobus solfataricus* to UV damage. *J. Bacteriol.*, **2007**, *189*, 8708–8718.

Gabbianelli, R., Battistoni, A., Polizio, F., Carri, M. T., Martino, A., Meier, B., & Desideri, R. G. Metal uptake of recombinant cambialistic superoxide dismutase from *Propionibacterium shermanii* is affected by growth conditions of host *Escherichia coli* cells. *Biochem. Biophys. Research Commun.*, **1995**, *216*, 841–847.

Gao, G., Le, D., Huang, L., Lu, H., Narumi, I., & Hua, Y. Internal promoter characterization and expression of the *Deinococcus radiodurans pprI-folP* gene cluster. *FEMS Microbiol. Lett.*, **2006**, *257*, 195–201.

Gehlani, A., Patel, R., Mangrola, A., & Dudhagara, P. Cultivation-independent comprehensive survey of bacterial diversity in Tulsi Shyam hot springs, India. *Genomics Data*, **2015**, *4*, 54–56.

Gérard, E., Jolivet, E., Prieur, D., & Forterre, P. DNA protection mechanisms are not involved in the radioresistance of the hyperthermophilic archaea *Pyrococcus abyssi* and *P. furiosus. Mol. Genet Genomics*, **2001**, *266*(1), 72–78.

Ghosal, D., Omelchenko, M. V., Gaidamakova, E. K., Matrosova, V. Y., Vasilenko, A., Venkateswaran, A., et al. How radiation kills cells: Survival of *Deinococcus radiodurans* and *Shewanella oneidensis* under oxidative stress. *FEMS Microbiol. Rev.*, **2005**, *29*, 361–375.

Ghosh, D., Bal, B., Kashyap, V. K., & Pal, S. Molecular phylogenetic exploration of bacterial diversity in a Bakreshwar (India) hot spring and culture of *Shewanella* related thermophiles. *Appl. Environ. Microbiol.*, **2003**, *69*, 7, 4332–4336.

Götz, D., Paytubi, S., Munro, S., Lundgren, M., Bernanderand, R., & White, M. F. Responses of hyperthermophilic crenarchaea to UV irradiation. *Genome Biol.*, **2007**, *8*, R220, doi: 10.1186/gb-2007–8–10-r220.

Graf, R., Anzali, S., Buenger, J., Pfluecker, F., & Driller, H. H. The multifunctional role of ecotine as a natural cell protectant. *Clinics in Dermatology*, **2008**, *26*, 326–333.

Grogan, D. W. The question of DNA-repair in hyperthermophilic archaea. *Trends Microbiol.*, **2000**, *8*, 180–185.

Harris, D. R., Pollock, S. V., Wood, E. A., Goiffon, R. J., Klingele, A. J., Cabot, E. L., et al. Directed evolution of ionizing radiation resistance in *E. coli. J. Bacteriol.*, **2009**, *191*, 16, 5240–5252.

Harris, D. R., Tanaka, M., Saveliev, S. V., Jolivet, E., Earl, A. M., Cox, M. M., & Battista, J. R. Preserving genome integrity: The DdrA protein of *Deinococcus radiodurans* R1. *PLoS Biol.*, **2004**, *2*(10), e304, doi: 10.1371/journal.pbio.0020304.

Hecker, M., & Völker, U. Non-specific, general and multiple stress resistance of growth-restricted *Bacillus subtilis* cells by the expression of the σB regulon. *Mol. Microbiol.*, **1998**, *29*, 1129–1136.

Henne, A., Briggemann, H., Raasch, C., Wiezer, A., Hartsch, T., Liesegang, H., Johann, A., et al. The genome sequence of the extreme thermophile *Thermus thermophilus*. *Nat. Biotechnol.*, **2004**, *22*, 547–553.

Hoiczyk, E., & Hansel, A. Cyanobacterial cell walls: News from an unusual prokaryotic envelope. *J. Bacteriol.*, **2000**, *182*, 1191–1199.

Horneck, G., Rettberg, P., Reitz, G., Wehner, J., Eschweiler, U., Strauch, K., Panitz, C., Starke, V., & Baumstark-Khan, C. Protection of bacterial spores in space, a contribution to the discussion on panspermia. *Orig. Life Evol. Biosph.*, **2001**, *31*, 527–547.

Hou, B., Xu, Z. W., Yang, C. W., Gao, Y., Zhao, S. F., & Zhang, C. G. Protective effects of inosine on mice subjected to lethal total-body ionizing irradiation. *J. Radiat. Res. (Tokyo)* **2007**, *48*, 57–62.

Hou, W., Wang, S., Dong, H., Jiang, H., Briggs, B. R., Peacock, J. P., Huang, Q., et al. A comprehensive census of microbial diversity in hot springs of Tengchong, Yunnan province China using 16S rRNA gene pyrosequencing. *PLoS One*, **2013**, *8*, e53350, 10.1371/journal.pone.0053350.

Hua, Y., Narumi, I., Gao, G., Tian, B., Satoh, K., Kitayama, S., & Shen, B. PprI: A general switch responsible for extreme radioresistance of *Deinococcus radiodurans*. *Biochem. Biophys. Res. Commun.*, **2003**, *306*, 354–360.

Huang, L., Hua, X., Lu, H., Gao, G., Tian, B., Shen, B., & Hua, Y. Three tandem HRDC domains have synergistic effect on the RecQ functions in *Deinococcus radiodurans*. *DNA Repair*, **2007**, *6*, 167–176.

Imlay, J. A. Cellular defenses against superoxide and hydrogen peroxide. *Annu. Rev. Biochem.*, **2008**, *77*, 755–776.

Imlay, J. A. Iron-sulfur clusters and the problem with oxygen. *Mol. Microbiol.*, **2006**, *59*, 4, 1073–1082.

Jakubovich, N. S., & Jenkinson, H. F. Out of iron age: New insights into the critical role of manganese homeostasis in bacteria. *Microbiol.*, **2001**, *147*, 1709–1718.

Jenney, F. E., & Adams, M. W. W. Hydrogenases of the model hyperthermophiles. *Annals of the New York Academy of Sciences*, **2008**, *1125*, 252–266.

Jenney, F. E., Verhagen, M. F. J. M., Cui, X., & Adams, M. W. W. Anaerobic microbes: Oxygen detoxification without superoxide dismutase. *Science*, **1999**, *286*(5438), 306–309.

Jolivet, E., Corre, E., L'Haridon, S., Forterre, P., & Prieur, D. *Thermococcus marinus* sp. nov., and *Thermococcus radiotolerans* sp. nov., two hyperthermophilic archaea from deep-sea hydrothermal vents that resist ionizing radiation. *Extremophiles*, **2004**, *8*, 219–227.

Jolivet, E., L'Haridon, S., Corre, E., Forterre, P., & Prieur, D. *Thermococcus gammatolerans* sp. nov., a hyperthermophilic archaeon from a deep-sea hydrothermal vent that resists ionizing radiation. *Int. J. Syst. Evol. Microbiol.*, **2003**, *53*, 847–851.

Jurado, V., Miller, A. Z., Alias-Villegas, C., Laiz, L., & Saiz-Jimenez, C. *Rubrobacter baracarensis* sp. nov., a novel member of the genus *Rubrobacter* isolated from a biodeteriorated monument. *Syst. Appl. Microbiol.*, **2012**, *35*, 5, 306–309.

Kecha, M., Benallaoua, S., Touzel, J. P., Bonaly, R., & Duchiron, F. Biochemical and phylogenetic characterization of novel terrestrial hyperthermophillic archaeon pertaining to the genus *Pyrococcus* from an Algerian hydrothermal hot spring. *Extremophiles*, **2007**, *11*, 65–73.

Kim, J. I., Sharma, A. K., Abbott, S. N., Wood, E. A., Dwyer, D. W., Jambura, A., Minton, K. W., Inman, R. B., Daly, M. J., & Cox, M. M. RecA Protein from the extremely radioresistant bacterium *Deinococcus radiodurans*: Expression, purification, and characterization. *J. Bacteriol.*, **2002**, *184*, 1649–1660.

Kish, A., Kirkali, G., Robinson, C., Rosenblatt, R., Jaruga, P., Dizdaroqlu, M., & DiRuggiero, J. Salt shield: intracellular salts provide cellular protection against ionizing radition in the halophilic archeon, *Halobacterium salinarum* NRC-1. *Environmental Microbiology*, **2009**, *11*(5), 1066–1078.

Komori, K., Miyara, T., DiRuggiero, J., Holley-Shanks, R., Hayashi, I., Cann, I. K., Mayanagi, K., Shingawa, H., & Ishino, Y. Both RadA and RadB are involved in homologous recombination in *Pyrococcus furiosus*. *J. Biol. Chem.*, **2000**, *275*(43), 33782–33790.

Kongpol, A., Kato, J., & Vangnai, A. S. Isolation and characterization of *Deinococcus geothermalis* T27, a slightly thermophilic and organic solvent-tolerant bacterium able to survive in the presence of high concentrations of ethyl acetate. *FEMS Microbiol. Lett.*, **2008**, *286*, 227–233.

Kottemann, M., Kish, A., Iloanusi, C., Bjork, S., & DiRuggiero, J. Physiological responses of the halophilic archaeon *Halobacterium* sp. strain NRC1 to desiccation and gamma irradiation. *Extremophiles*, **2005**, *9*, 219–227.

Krapp, A. R., Rodriguez, R. E., Poli, H. O., Paladini, D. H., Palatnik, J. F., & Carrillo, N. The flavoenzyme ferredoxin (flavodoxin)-NADP(H) Reductase modulates NADP(H) homeostasis during the *soxRS* response of *Escherichia coli*. *J. Bacteriol.*, **2002**, *184*, 1474–1480.

Krauss, I. R., Merlino, A., Pica, A., Rullo, R., Bertoni, A., Capasso, A., Amato, M., Riccitiello, F., De Vendittis, E., & Sica, F. Fine tuning of metal–specific activity in Mn-like group of cambialistic superoxide dismutases. *RSC Adv.*, **2015**, *107*, 87876–87887.

Kumar, R., Acharya, C., & Joshi, S. R. Isolation and analyses of uranium tolerant *Serratia marcescens* strains and their utilization for aerobic uranium U(IV) biosorption. *The Journal of Microbiology*, **2011**, *49*, 4, 568–574.

Kunkel, T. A., & Erie, D. A. DNA mismatch repair. *Annu. Rev. Biochem.*, **2005**, *74*, 681–710.

Kuzminov, A. Recombinational repair of DNA damage in *Escherichia coli* and bacteriophage lambda. *Microbiol. Mol. Biol. Rev.*, **1999**, *63*, 751–813.

Lancaster, V. L., LoBrutto, R., Selvaraj, F. M., & Blankenship, R. E. A cambialistic super-oxide dismutase in thermophilic photosynthetic bacterium *Chloroflexus aurantiacus*. *J. Bacteriol.*, **2004**, *186*, 11, 3408–3414.

Lancaster, W. A., Praissman, J. L., Poole, II F. L., Cvetkovic, A., Menon, A. L., Scott, J. W., et al. A computational framework for proteome-wide pursuit and prediction of metalloproteins using ICP-MS and MS/MS data. *BMC Bioinformatics*, **2011**, *12*, 64, http://www.biomedcentral.com/1471–2105/12/64 (Accessed on 7 October 2019).

Lange, C. C., Wackett, L. P., Minton, K. W., & Daly, M. J. Engineering a recombinant *Deinococcus radiodurans* for organopollutant degradation in radioactive mixed waste environments. *Nat. Biotechnol.*, **1998/2011**, *16*, 929–933.

Lesniak, J., Barton, W. A., & Nikolov, D. B. Structural and functional features of the *Escherichia coli* hydroperoxide resistance protein OsmC. *Protein Sci.*, **2003**, *12*, 2838–2843.

Letzen, G., & Schwaz, T. T. Extremolytes: Natural compounds from extremophiles for versatile applications. *Appl. Microbiol. Biotechnol.*, **2006**, *72*, 4, 623–634.

Levin-Zaidman, S., Englander, J., Shimoni, E., Sharma, A. K., Minton, K. W., & Minsky, A. Ringlike structure of the *Deinococcus radiodurans* genome: A key to radioresistance? *Science*, **2003**, *299*, 254–256.

Liedert, C., Peltola, M., Bernhardt, J., Neubauer, P., & Salkinoja-Salonen, M. Physiology of resistant *Deinococcus geothermalis* bacterium aerobically cultivated in low-manganese medium. *J. Bacteriol.*, **2012**, *194*, 1552–1561.

Lin, L. H., Slater, G. F., Sherwood, L. B., Lacrampe-Couloume, G., & Onstott, T. C. The yield and isotopic composition of radiolytic H_2, a potential energy source for the deep subsurface biosphere. *Geochimica et Cosmochimica Acta*, **2005**, *69*, 4, 893–903.

Liochev, S. I., & Fridovich, I. Fumarase C, the stable fumarase of *Escherichia coli*, is controlled by the *soxRS* regulon. *Proc. Natl. Acad. Sci. USA*, **1992**, *89*, 5892–5896.

Little, J. W., Mount, D. W., & Yanisch-Perron, C. R. Purified LexA protein is a repressor of the recA and lexA genes. *Proc. Natl. Acad. Sci. USA*, **1981**, *78*, 4199–4203.

Liu, Y., Zhou, J., Omelchenko, M. V., Beliaev, A. S., Venkateswaran, A., Stair, J., et al. Transcriptome dynamics of *Deinococcus radiodurans* recovering from ionizing radiation. *Proc. Natl. Acad. Sci. USA*, **2003**, *100*(7), 4191–4196.

Macilwain, C. Science seeks weapons clean-up role. *Nature*, **1996**, *383*, 375–379.

Makarova, K. S., Aravind, L., Daly, M. J., & Koonin, E. V. Specific expansion of protein families in the radioresistant bacterium *Deinococcus radiodurans*. *Genetica*, **2000**, *108*, 25–34.

Makarova, K. S., Aravind, L., Grishin, N. V., Rogozin, I. B., & Koonin, E. V. A DNA repair system specific for thermophilic archaea and bacteria predicted by genomic context analysis. *Nucleic Acids Res.*, **2002**, *30*(2), 482–496.

Makarova, K. S., Aravind, L., Wolf, Y. I., Tatusov, R. L., Minton, K. W., Koonin, E. V., & Daly, M. J. Genome of the extremely radiation-resistant bacterium *Deinococcus radiodurans* viewed from the perspective of comparative genomics. *Microbiol. Mol. Biol. Rev.*, **2001**, *65*, 44–79.

Makarova, K. S., Omelchenko, M. V., Gaidamakova, E. K., Matrosova, V. Y., Vasilenko, A., et al. *Deinococcus geothermalis:* The pool of extreme radiation resistance genes shrinks. *PLoS One*, **2007**, *2*(9), e955, doi: 10.1371/journal.pone.0000955.

Mandelli, F., Franco, C. J. P., Citadini, A. P., Bŭchli, F., Alvarez, T. M., Oliveira, R. J., Leite, V. B., Paes, L. A. F., Mercadante, A. Z., & Squina, F. M. The characterization of a thermostable and cambialistic superoxide dismutase form *Thermus filiformis*. *Lett. Appl. Microbiol.*, **2013**, *57*(1), 40–46.

Martins, A., & Shuman, S. An RNA ligase from *Deinococcus radiodurans*. *J. Biol. Chem.*, **2004**, *279*(49), 50654–50661.

Mattimore, V., & Battista, R. Radioresistance of *Deinococcus radiodurans:* Functions necessary to survive ionizing radiation are also necessary to survive prolonged desiccation. *J. Bacteriol.*, **1996**, *78*, 633–637.

Mehetre, G. T., Paranjpe, A. S., Dastager, S., & Dharne, M. S. Complete metagenome sequencing based bacterial diversity and functional insights from basaltic hot spring of Unkeshwar, Maharashtra, India. *Genomics Data*, **2015**, *7,* 140–143.

Mostertz, J., Scharf, C., Hecker, M., & Homuth, G. Transcriptome and proteome analysis of *Bacillus subtilis* gene expression in response to superoxide and peroxide stress. *Microbiology*, **2004**, *150*, 497–512.

Nachin, L., Loiseau, L., Expert, D., & Barras, F. SufC: An unorthodox cytoplasmic ABC/ATPase required for [Fe-S] biogenesis under oxidative stress. *EMBO J.*, **2003**, *22*, 427–437.

Narumi, I., Satoh, K., Cui, S., Funayama, T., Kitayama, S., & Watanabe, H. PprA: A novel protein from *Deinococcus radiodurans* that stimulates DNA ligation. *Mol. Microbiol.*, **2004**, *54*, 278–285.

Narumi, I., Satoh, K., Kikuchi, M., Funayama, T., Yanagisawa, T., Kobayashi, Y., Watanabe, H., & Yamamoto, K. The LexA protein from *Deinococcus radiodurans* is not involved in RecA induction following gamma irradiation. *J. Bacteriol.*, **2001**, *183*(23), 6951–6956.

Nazina, T. N., Lebedeva, E. V., Poltaraus, A. B., Tourova, T. P., Grigoryan, A. A., Sokolova, D. S., Lysenko, A. M., & Osipov, G. A. *Geobacillus gargensis* sp. nov., a novel thermophile from a hot spring, and the reclassification of *Bacillus vulcani* as *Geobacillus vulcani* comb. nov. *Int. J. Syst. Evol. Microbiol.*, **2004**, *54*, 2019–2024.

Nicholson, W. L. Ancient micronauts: Interplanetary transport of microbes by cosmic impacts. *Trends Microbiol.*, **2009**, *17*, 243–250.

Nicholson, W. L., Munakata, N., Horneck, G., Melosh, H. L., & Setlow, P. Resistance of *Bacillus* endospores to extreme terrestrial and extraterrestrial environments. *Microbiol. Mol. Biol. Rev.*, **2000**, *64*, 548–572.

Niederberger, T. D., Ronimus, R. S., & Morgan, H. W. The microbial ecology of a high temperature near neutral spring situated in Rotoura, New Zealand. *Microbiol. Res.*, **2008**, *163*(5), 594–603.

Nisbet, E. G., & Sleep, N. H. The habitat and nature of the early life. *Nature*, **2001**, *409*, 1083–1091.

Nobre, A., Alarico, S., Fernandes, C., Empadinhas, N., & Da Costa, M. S. A unique combination of genetic systems for the synthesis of trehalose in *Rubrobacter xylanophilus*: Properties of a rare actinobacterial TreT. *J. Bacteriol.*, **2008**, *190*, 7939–7946.

Nowosielska, A., & Marinus, M. G. DNA mismatch repair-induced double-strand breaks. *DNA Repair*, **2008**, *7*, 48–56.

Ohta, T., Tokishita, S. C., Mochizuki, K., Jun, K., Masahide, S., & Hideo, Y. UV sensitivity and mutagenesis of the extremely thermophilic eubacterium *Thermus thermophilus* HB27. *Genes and Environment*, **2006**, *28*, 2, 56–61.

Omelchenko, M. V., Wolf, Y. I., Gaidamakova, E. L., Matrosova, V. Y., Vailenko, A., Zhai, M., Daly, M. J., Koonin, E. V., & Makarova, K. S. Comparative genomics of *Thermus thermophilus* and *Deinococcus radiodurans:* Divergent routes of adaptation to thermophily and radiation resistance. *BMC Evolutionary Biology*, **2005**, *5*, 57, doi: 10.1186/1471-2148-5-57.

Pavlopoulou, A., Savva, G. D., Louka, M., Bagos, P. G., Vorgias, C. E., Michalopoulos, I., & Georgakilas, A. G. Unraveling the mechanisms of extreme radioresistance in prokaryotes: Lessons from nature. *Mutation Research/Reviews in Mutation Research*, **2016**, *767*, 92–107.

Potts, M. Mechanisms of desiccation tolerance in cyanobacteria. *Eur. J. Phycol.*, **1999**, *34*, 319–328.

Prestrelski, S. J., Tedeschi, N., Arakawa, T., & Carpentert, J. F. Dehydration-induced conformational transitions in proteins and their inhibition by stabilizers. *Biophys. J.*, **1993**, *65*, 661–671.

Rampelotto, P. H. Resistance of microorganisms to extreme environmental conditions and its contribution to astrobiology. *Sustainability*, **2010**, *2*(6), 1602–1623.

Rampelotto, P. H., Rosa, M. B., Schuch, A. P., Pinheiro, D. K., Schuch, N. J., & Munakata, N. Exobiology research in the south of Brazil aiming the monitoring of the biogenically-effective solar radiation. *Astrobiology*, **2007**, *7*, 502–540.

Rampelotto, P. H., Rosa, M. B., Schuch, N. J., & Munakata, N. Exobiological application of spore dosimeter in studies involving solar UV radiation. *Orig. Life Evol. Biosph.*, **2009**, *39*, 373–374.

Ranawat, P., & Rawat, S. Radiation resistance in thermophiles: Mechanisms and applications. *World J. Microbiol. Biotechnol.*, **2017b**, *33*(112), 1–22.

Ranawat, P., & Rawat, S. Stress response physiology of thermophiles. *Arch. Microbiol.*, **2017a**, *199*, 391–414.

Reysenbach, A. L., Ehringer, M., & Hershberger, K. Microbial diversity at 83°C in the calcite springs, Yellowstone National Park: Another environment where the Aquificales and "Korarchaeota" coexist. *Extremophiles*, **2000**, *4*, 61–67.

Reysenbach, A. L., Wickham, G. S., & Pace, N. R. Phylogenetic analysis of the hyperthermophilic pink filament community in Octopus Spring, Yellowstone National Park. *Appl. Environ. Microbiol.*, **1994**, *60*(6), 2113–2119.

Robinson, C. K., Webb, K., Kaur, A., Jaruga, P., Dizdaroglu, M., Baliga, N. S., Place, A., & Di Ruggiero. A major role for nonenzymatic antioxidant processes in the radioresistance of *Halobacterium salinarum*. *J. Bacteriol.*, **2011**, *193*(7), 1653–1662.

Rosa, M. B., Rampelotto, P. H., Schuch, N. J., Schuch, A. P., & Munakata, N. Spore dosimetry: *Bacillus subtilis* TKJ6312 as biosensor of biologically effective solar radiation. *Quim. Nova.*, **2009**, *32*, 282–285.

Sar, P., Kazy, S., Paul, B., & Sarkar, A. Metal bioremediation by thermophilic microorganisms. In: Satyanarayan, T., Littllechild, J., & Kawarabayasi, Y., (eds.), *Thermophilic Microbes in Environment and Industrial Biotechnology: Biotechnology of Thermophiles* (pp. 171–201). Springer, New York, **2013**.

Schaffer, C., Franck, W. L., Scheberl, A., Kosma, P., McDermott, T. R., & Messner, P. Classification of isolates from locations in Austria and Yellowstone National Park as *Geobacillus tepidamans* sp. nov. *Int. J. Syst. Evol. Microbiol.*, **2004**, *54*, 6, 2361–2368.

Schut, G. J., Nixon, W. J., Lipscomb, G. L., Scott, R. A., & Adams, M. W. Mutational analyses of the enzymes involved in the metabolism of hydrogen by the hyperthermophilic archaeon *Pyrococcus furiosus*. *Frontiers in Microbiology*, **2012**, *3*, 163, doi: 10.3389/fmicb.2012.00163.

Sharma, A., Jani, K., Souche, Y. S., & Pandey, A. Microbial diversity of Soldhar, hot spring, India, assessed by analyzing 16S rRNA and protein coding genes. *Annl. Microbiol.*, **2015**, *65*(3), 1323–1332.

Singh, A., & Subudhi, E. Structural insights of microbial community of Deulajhari (India) hot spring using 16S-rRNA based metagenomic sequencing. *Genomics Data*, **2016**, *7*, 101–102.

Singh, O. V., & Gabani, P. Extremophiles: Radiation resistance microbial reserves and therapeutic implications. *J. Appl. Microbiol.*, **2011**, *110*, 4, 851–861.

Slade, D., & Radman, M. Oxidative stress resistance in *Deinococcus radiodurans*. *Microbiol. Mol. Biol. Rev.*, **2011**, *75*(1), 133–191.

Slobodkin, A., Reysenbach, A. L., Strutz, N., Dreier, M., & Wiegel, J. *Thermoterrabacterium ferrireducens* gen. nov., sp. nov., a thermophilic anaerobic dissimilatory Fe(III)-reducing bacterium from a continental hot spring. *Int. J. Syst. Bacteriol.*, **1997**, *47*(2), 541–547.

Soppa, J. Evolutionary advantages of polyploidy in halophilic archaea. *Biochem. Soc. Trans.*, **2013**, *41*, 339–343.

Soppa, J. Polyploidy in archaea and bacteria: About desiccation resistance, giant cell size, long-term survival, enforcement by a eukaryotic host and additional aspects. *J. Mol. Microbiol. Biotechnol.*, **2014**, *24*, 409–419.

Stadtman, E. R., & Levine, R. L. Free radical-mediated oxidation of free amino acids and amino acid residues in proteins. *Amino Acids*, **2003**, *25*(3/4), 207–218.

Stepanov, V. G., Vaishampayan, P., Venkateswaran, K., & Fox, G. E. Draft genome sequence of *Deinococcus phoenicis*, a novel strain isolated during the phoenix lander spacecraft assembly. *Genome Announc.*, **2014**, *2*(2), e00301–14, doi: 10.1128/genomeA.00301–14.

Strand, K. R., Sun, C., Li, T., Jenney, F. E., Schut, G. J., & Adams, M. W. W. Oxidative stress protection and the repair response to hydrogen peroxide in the hyperthermophilic archaeon *Pyrococcus furiosus* and in related species. *Arch Microbiol.*, **2010**, *192*, 6, 447–459.

Tanaka, M., Earl, A. M., Howell, H. A., Park, M. J., Eisen, J. A., Peterson, S. N., & Battista, J. R. Analysis of *Deinococcus radiodurans*'s transcriptional response to ionizing radiation and desiccation reveals novel proteins that contribute to extreme radioresistance. *Genetics*, **2004**, *168*, 21–33.

Terato, H., Suzuki, K., Nishioka, N., Okamoto, A., Shimazaki-Tokuyama, Y., Inoue, Y., & Saito, T. Characterization and radio-resistant function of manganese superoxide dismutase of *Rubrobacter radiotolerans*. *J. Radiat. Res. (Tokyo)* **2011**, *52*, 735–742.

Touati, E., Laurent-Winter, C., Quillardet, P., & Hofnung, M. Global response of *Escherichia coli* cells to a treatment with 7-methoxy-2-nitronaphtho[2,1-*b*]furan (R7000), an extremely potent mutagen. *Mutat. Res.*, **1996**, *349*, 193–200.

Uemori, T., Ishino, Y., Toh, H., Asada, K., & Kato, I. Organization and nucleotide sequence of the DNA polymerase gene from *Pyrococcus furiosus*. *Nucl. Acid Res.*, **1993**, *21*, 259–265.

Urmania, V. V. Bioremediation of toxic heavy metals using acidothermophilic autotrophies. *Bioresour. Technol.*, **2006**, *97*(10), 1237–1242.

Venkateswaran, A., McFarlan, S. C., Ghosal, D., Minton, K. W., Vasilenko, A., Makarova, K., Lawrence, P. W., & Daly, M. J. Physiologic determinants of radiation resistance in *Deinococcus radiodurans*. *Appl. Environ. Microbiol.*, **2000**, *66*, 2620–2626.

Visick, J. E., & Clarke, S. Repair, refold, recycle: how bacteria can deal with spontaneous and environmental damage to proteins. *Mol. Microbiol.*, **1995**, *5*, 835–845.

Von Sontaag, C. *The Chemical Basis of Radiation Biology*. Taylor and Francis, London, **1987**.

Vreeland, R. H., Rosenzweig, W. D., & Powers, D. W. Isolation of a 250 million year-old halotolerant bacterium from a primary salt crystal. *Nature*, **2000**, *407*, 897–900.

Wanarska, M., Krawczyk, B., Hildebrandt, P., & Kur, J. RecA proteins from *Deinococcus geothermalis* and *Deinococcus murrayi*-cloning, purification and biochemical characterization. *BMC Molecular Biology*, **2011**, *12*(17), 1–13.

Webb, K. M., & DiRuggiero, J. Role of Mn^{2+} and compatible solutes in the radiation resistance of thermophilic bacteria and archaea. *Archaea*, **2012**, 845756, http://dx.doi.org/10.1155/2012/845756 (Accessed on 7 October 2019).

Weisburg, W. G., Giovannoni, S. J., & Woese, C. R. The *Deinococcus-Thermus* phylum and the effect of rRNA composition on phylogenetic tree construction. *Syst. Appl. Microbiol.*, **1989**, *11*, 128–134.

Williams, E., Lowe, T. M., Savas, J., & DiRuggiero, J. Microarray analysis of the hyperthermophilic archaeon *Pyrococcus furiosus* exposed to gamma irradiation. *Extremophiles*, **2007**, *11*(1), 19–29.

Wright, M. H., Patel, B. K. C., & Greens, A. C. Thermophilic bacteria from Paralana hot springs in the Northern Flinders ranges of South Australia. *Australian Society for Microbiology Annual Scientific Meeting*, Brisbane, Australia, **2012**.

Youn, H. D., Kim, E. J., Roe, J. H., Hah, Y. C., & Kang, S. O. A novel nickel containing superoxide dismutase from *Streptomyces* spp. *Biochem. J.*, **1996**, *318*, 889–896.

Yuan, M., Chen, M., Zhang, W., Lu, W., Wang, J., Yang, M., et al. Genome sequence and transcriptome analysis of the radioresistant bacterium *Deinococcus gobiensis*: Insights into the extreme environmental adaptations. *PLoS One*, **2012**, *7*(3), e34458, doi: 10.1371/journal.pone.0034458.

Zhang, Y. M., Liu, J. K., & Wong, T. Y. The DNA excision repair system of the highly radioresistant bacterium *Deinococcus radiodurans* is facilitated by the pentose phosphate pathway. *Mol. Microbiol.*, **2003**, *48*, 1317–1323.

Zhang, Y. M., Wong, T. Y., Chen, L. Y., Lin, C. S., & Liu, J. K. Induction of a futile Embden-Mayerhof-Parnas pathway in *Deinococcus radiodurans* by Mn: Possible role of the pentose phosphate pathway in cell survival. *Appl. Environ. Microbiol.*, **2000**, *1*, 105–112.

Zivanovic, Y., Armenqaud, J., Laqorce, A., Leplat, C., Guérin, P., Dutertre, M., Anthouard, V., Forterre, P., Wincker, P., & Confalonieri, F. Genome analysis and genome-wide proteomics of *Thermococcus gammatolerans*, the most radioresistant organism known amongst the archaea. *Genome Biol.*, **2009**, *10*, 6, R70, doi: 10.1186/gb-2009-10-6-r70.

CHAPTER 11

DYNAMIC POTENTIAL OF INDIGENOUS AND EFFECTIVE MICROBES IN WASTEWATER TREATMENT PROCESSES

JEYABALAN SANGEETHA,[1]
SHIVASHARANA CHANDRABANDA THIMMAPPA,[2]
DEVARAJAN THANGADURAI,[3]
MEGHA RAMACHANDRA SHINGE,[2] ABHISHEK MUNDARAGI,[4]
RAVICHANDRA HOSPET,[3] SIMMI MAXIM STEFFI,[1]
MOHAMMED ABDUL MUJEEB,[2] and
PRATHIMA PURUSHOTHAM[3]

[1]Department of Environmental Science, Central University of Kerala, Periye, Kasaragod–561716, Kerala, India

[2]Department of Microbiology and Biotechnology, Karnatak University, Dharwad–580003, Karnataka, India

[3]Department of Botany, Karnatak University, Dharwad–580003, Karnataka, India

[4]Department of Microbiology, Davangere University, Davangere–577002, Karnataka, India

11.1 INTRODUCTION

As the world population grows and many developing countries modernize, the significance of water supply and wastewater treatment becomes a much greater area in the welfare of nations. Nowadays, the competition for water sources combines with adverse amalgamates of wastewater releases with freshwater supplies makes subsidiary challenge on wastewater treatment systems (Aziz and Mojiri, 2014). Recently, scientists focus on wastewater treatment by various methods with minimal cost and maximum efficiency. Rapid industrialization and urbanization release large volumes of waste effluents, which are drastically utilized as a potent resource for agriculture and irrigation in urban and rural areas. Wastewater treatment leads significant economic growth, and a healthy environment assists countless livelihoods mainly of farmers and substantially improves the quality of natural water bodies (Marshall et al., 2007; Rajasulochana and Preethy, 2016). It is noted that highly polluted and drastic effects of consumption of such polluted water and its sanitation problem are increasing gradually in most developing countries. This leads to water scarcity and makes a negative influence on human livelihoods, economic development, and environmental sustainability around the world. Hence, there is an urgent need to address the seriousness of the issue to protect water bodies from getting polluted and to develop advanced, cost-effective techniques for the protection and sustainable utilization of natural water resources (Kaur et al., 2012; Shivajirao, 2012; Rajasulochana and Preethy, 2016).

The importance and necessity of industrial and municipal wastewater treatment have become more apparent in program to conserve and protect vital water resources. A current challenge to professionals in the wastewater treatment field is the potential utilization of effective microbial sources for municipal and industry-based wastewater treatment in the soundest and cost-effective manner. It is investigated that nearly 1.1 billion people drink unsafe water all over the world (Rajasulochana and Preethy, 2016). As per the reports of the World Bank, around 21% of the communicable diseases in India, are from waterborne (Marshall et al., 2007; Aziz and Mojiri, 2014). The majority of microbial pathogens observed in wastewater treatment plants (WWTPs) are bacteria, viruses, protozoa, algae, fungi, and helminths. The existence of these pathogens in water leads to the spread of various waterborne diseases. The diverse microbial pathogens in wastewater can cause severe chronic diseases with long-term effects, such as stomach ulcer, typhoid, gastroenteritis, and degenerative heart disease. The population density and diversity of these microbial pollutants can vary depending on the intensity and prevalence of infection in sewered community (Akpor et al., 2014). The detection, isolation, and identification of the different types of microbial pathogens in wastewater are always laborious, time-consuming, and expensive. To avoid this, indicator organisms are always used to analyze the relative risk of the possible presence of specific pathogen in wastewater (Paillard et al., 2005; Akpor et al., 2014). The major chemical pollutants constitute in wastewater are phosphorus and nitrogen. It is observed that pesticides, detergents, and heavy metals are the most persistent limiting agents in the process of eutrophication (Schultz, 2005; Thawale et al., 2006).

Generally, the process for removal of impurities in wastewater is characterized into chemical and biological process. Chemical removal techniques include coagulation/flocculation, chloramination, chlorination, ozonation, and ultraviolet (UV) light treatment. These techniques involve addition of chemicals to form particles which settle and remove contaminants. The treated water is then decanted and appropriately reused or disposed of after resultant sludge is dewatered to reduce the volume. In the case of biological-wastewater treatment processes, the potential of the microbial community utilized for various wastewater constituents to enhance the microbial metabolism and cell synthesis must be removed before discharge. This microbial metabolic process can effectively eliminate contaminants that are as diverse as raw substances and by-products (Schultz, 2005; Akpor et al., 2014). Many conventional technologies available for wastewater treatments are present for decades (Narmadha and Mary Selvam Kavitha, 2012; Rajasulochana and Preethy, 2016), but the effectiveness and flexibility of the technology are still challenged. The advancement in new green methods and usage of microbial populations are being popularized to overcome the less effective traditional methods of wastewater treatment (Shivajirao, 2012; Kumar and Sai Gopal, 2015).

Sustainable environmental utilization and conservation have the foremost significance in the present life of mankind. Scientists have been developing new technologies and techniques for the improvement of wastewater management in agricultural and industrial waste. Indigenous microorganisms (IMOs) and effective microorganisms (EMOs) based on technology, which influenced effectively in the eastern part of the world for the enhancement of wastewater management. IMOs are the class of an innate microbial consortium which has potent ability in biodegradation, biocomposting, bioleaching, and nitrogen fixation. EMOs shows specificity in their functional activity and engineered with respect to specific needs for the effective applications in various areas of biological sciences. The advancement in the wastewater treatment technology leads to usage of these IMOs and EMOs as potential microbial tools for wastewater treatment (Aziz and Mojiri, 2014). Industrial wastewater treatment standards are concerned with the removal of suspended solids, biodegradable organics, and pathogens. Any of the more stringent standards that have been developed recently deal with the removal of nutrients and opportunist pollutants. The significant physical factors of the wastewater are the total solids content, which include colloidal matter, settleable matter, and floating matter (Paillard et al., 2005; Kumar and Sai Gopal, 2015). The municipal wastewater is composed mainly of anthropogenic and agricultural wastes that are rich in nutritional supplements such as carbon, nitrogen, and phosphorus. In the meanwhile, it is noted that the cost of biological treatment of wastewater is increasing worldwide due to the population growth in urban and semi-urban areas, such problems have to address by adopting advanced, cost-effective wastewater treatment systems (Aziz and Mojiri, 2014).

The elemental reason for the wastewater treatment is to avoid the impact of water pollution and to protect natural water bodies to get contaminated by hazardous pollutants, thereby safeguarding public health through effective wastewater management against the spread of diseases. This is performed through series of wastewater treatment methods such as activated sludge, stabilization ponds, trickling filters, constructed wetlands, and membrane bioreactors wastewater treatment systems (Haandel and Lubbe, 2007; Akpor et al., 2014). The activated sludge process is one of the efficient bio-nutrient removal methods incorporated in WWTPs. Activated sludge is comprised of various microbial consortia, among which bacterial populations exhibit potent activities in the wastewater treatment process (Aziz and Mojiri, 2014). Microalgae or microphytes are the microscopic algae able to perform photosynthesis similar to plants and reproduce quickly with available nutrients such as phosphorus, nitrogen, and CO_2 from their environment. As the outcome of photosynthesis, released oxygen is utilized by activated sludge bacteria through the microalgae activated sludge (MAAS) process (Anbalagan et al., 2016; Huijun and Qiuyan, 2016).

The quality and reproducibility of wastewater effluents are responsible for the degradation of receiving water sources, such as rivers, lakes, and streams. Microbial populations are of major importance in treating municipal and industrial wastewater. The application of microorganisms has positive effects on the outcome of aquaculture and irrigation operations. The efficient microbial activities include the removal of toxic materials such as nitrite, ammonia, hydrogen sulfide, and degradation of uneaten feed (Lu et al., 2009; Huijun and Qiuyan, 2016). These and other functions make microorganisms the key players in the health and sustainability of water bodies. The significant role of various microbial consortium application in the wastewater treatment with specific importance of bacteria and protozoa in removal of nitrogen, phosphorus, and other pollutants observed that microbial mediated wastewater treatment system is very effective and essential to challenge water pollutants especially from municipal and industrial origins (Thawale et al., 2006; Templeton and Butler, 2011; Huijun and Qiuyan, 2016).

11.2   INDUSTRIAL AND MUNICIPAL WASTEWATER: SOURCES AND TREATMENT PROCESSES

Environmental pollution due to direct discharge of wastewater into water bodies is a major concern as the components and contaminants of waste are having a hazardous effect on public health and the ecological system. Rapid industrialization and anthropogenic activities are the major sources of contaminants of wastewater (Abdel-Raouf et al., 2012). However, as the industrial wastewater and municipal wastewater have different sources and also various contaminants, wastewater treatment strategies are also different, depends on the properties of wastewater to be treated (Templeton and Butler, 2011). Energy required for the treatment process, sludge generation and disposal strategies, complexity of the process, strength, and adaptability of the treatment, cost-effectiveness, and need of manpower are the important factors directed at the effective process of selection and designing. Though the sources and pollutants of industrial and municipal wastewater are different, they involve overall principles of unit operations and unit processes for the removal of contaminants from wastewater (EPA, 1997; Akpor et al., 2014).

11.2.1   UNIT OPERATIONS IN WASTEWATER TREATMENT

The unit operation consists of physical operations required to eliminate the contaminants present in the water (Kapagiannidis et al., 2012). The screening process involves uses of different types of coarse, medium, and fine screens meant for removal of larger substances from wastewater like rags, paper, plastics, and metals. The screening method is essential in preventing clogging and damage of downstream equipment and processes (Templeton and Butler, 2011). Grit removal, also called primary clarification for both industrial and municipal wastewater, contains different types of grit as domestic and food waste, organic, inorganic, and heavier solid materials (Narmadha and Mary Selvam Kavitha, 2012). Grit removal facilities usually precede primary clarification and monitor screening.

Quantity and characteristics of grit are crucial factors in the selection of the existing type of grit removal system, for example, aerated grit chambers, vortex or paddle grit, detritus tanks, horizontal flow grit chamber and hydrocyclones (Tchobanoglous et al., 2014).

11.2.2   UNIT PROCESSES IN WASTEWATER TREATMENT

11.2.2.1   MICROBE MEDIATED BIOLOGICAL WASTEWATER TREATMENT PROCESS

These set of processes involves biological and chemical conversion of waste. Furthermore, secondary sewage treatment process includes series of unit processes that facilitates biological degradation of sewage with the use of microorganism. In the biological waste treatment process, microorganisms usually use aerobic metabolism to degrade organic pollutants in the liquid sludge (Nielsen et al., 2009, 2010). The objective of microbial degradation of sewage aimed at forcing the resident microbial flora of sewage to degrade organic pollutants for safe treated effluent discharge. The resident microbial flora of sewage plays an important role in effective biological treatment process (Siezen and Galardini, 2008). Understanding the ecology of sewage to explore variety of microorganism's resident of sewage can play crucial role in designing sewage treatment plants and to carryout effective biodegradation of sewage. Raw sewage contains high biodegradable organic compounds and rich in nutrients, signifies excellent medium to favor the growth of microorganisms (Frigon et al., 2013). Variety of microbes have a different mode of metabolism and decomposing the ability of different pollutants, which allows degradation of different organic waste at once in a treatment plant. Microbes may follow the anaerobic or aerobic mode of waste degradation. The rate of microbial growth in a particular reactor directly depends on the amount of organic waste present in sewage, and BOD indicates the quality of treated wastewater (Chudoba et al., 1992).

11.2.2.2   ACTIVATED SLUDGE PROCESS

The activated sludge process includes the treatment of sewage using biological flocs of bacteria and protozoa and aeration. In this process, dissolved organic compounds are oxidized for nutrient uptake. Sludge is aerated in aeration tank where microorganisms metabolize carbonaceous waste, part of this used to synthesize new cells, and part is oxidized into CO_2 and water to derive energy (Ahansazan et al., 2014). Cell biomass formed in the reaction are separated from the liquid stream in the form of flocculent sludge while settled biomass in the form of activated sludge returned to aeration tank, remaining forms waste or surplus sludge. A part of activated sludge, the nitrogenous matter is mainly oxidized into ammonia, and nitrogen also added in removing nitrogen and phosphorous (Tadkaew et al., 2010).

11.2.2.3   ATTACHED GROWTH PROCESSES

Hybridized growth processes, in combination with activated sludge and attached growth system, are incorporated into unit processes in the treatment of sewage. The attached growth system includes rotating biological contactors (RBC), packed beds, or suspended carrier materials. In general, these systems contain fixed-film components, which safeguards disturbance in the process systems (Gavrilescu and Macoveanu, 2000). An increase in hydrostatic pressure, expulsion of toxic compounds into the system, or breakdown in the aeration system are the demerits of process disturbance. Process disturbance takes account of the principle types of attached growth membranes (Marti et al., 2011).

11.2.2.3.1   Continuously Flowing Suspended Systems With Fixed Film Growth Membrane

These systems are composed of plastic material and are possible to unite with fixed packing in many cases to maintain unique flow patterns. It enhances microbial growth in the fixed film, which is in contact with wastewater under treatment (Han et al., 2005).

11.2.2.3.2   Continuous Flow Suspended With Suspended Internal Packing

These materials designed to float or sink depending on the specificity of the process. These systems enhance the overall interface of attached microbial growth by circulating it into the water treatment column (Marti et al., 2011).

11.2.2.3.3   Sequencing Batch Reactors (SBRs) With Internal Packing

In this reactor system, sequential batches of wastewater stored for a period and treated. Batch under treatment seeded with dynamic bacterial culture and oxygenized from the treatment period. This reaction ends with the settling of flocculated bacteria and other solids. The supernatant transferred to the next chamber, the next batch begins, and the cycle repeats continuously (Fang et al., 1993).

11.2.2.3.4   Rotating Biological Contactors (RBC)

It consists of a rotating disc, hence part of fixed film partially submerged per rotation into the flow to react with wastewater under treatment. Process optimization is possible to achieve by adjusting the speed per rotation and depth of submergence (Tawfik et al., 2006; Hassard et al., 2014).

11.2.3   CHEMICAL AIDED UNIT PROCESSES

Chemical treatment processes are utilized to eliminate dissolved inorganic pollutants leftover in secondary biological treatment processes. These processes aimed at the removal of inorganic pollutants from wastewater and maintaining water quality for its safe discharge into the environment (Schuler et al., 2001). Chemical contaminants of water can be removed by changing the temperature or by precipitation as solids carried out by the addition of acid or alkali in the process of chemical precipitation. Maintaining the required pH of the water is the main objective of neutralization, addition of lime, calcium hydroxide, sodium hydroxide, and sodium carbonate are the most common chemicals in use to adjust pH at an acceptable range. Disinfection is the last process of unit processes required for wastewater treatment, and this treatment is conducted by treating the effluent with the disinfectant to inactivate pathogens such as microbes, viruses, and protozoan and to meet the wastewater discharge standards. The ideal disinfectant should be less harmful, and with bacterial toxicity, and should have reliable means of detecting the presence of residues. Chemical disinfectant includes chlorine, ozone, ultraviolet radiation (UV-R), chlorine dioxide, and bromine (Fayza et al., 2007) (Figure 11.1).

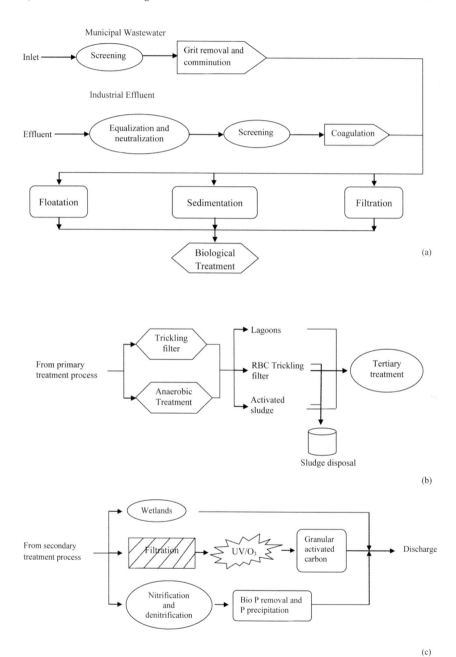

FIGURE 11.1 General wastewater treatment process: (a) Primary wastewater treatment, (b) Secondary wastewater treatment, and (c) Tertiary wastewater treatment.

11.3   INDIGENOUS AND EFFECTIVE MICROORGANISMS (EMOs) IN WASTEWATER TREATMENT

The waste management either solid or liquid is of paramount necessity. Due to increased urbanization and migration of people to cities in search for job has resulted in increased consumption of natural resources and accumulation of enormous waste. In addition, intensive agriculture practices to fulfill food requirement has enabled forced usage of synthetic compounds without understanding its adverse impact on the environment. Several industries among developing nations dispose waste into water bodies and this has raised concerns among large population. In recent years, biochemical or biological degradation of pollutants has received a great attention because of the sole reason that it has a potential to degrade wide arrays of recalcitrant pollutants including xenobiotics (Mueller et al., 1991; Barbeau et al., 1997; Cameron et al., 2000; Tripathi and Garg, 2013). Further, it is considered to be more safe and economic process when compared to other bioremediation processes.

Nevertheless, extensive studies on the biodegradation have now enabled to use a group of particular microorganisms or microbial consortia in removal of complex recalcitrant in the environment. Earlier findings suggest that indigenous microorganism own a potential to degrade wide range of pollutants (Cai et al., 2013; Kumar and Gopal, 2015). However, time involved is essentially more as these develop a mechanism of degradation naturally, under the favorable conditions which are termed as natural attenuation. Further, one of the key constraints is fluctuating environmental factors as these microorganisms are very sensitive to environmental conditions (pH, temperature, and type of pollutant). Nevertheless, approaches such as cell-free enzymes from such sensitive organisms have been found to address this issue. Several microorganisms have been isolated from the wide array of contaminated sites that are known to effectively degrade the pollutants under laboratory conditions. *In situ* bioremediation techniques especially bioaugmentation and biostimulation are more interesting than that of those *ex-situ* processes (Tyagi et al., 1993). The bioaugmentation approach is receiving a great deal of attention as it involves the usage of exogenous microorganisms for bioremediation of pollutants. Additionally, it can involve either the use of an individual group or consortia of microorganisms. However, recent studies suggest that consortia of microorganisms are comparatively more efficient on contrary to individual group of microorganisms.

In 1970s, for the first time, Teruo Higa at the University of Ryukyus, Okinawa, Japan, proposed the concept of effective microorganism for bioremediation of a wide range of pollutants (Higa, 1989, 1991, 1994; Higa and Chinen, 1998). The EMOs involve consortia of microorganisms such as lactic acid bacteria (*Lactobacillus plantarum*, *Lactobacillus casei*, and *Streptococcus lactis*), photosynthetic bacteria (*Rhodopseudomonas palustrus* and *Rhodobacter spaeroides*) and yeasts (*Saccharomyces cerevisiae* and *Candida utilis*), actinomycetes (*Streptomyces albus* and *Streptomyces griseus*). Teruo and James (1994) identified more than 80 beneficial species of microorganisms. The selected potential microorganisms for efficient degradation allow degrading a wide array of pollutants. The approach of EMOs has now gained much appreciation and is been followed across globe for removal of pollutants among different environments. These microorganisms have a tendency to grow under both aerobic as well as anaerobic conditions. Additionally, the combination of microorganisms used in developing this mixture is in such a manner that they are symbiotic against each other and work efficiently under different substrates. Further, the medium used for cultivation of these EMO is very economic and has shown to be efficient after storing for long duration. EMOs are widely used as insect repellent, foliar spray and as compost (PSDC, 2009; Zakaria et al., 2010).

On contrary, IMOs are mixture of naturally occurring group of microorganisms present over the contaminated sites. The efficiency of IMO technology was proven to be reliable and ecofriendly when compared to intensive or chemical farming. The IMO technology improves soil and plant health, and benefits other soil-inhabiting organisms. This approach has been largely followed by farmers of many Southeast Asian countries and has been employed for home gardening as well as commercial farming. IMO has also been reported to facilitate the removal of unpleasant odor among the poultry and piggery (Kumar and Gopal, 2015) (Tables 11.1 and 11.2).

11.4   POTENTIALITY OF IMOs AND EMOs

Wastewater generated from different sources shows wide variations. Generally, industrial wastewaters comprise suspended, colloidal, and dissolved particles, including complex organic matter. The chemical composition may significantly contribute for increase in the acidity or alkalinity (Cheryan and Rajagopalan, 1998; Verma et al., 2012). These

TABLE 11.1 Application of Indigenous Microorganisms in Wastewater Treatment Process

Microorganism	Compound Treated	Sources	References
Bacteria			
Haliscomenobacter hydrossis	Sludge bulking	Domestic and municipal	Mielczarek et al., 2012; Kowalska et al., 2015
Acidithiobacillus ferrooxidans, Acidithiobacillus thiooxidans	Metals	Sewage	Tyagi et al., 1993; Chan et al., 2003
Burkholderia pickettii	Quinoline	Coke effluent	Jianlong et al., 2002
Nitrospira oligotropha, Nitrosomonas oligotropha	Organic matter	Municipal	Harms et al., 2003; Siripong and Rittmann, 2007
Rhodopseudomonas blastica	Reduction of COD and BOD	Rubber sheet	Kantachote et al., 2005
Ion oxidizing bacteria	Heavy metals	Sewage	Xiang et al., 2000
Nitrosomonas sp., *Nitrobacter* sp.	Nutrients, ammonia	Sewage	Wagner, 1996
Pseudomonas sp.	Nutrients	Sewage	Salla et al., 1989
Achromobacter sp.	Pyridine	Industry	Deng et al., 2011
Alcaligens sp.	Phenol	Municipal	Wang et al., 2016
Zooglea ramigera	Lower BOD	Municipal	Rossello-Mora et al., 1995
Acinetobacter sp.	Phenol	Municipal	Wang et al., 2016
Candidatus sp., *Nitrotoga arctica*	Nutrient, ammonia	Municipal	Lücker et al., 2015; Saunders et al., 2016
Microthrix parvicella	Sludge bulking, phosphorus removal	Municipal	Wang et al., 2014
Dechloromonas sp.	Nutrients	Activated sludge	Zhang et al., 2012
Prosthecobacter sp.	Nutrients	Activated sludge	Zhang et al., 2012
Caldilinea sp.	Nutrients	Activated sludge	Zhang et al., 2012
Tricoccus sp.	Nutrients	Activated sludge	Zhang et al., 2012

TABLE 11.1 *(Continued)*

Microorganism	Compound Treated	Sources	References
Thiobacillus sp.	Nutrients	Municipal	Ma et al., 2015
Comamonas sp.	Nutrients	Municipal	Ma et al., 2015
Thauera sp.	Nutrients	Municipal	Jiang et al., 2008
Azoarcus sp.	Nutrients	Municipal	Ma et al., 2015
Rhodoplanes sp.	Sludge bulking	Municipal	Ma et al., 2015
Nostocola limicola	Sludge bulking	Municipal	Guo and Zhang, 2012
Mycobacterium fortuitum	Sludge bulking	Activated sludge	Guo and Zhang, 2012
Accumulibacter sp.	Phosphorus	Domestic and municipal	Oehmen et al., 2007; He et al., 2008
Tetrasphaera sp.	Phosphorus	Domestic and municipal	Nielsen et al., 2010; Nguyen et al., 2011
Pseudomonas paucimobilis	Bisphenol A	Municipal	Ike et al., 1995
Sphingomonas bisphenolicum	Bisphenol A	Municipal	Oshiman et al., 2007
Sinorhizobium meliloti	Bisphenol A	Sewage sludge	Mohapatra et al., 2010
Microbacterium oxydans	Phenol	Effluent	Wang et al., 2016
Pseudomonas aeruginosa	Heavy metal	Effluent	Olawale, 2014
Fungi			
Penecillium corylophilum, P. waksmanii, P. citrinum	Organic matter	Domestic and municipal wastewater	Fakhrul-Razi et al., 2002
Aspergillus terrius, A. flavus	Organic matter	Domestic and municipal wastewater	Fakhrul-Razi et al., 2002
Trichoderma harzianum	Organic matter	Domestic and Municipal wastewater	Fakhrul-Razi et al., 2002

TABLE 11.2   Applications of Effective Microorganisms in Wastewater Treatment Process

Microorganism	Compound treated	Sources	References
Algae			
Chlorella vulgaris	Nutrient, ammonium, and phosphorus ions, metal recovery	Municipal, synthetic wastewater, effluent	Lau et al., 1995; Gomes et al., 1998; De-Bashan et al., 2002
Bacteria			
Aquabacterium sp.	Nutrients	Municipal	Zhang et al., 2016
Thauera sp.	Nitrogen	Domestic	Peng et al., 2014
Kuenenia stuttgartiensis, Brocadia anammoxidans	Ammonia	Domestic, municipal	Van Der Star et al., 2007
Arthrobacter sp.	COD removal	Refinery effluent	Sahoo et al., 2014
Lactobacillus plantarum, Lactobacillus casei, Streptoccus lactis, Rhodopseudomonas palustrus, Rhodobacter spaeroides	Solid liquid separation, reducing sludge volume, decompose organic matter	Municipal	Higa and Chinen, 1998; Zhao et al., 2006; Ke et al., 2009; Shalaby, 2011
Pseudomonas sp.	Detoxifies the metals	Sewage	Monica et al., 2011
Cyanobacteria			
Oscillatoria sp., *Phormidium tenue, Phormidium bohneri, Phormidium tenue*	Nitrogen, phosphorus	Domestic, agriculture	Chevalier et al., 2000
Fungi			
Bjerkandera adust	Decolonization	Textiles	Spina et al., 2012
Aspergillus terreus, Rhizopus sexualis	COD, BOD, TDS, oil, and grease, heavy metals	Effluent	Dias et al., 2002; Jogdand et al., 2012

TABLE 11.2 (Continued)

Microorganism	Compound treated	Sources	References
Penicillium chrysogenum, Rhizopus oryzae	Metal recovery	Effluent	Gomes et al., 1998
Aspergillus oryzae, Mucor hiemalis	Breakdown of organic matters	Municipal	Zakaria et al., 2010
Yeast			
Rhodotorula mucilaginosa	COD, phenol	Municipal	Jarboui et al., 2012
Saccharomyces cerevisiae, Candida utilis, Streptomyces albus, Streptomyces griseus	Metabolites used as substrate for lactic acid bacteria, heavy metals	Municipal, effluent	Gomes et al., 1998; Higa and Chinen, 1998; Zhao et al., 2006; Ke et al., 2009; Shalaby, 2011

factors have profound effects on the growth and survivability of IMO. The potential of EM technology is that it can effectively degrade complex compounds such as xenobiotics and organic matter present in the waste-water. As aforementioned, EM has been also found to subsidize the growth of pathogenic microorganisms in wastewater, such as fecal coli-form bacteria. It reduces the BOD, COD, TSS, total, and fecal coliforms and significantly minimizes the usage of chemicals such as alum, lime, chlorine for treatment, and contributes in cost reduction. Sludge volume greatly reduces up to 20%. It may also aid in the maintenance of the right pH, which greatly influences the growth of wide bacteria and largely subsidizes the foul smell. No design alteration in the ETP is required, and it can be directly applied to natural streams. More interestingly, this technology is found to be simple, natural, economic, and environment friendly (El-Sherbiny et al., 2001; Shalaby, 2011).

A study by Okuda and Higa (1995) demonstrated that the application of EM to the sewage water significantly reduced the BOD and COD up to 93% and 20%, respectively. Further, a decrease in suspended solids up to 94% was also reported on contrary to untreated water. Additionally, the decrease in nitrogen and phosphorus content was observed. This treated water when evaluated for the cucumber cultivation they found some prom-ising results; EM treated water showed increased plant survival, leaf size, and dry weight. In addition, they observed the increase in vitamin C, and chlorophyll content, and root activity was also found to be enhanced among EM treated water. The sewage sludge after wastewater treatment was also evaluated for use as fertilizer for tomato cultivation and their study indi-cated that EM treated sludge of 500 g and 1 Kg resulted in increased plant height, leaf number and fresh weight on contrary to untreated and bare soil. Thus their study indicated that application of EM for sewage treat-ment could offer multiple facets and can be exploited for recycling the water as well as the sludge.

Studies also show that the increased concentrations of heavy metals in the wastewater have a significant effect on the EM activity. A study of Sheng et al. (2008) reported the adverse effects of heavy metals on EM and their study highlighted that various heavy metals found across the wastewater severely affect the DNA of EM. Wherein their study mainly focused on the assessment of heavy metals (As^{3+}, Cd^{2+}, Cr^{3+}, Cu^{2+}, Hg^{2+}, Pb^{2+}, and Zn^{2+}) on EM DNA using *in vitro* assay (Comet assay). Their study demonstrated that the damage of the DNA of EM were nega-tively correlated with their treatment capability and that EM bacteria can withstand up to maximum concentrations of 0.05 mg/L for As^{3+}, 0.2 mg/L for Hg^{2+}, 0.5 mg/L for Cd^{2+}, Cr^{3+}, and Cu^{2+}, and 1 mg/L for Pb^{2+} and Zn^{2+}. Further concluded that As^{3+} at 0.05 mg/L concentration and Hg^{2+} at 0.20 mg/L concentration DNA damage was more and their ability of waste-water treatment reduced drastically.

11.5   STRATEGY TO ENHANCE THE IMOs IN WASTEWATER TREATMENT PROCESS

Laboratory scale experiments using indigenous strains of *Bacillus* sp., *Brevundimonas* sp. and *Shewanella* sp. in the removal of alarming concentrations of calcium, iron, and magnesium in surface waters from mined lands and petroleum contaminated sites is well documented (Gupta et al., 2000; Philip and Venkobachr, 2001; Srinath et al., 2003; Kim et al., 2007; Fosso-Kankeu et al., 2009; Karamalidis et al., 2010). Experimental batch reactor studies by Thakur (2004) under controlled conditions showed that IMO were able to decolorize kraft pulp bleached effluent. In addition, the study also highlighted on the reduction of adsorbable organic halogens and removal of color and lignin in pulp and paper mill effluent using fungi and bacteria. The study involved identification of eight potential fungal and three bacterial isolates from various sources such as decomposed wood, sediment core and pulp and paper mill effluent. Fungi such as *Paecilomyces* sp. (F3) was found to be more effective as decolorizing strain followed by F5 (*Phoma* sp.) and F7 (*Paecilomyces varioti*). Additionally, *Paecilomyces* sp. (F3) was able to reduce 80% color and lignin. Whereas bacteria such as *Pseudomonas aeruginosa*, *Acinetobacter calcoaceticus*, and *Klebsiella pneumoniae* removed color up to 45%, 39%, and 25%, respectively. Furthermore, the study indicated that percentage reduction of color and lignin enhanced when both fungal and bacterial strains were simultaneously applied.

Haq et al. (2016) isolated the novel bacterial strain *Serratia liquefaciens* (LD-5) from pulp and paper mill effluent, capable of decolorizing Azure B dye. Wherein they found that under optimal conditions of temperature (30°C), pH (7.6) and constant shaking (120 rpm) for 144 h, *Serratia liquefaciens* (LD-5) was able to decolorize up to 72%, and further degrade lignin and phenol up to 58% and 95%, respectively. In addition, residual toxicity of the treated effluent was evaluated by alkaline single cell gel electrophoresis assay wherein toxicity reduction to treated effluent was 49.4%. Thus, their study demonstrated the potentiality of *Serratia liquefaciens* as novel bacterium for bioremediation of pulp and paper mill effluent.

Abdelouas et al. (1998) reported the reduction and immobilization of Uranium (VI) species in uranium-contaminated groundwater by indigenous bacteria *Pseudomonas aeruginosa*, *P. stutzeri* and *Shewanella putrefaciens*. Indigenous iron-oxidizing bacteria, viz. *Acidithiobacillus ferrooxidans* and *Acidithiobacillus thiooxidans* are reported to be potential bioleaching agents for heavy metals across sewage sludge (Tyagi et al., 2013). A study by Xiang et al. (2000) suggested that indigenous iron-oxidizing bacteria significantly removed heavy metals such as Cr, Cu, Zn, and Ni from the anaerobically digested sewage sludge. In another study, Pathak et al. (2009) reported the potential of indigenous iron-oxidizing microorganisms enriched with ammonium ferrous sulfate, and ferrous sulfate as an energy source could yield reduced concentrations of heavy metals across sludge. A recent study by Garcha et al. (2016) reported the isolation and characterization of ten potential indigenous bacterial strains belonging to genera *Bacillus* sp., *Escherichia* sp., *Lysinibacillus* sp. and *Brevibacillus* sp. from dairy wastewater and sludge. Their study indicated that bioaugmentation of these potential isolates reduced the BOD, TSS, and oil and grease content up to 89.8, 88.6, and 96.9%, respectively.

Studies have shown that the application of mixed cultures to the waste treatment is generally more pronounced and effective than that of using a single microbial culture. Bala et al. (2015) reported the reduction of organic load palm oil mill effluent using indigenous bacteria such as *Micrococcus luteus*, *Stenotrophomonas maltophilia*, *Bacillus cereus*, *Providencia vermicola*, *Klebsiella pneumoniae*, and *Bacillus subtilis*. These bacteria showed an increased reduction of BOD and COD in the wastewater. Further, their study indicated that mixed cultures of selected bacteria *Bacillus cereus* and *Bacillus subtilis* were more effective treatment for palm oil mill effluent. Several field application studies have been conducted previously for assessment of EM and IM technologies as fertilizer from food processing effluent (Wood et al., 2002). Mbouobda et al. (2013) carried an experiment that application of EMO and IMO manure on *Colocasia esculenta* crop. Their study showed that upon EMO and IMO manure treatment in a randomized complete block design (RCBD) design for five months, EMO manure significantly influenced the growth of Taro (*Colocassia esculenta*) followed by IMO manure. Further, the phenolic content, peroxidase, and polyphenol oxidase enzyme activities were higher with respect to EM treated plants. However, their study indicated that EM and IMO manure was ineffective controlling the leaf blight disease. In another study by Mbouobda et al. (2014) reported that Irish potato (*Solanum tuberosum*) treated with EM and IMO Bokashi resulted in increased growth and productivity. Additionally, phenol content, PME, PPO, and POX activities were enhanced among the treated plants when compared to control plants.

11.6 EFFECT OF EMO ON THE QUALITY OF TREATED WASTEWATER

The use of EMOs in the biochemical decomposition of wastewaters mostly depends on bacteria, produce stable end products, and waste is converted into carbon dioxide, water with other nontoxic end products. More and more microorganisms are used in these days for the successful treatment of wastewater that originates from industries such as petroleum, dyes, leather, or sewage water from the municipal sources (Zhao et al., 2006). The effect of EMOs in the treatment of wastewater showed a considerable increase in BOD and decreased pH levels, and also the settlement of sludge in the treatment tank was seen to have improved. However, when compared with the treated tanks, significant solids remain as it is. The level of suspended solids in the effluent was not reduced considerably in some of the experiments conducted for the treatment of the wastewater. After EMOs application, some other parameters also showed similar results, except total suspended solids. Highly variable conditions are observed in the septic tanks during the monitoring period. Following the use of EMOs for a considerable period, the result shows the decreased alkalinity and electric conductance (EC) of the treated wastewater. Results have also revealed that the sludge volumes may considerably get reduced due to the EMO application. In the case of solids, the breakdown of the organic matter leads to an initial increase in entrained solids of the septic tank. When EMO added, there is an increase in the reliability of the microbial ecosystems, which are 'notoriously fragile' and pathogenic microorganisms get excluded due to competition, which in turn favors microbes that are beneficial and already present in the process. Due to the chemically stirring of the sludge, the septic tanks show variance in the results, which leads to the changes in the EC, total alkalinity, and pH of the water undergoing treatment. Certain parameters such as pH can be influenced, if the microbial populations are disturbed at any trophic level, and the impact can also be seen on the other intermediates such as the complete microbial community and overall efficiency of the treatment tank (Linich, 2001).

A study was conducted by Melamane et al. (2007) and Manyuchi and Ketiwa (2013) on petroleum wastewater to remove the organic pollutants by treating with various biological methods such as activated sludge reactors or biofilm-based reactor. The wastewater from petroleum industries and refineries mainly contains organic matter, oil, and other compounds. The wastewater arising from the petroleum refinery contains compounds that are hazardous with adverse effects on the ecosystem. The treatment occurs in two stages, firstly, pre-treatment stage where oil, suspended materials, and grease are reduced. Secondly, degrade, and decrease the pollutants by an advanced method to acceptable values. Based on the availability of dissolved oxygen (DO), biological treatment methods are generally divided into aerobic and anaerobic methods (Zhao et al., 2006). The products of chemical and biochemical reactions in the anaerobic system produce displeasing odors and colors in water. Thus, to reduce this, oxygen availability is vital (Attiogbe et al., 2007). The organic compounds and recalcitrant components in an aerobic biological process are converted into CO_2, water, and solid biological products. Anaerobic biological technology is highly efficient and is been widely applied in wastewater treatment (Lettinga et al., 2001).

In a study conducted by Zhao et al. (2006) on petroleum refinery wastewater using a reactor immobilized with EMOs showed the degradation efficiencies of 78% total organic carbon (TOC) removal and 94% oil removal. Similarly, Satyawali and Balakrishnan (2008) showed an improved chemical oxygen demand (COD) removal by the aerobic biological process. Recent advances have lead to the building of a reactor called as up-flow anaerobic sludge bed reactor for the treatment of petroleum wastewater design which is simple with easy maintenance and construction (Rastegar et al., 2011). For efficient removal of COD up to 81.07%, an anaerobic packed-bed biofilm reactor was used in combination with an up-flow anaerobic sludge blanket reactor (UASB) in the treatment of petroleum refinery effluents (Nasirpour et al., 2015). Several types of bioreactors with EMOs are used for the successful treatment of petroleum refinery wastewater such as the anaerobic treatment process (a UASB reactor) by Gasim et al. (2013) with the efficiency of removing COD levels as high as of 82% and conversion of half of the organic compounds into biogas. Another experiment conducted by Zou (2015) showed that the oil in the heavy oil wastewater, COD, and ammonia nitrogen (NH_3-N) were removed by 86.5,

90.8, and 90.2%, respectively when biological aerated filter (BAF) system was used in combination with UASB. Temperature plays a vital role in the biotransformation of total naphthenic acids (NAs) through the activated sludge system (Wang et al., 2014, 2015, 2016). The study revealed that the percentage of removal efficiency in summer for total NAs was 73% higher than winter 53% because in the activated sludge system, the microbial biotransformation activities are high in summer.

A. terreus and *R. sexualis* are two fungi whose enzymatic productivity was checked by growing in non-composted sawdust and municipal sludge enzymatic contents and were tested after14 days of incubation. It was seen that the concentration of α-amylase and cellulase were low in the municipal sludge, and also the non-composted sawdust. Similarly, protease and α-amylase enzymes were found to be in lower concentrations in non-composted sawdust. The enzymes, such as cellulase and α-amylase, are derived from the composted sawdust due to the luxuriant fungal growth serving as a store of enzymes. The sawdust attaches to the surface working together with the microbial film, and it acts as a good supporting matrix. Apart from this, it has several other benefits, such as a good nutrient source and slow biodegradable material that can be effectively used in the treatment of wastewater. A drastic effect on the quality of wastewater was seen with the application of non-composted sawdust with respect to COD and BOD. The use of pure fungal growth is seen as highly efficient in the process of wastewater treatment. Significant results were obtained during the biological treatment processes for the total dissolved solids (TDS) ratio by pure fungal pellets or fungal-sawdust compost (bio-mixture) after 8 h of incubation (retention time) could be reduced by 67.5–74.6%. Starch decomposition up to 90% was achieved in the treatment process of starch-wastewater by fungal pellets of *Aspergillus niger* when applied for 16 days (Fujita et al., 1993).

In another treatment method to biodegrade, the dairy effluent waste-water, yeast isolate, and two bacterial isolates were isolated from the dairy sludge with a model consisting of layers of sawdust and activated charcoal was used as a filtering media. A mixed culture of yeast and a microbial isolate was prepared by taking 1:1 ratio. Each isolate and the mixed culture were used as an inoculum to treat the diary wastewater. Among all the isolates, a mixed culture of dairy sludge proved to be most efficient in the treatment of wastewater after 48h aeration period. The reduction efficiency of the mixed culture was highest by 47.52% in BOD when compared with other isolates. Studies suggested that the use of mixed culture after the

48h aeration period is more effective in the treatment of wastewaters. It can reduce some of the physicochemical parameters of wastewater. The reduction in the total solid content, BOD, TDS, COD, EC, chlorides, and sulfates were caused by the bacterial isolate being the second most effective in the treatment. However, for the reduction in the turbidity and oil and gas content, yeast isolate was found to be a more helpful and effective single culture whereas, the final conclusion is that the mixed culture would prove to be more beneficial and effective than a single culture.

11.7 FACTORS AFFECTING WASTEWATER TREATMENT PROCESSES

It is very important to maintain active and mixed microorganisms in a biological wastewater treatment system because it influences the successful working of the biological treatment process (Dara, 1993). In rotating biological filter and trickling filter, microbes are found attached to certain types of medium that support their growth, and in activated sludge process and anaerobic digestion, they are found as suspended growth (Cohen, 2000). Microbes utilize the organic waste matter as their food resource, which helps them in the synthesis of new cell material, and the degradation of certain organic matter to more simpler compounds provides energy for synthesis and functioning of cell maintenance (Waites et al., 2001). Biological growth includes both the processes of biodegradation and cell synthesis. Biological growth is influenced by environmental factors like the presence of toxic agents, pH, temperature, and mixing intensity (Grady et al., 2011). The deficiency of these factors will affect the biological growth, which leads to the inefficiency of the process. In order to maximize the process efficiency, all the operational conditions should be maintained constant (Grady et al., 2011). Wastewater treatment using microorganisms are influenced by factors like pH and alkalinity, temperature, nutrient availability and oxygen demand (Gerardi, 2006; Fulekar, 2010)

11.7.1 pH AND ALKALINITY

Microbes used for wastewater treatment require optimum pH for proper growth, and most of them prefer pH value of 6 to 9 (Fulekar, 2010). If the pH value becomes less than 6 then it will affect the enzymatic activity (reduction), hydrogen sulfide production (increase), formation of floc and uncontrollable growth of certain other microorganisms. When pH value exceeds 9, it affects enzyme activity, ammonia production, nitrification process, and formation of floc. The pH regulates the degree of substrate ionization, waste, and nutrients. Substrate utilization and waste production by microbes cause a change in the pH of the biological treatment plant, which in turn affects the ability of microbes to treat wastewater (Gerardi, 2006). Alkalinity acts as a buffer to maintain the pH at a normal level. Urea, protein, and carbonates of calcium and sodium are the source of alkalinity. When acids are produced as a result of metabolism by the microbes, alkalinity prevents the sudden change in pH (Fulekar, 2010).

11.7.2   TEMPERATURE

Enzymatic activity, substrate, and nutrient diffusion into cells of microbes are related to temperature change. These processes increase with an increase in temperature. If the temperature is very high, it will result in the denaturation of an enzyme that is related to the catalysis of biochemical reactions (Gerardi, 2006). Psychrophilic microbes prefer less temperature (15–20°C) for their growth compared to thermophilic and mesophilic microbes (Russell and Fukunaga, 1990; D'Amico et al., 2006; Gerardi, 2006; Fulekar, 2010), thermophilic bacteria prefer high-temperature range between 45–65°C (Russell and Fukunaga, 1990; Gerardi, 2006) and mesophilic prefer a temperature range of 25–45°C (Bergey, 2014). If the temperature goes beyond 45°C, it will cause the death of mesophilic microbes, which results in the growth of thermophilic microbes.

11.7.3   NUTRIENT AVAILABILITY

Macronutrients such as magnesium, calcium, potassium, sulfur, phosphorus, nitrogen, and micronutrients such as cobalt, chromium, zinc, manganese, copper, iron, and boron are essential for the growth of microbes because they promote the growth and helps in enzyme synthesis which is responsible for biochemical reactions. Phosphorus and nitrogen play an important role in the synthesis of cell mass (Fulekar, 2010). Nitrogen helps to breakdown the carbonaceous substrates (Bergey, 2014). Another important nutrient is carbon, which is an important component of carbohydrates, proteins, and lipids that promotes cell growth of the microbes (Mara and Horan, 2003).

11.7.4 OXYGEN DEMAND

Oxygen is an important requirement for microbes because aerobic microbes decompose waste material using oxygen. Biological oxygen demand (BOD), COD, and DO are important parameters that determine the strength of pollutants present in wastewater (Dara, 1993; Fulekar, 2010; Bergey, 2014). BOD refers to the amount of oxygen required by the microbes to biochemically degrade and transform the organic content of the wastewater aerobically (Dara, 1993). American Society of Testing and Materials (ASTM) defines COD as the amount of oxygen (expressed in mg/l) consumed under specified conditions in the oxidation of organic and oxidizable inorganic matter, corrected for the influence chlorides (Dara, 1993). The level of DO is important to understand the efficiency of a biological wastewater plant, and a low level of DO content will lead to anaerobic conditions, thus resulting in an inefficient biological treatment plant (Bergey, 2014). Microbes require DO in the range of 1 to 3 mg/l for decomposing the waste. When the pollutant content is high, a huge amount of microbes will be required to decompose the waste, which requires a large amount of oxygen content, thus resulting in high demand for oxygen (Fulekar, 2010).

11.8 CONCLUSION AND FUTURE PERSPECTIVES

Despite the high relative proportion of the population served by wastewater treatment facilities, discharge of municipal and industrial effluents continues to have significant adverse impacts on receiving water bodies. The environmental consequences of wastewater discharge into the environment are, however, difficult to generalize largely due to regional variations in the level of wastewater treatment and the nature of receiving water bodies. The significance of wastewater treatment is to avoid the spread of diseases by protecting water resources against pollutants. Treatment of wastewater is one of the strategies for the management of water quality and its sustainable utilization. Due to some limitations on the use of chemical treatment in wastewater treatment, biological treatment is now employed to avoid the unnatural conditions in water resources. The incidence of nitrogen and phosphorus compounds in wastewater treatment and their effective removal from the wastewater has been extensively discussed. Most of the nutrient removal studies happen in the presence of bacteria and protozoa, and their roles in the removal of nutrients have been well documented. Fungi have also reported to increase the settleability, degradability of wastewater sludge, and contribute to the sludge management strategy. Our understanding of microbial community structure in wastewater treatment systems continues to advance rapidly owing to ongoing development and application of molecular methods. Nowadays, for most of the major processes in wastewater treatment systems, the application of culture independent eukaryotes and prokaryotic microorganisms has been analyzed, which provides a great opportunity for the exploration of novel microorganisms in the studies of wastewater treatment. The significant investigations have shown that the functional diversity of important prokaryotic groups in wastewater treatment systems can be influenced by the plant design and also by the changes in the process stability. Hence, it is necessary to look towards dynamic, safe, and sustainable wastewater treatment plans and policies confining low-cost decentralized microbial mediated wastewater treatment technologies.

KEYWORDS

- effective microorganisms
- indigenous microorganisms
- oxygen demand
- sustainability
- urbanization
- wastewater treatment

REFERENCES

Abdelouas, A., Lu, Y., Lutze, W., & Nuttall, H. E. Reduction of U(VI) to U(IV) by indigenous bacteria in contaminated groundwater. *Journal of Contaminant Hydrology*, **1998**, *35*(1–3), 217–233.

Abdel-Raouf, N., Al-Homaidan, A. A., & Ibraheem, I. B. M. Microalgae and wastewater treatment. *Saudi J. Biol. Sci.*, **2012**, *19*, 257–275.

Ahansazan, B., Afrashteh, H., Ahansazan, N., & Ahansazan, Z. Activated sludge process overview. *Inter J. Environ. Sci. Devt.*, **2014**, *5*(1), 81–85.

Akpor, O. B., Ogundeji, M. D., Olaolu, T. D., & Aderiye, B. I. Microbial roles and dynamics in wastewater treatment systems: An overview. *Int. J. Pure Appl. Biosci.*, **2014**, *2*(1), 156–168.

Anbalagan, A., Schwede, S., Lindberg, C. F., & Nehrenheim, E. Influence of hydraulic retention time on indigenous microalgae and activated sludge process. *Water Res.*, **2016**, *91*, 277–284.

Attiogbe, F. K., Glover-Amengor, M., & Nyadziehe, K. T. Correlating biochemical and chemical oxygen demand of effluents, a case study of selected industries in Kumasi, Ghana. *W Afr. J. Appl. Ecol.*, **2007**, *11*, 110–118.

Aziz, H. A., & Mojiri, A. *Wastewater Engineering: Advanced Wastewater Treatment Systems*. IJSR Publications, Penang, Malaysia, **2014**.

Bala, J. D., Lalung, J., & Ismail, N. Studies on the reduction of organic load from palm oil mill effluent (POME) by bacterial strains. *International Journal of Recycling of Organic Waste in Agriculture*, **2015**, *4*(1), 1–10.

Barbeau, C., Deschenes, L., Karamanev, D., Comeau, Y., & Samson, R. Bioremediation of pentachlorophenol-contaminated soil by bioaugmentation using activated soil. *Applied Microbiology and Biotechnology*, **1997**, *48*(6), 745–752.

Bergey, D. H. *Wastewater Microbiology* (pp. 282–283). Scientific International Private Limited, New Delhi, India, **2014**.

Cai, M., Yao, J., Yang, H., Wang, R., & Masakorala, K. Aerobic biodegradation process of petroleum and pathway of main compounds in water flooding well of Dagang oil field. *Bioresource Technology*, **2013**, *144*, 100–106.

Cameron, M. D., Timofeevski, S., & Aust, S. D. Enzymology of *Phanerochaete chrysosporium* with respect to the degradation of recalcitrant compounds and xenobiotics. *Applied Microbiology and Biotechnology*, **2000**, *54*(6), 751–758.

Chan, L., Gu, X., & Wong, J. Comparison of bioleaching of heavy metals from sewage sludge using iron-and sulfur-oxidizing bacteria. *Adv. Environ. Res.*, **2003**, *7*, 603–607.

Cheryan, M., & Rajagopalan, N. Membrane processing of oily streams: Wastewater treatment and waste reduction. *Journal of Membrane Science*, **1998**, *151*(1), 13–28.

Chevalier, P., Proulx, D., Lessard, P., Vincent, W. F., & De la Noüe, J. Nitrogen and phosphorus removal by high latitude mat-forming cyanobacteria for potential use in tertiary wastewater treatment. *Journal of Applied Phycology*, **2000**, *12*(2), 105–112.

Chudoba, P., Morel, A., & Capdeville, B. The case of both energetic uncoupling and metabolic selection of microorganisms in the OSA activated sludge system. *Environ. Technol.*, **2000**, *13*, 761–770.

Cohen, Y. Biofltration – the treatment of fluids by microorganisms immobilized into the filter bedding material: A review. *Bioresource Technology*, **2000**, *77*(2001), 2–4.

D'Amico, S., Collins, T., Marx, J., Feller, G., & Gerday, C. Psychrophilic microorganisms: Challenges for life. *EMBO Rep.*, *7*(4), 385–389.

Dara, S. S. A. *Textbook of Environmental Chemistry and Pollution Control* (pp. 70–72). S. Chand & Company, New Delhi, India, **1993**.

De-Bashan, L. E., Moreno, M., Hernandez, J. P., Bashana Y., Removal of ammonium and phosphorus ions from synthetic wastewater by the microalgae *Chlorella vulgaris* co-immobilized in alginate beads with the microalgae growth-promoting bacterium *Azospirillum brasilense. Water Research,* **2002**, *36*(12), 2941–2948.

Deng, X., Wei, C., Ren, Y., & Chai, X. Isolation and identification of *Achromobacter* sp. DN-06 and evaluation of its pyridine degradation kinetics. *Water Air Soil Pollution,* **2011**, *221,* 365–375.

Dias, M. A., Lacerda, L. C. A., Pimentel, P. F., De Castro, H. F., & Rosa, C. A. Removal of heavy metals by an *Aspergillus terreus* strain immobilized in a polyurethane matrix. *Letters in Applied Microbiology,* **2002**, *34,* 46–50.

El-Sherbiny, M., Al-Sarawey, A., & Elmitwalli, T. Aerobic biodegradability of Egyptian domestic sewage. *Proceeding "Environment," Egyptian Ministry for Environmental Affairs, Cairo, Egypt,* **2001**.

EPA. *Wastewater Treatment Manuals: Primary, Secondary and Tertiary Treatment* (pp. 11–100). Environmental Protection Agency, Wexford, Ireland, **1997**.

Fakhrul-Razi, A., Alam, M. Z., Idris, A., Abd-Aziz, S., & Molla, A. H. Filamentous fungi in indah water konsortium (iwk) sewage treatment plant for biological treatment of domestic wastewater sludge. *Journal of Environmental Science and Health Part A,* **2002**, *37*(3), 309–320.

Fang, H. H. P., Yeong, C. L. Y., Book, K. M., & Chiu, C. M. Removal of COD and nitrogen in wastewater using sequencing batch reactor with fibrous packaging. *Wat. Sci. Tech.,* **1993**, *27*(7), 125–131.

Fayza, A. N., Doma, H. S., Abdel-Halim, H. S., & El-Shafai, S. A. Chemical industry wastewater treatment. *The Environmentalist,* **2007**, *27,* 275–286.

Fosso-Kankeu, E., Mulaba-Bafubiandi, A. F., Mamba, B. B., & Barnard, T. G. Mitigation of Ca, Fe, and Mg loads in surface waters around mining areas using indigenous microorganism strains. *Physics and Chemistry of the Earth, Parts A/B/C,* **2009**, *34*(13/16), 825–829.

Frigon, D., Biswal, B. K., Mazza, A., Masson, L., & Gehra, R. Biological and physico-chemical wastewater treatment processes reduce the prevalence of virulent *Escherichia coli. Appl. Environ. Microbiol.,* **2013**, *79*(3), 835–844.

Fujita, M., Iwahori, K., & Yamakawa, K. Pellet formation of fungi and its application to starch wastewater treatment. *Proceedings of the IAWQ Symposium on Waste Management Problems in Agro-Industries* (pp. 267–274). Pergamon Press, New York, **1993**.

Fulekar, M. H. *Environmental Biotechnology* (pp. 227–229). Science Publishers, Enfield, USA, **2010**.

Garcha, S., Verma, N., & Brar, S. K. Isolation, characterization and identification of microorganisms from unorganized dairy sector wastewater and sludge samples and evaluation of their biodegradability. *Water Resources and Industry,* **2016**, *16,* 19–28.

Gasim, H. A., Kutty, S. R. M., Hasnain-Isa, M., & Alemu, L. T. Optimization of anaerobic treatment of petroleum refinery wastewater using artificial neural networks. *Res. J. Appl. Sci. Eng. Tech.,* **2013**, *6,* 2077–2082.

Gavrilescu, M., & Macoveanu, M. Attached-growth process engineering in wastewater treatment. *Bioprocess Biosyst. Eng.,* **2000**, *23*(1), 95–106.

Gerardi, M. H. *Wastewater Bacteria* (pp. 27–30). John Wiley & Sons, Canada, **2006**.

Gomes, N. C. M., MendoncËa-Hagler, L. C. S., & Savvaidis, I. Metal bioremediation by micro-organisms. *Brazilian Journal of Microbiology,* **1998**, *29,* 85–92.

Grady, C. P., Daiger, G. T., Love, N. G., & Filipe, C. D. M. *Biological Wastewater Treatment* (pp. 580–591). CRC Press, Boca Raton, USA, **2011**.

Guo, F., & Zhang, T. Profiling bulking and foaming bacteria in activated sludge by high throughput sequencing. *Water Res.*, **2012**, *46*(8), 2772–2782.

Gupta, R., Ahuja, P., Khan, S., Saxena, R. K., & Mohapatra, H. Microbial biosorbents: Meeting challenges of heavy metal pollution in aqueous solutions. *Current Science*, **2000**, 967–973.

Haandel, V. A., & Lubbe, V. J. *Handbook Biological Wastewater Treatment.* Quist Publishing, Leidschendam, **2007**.

Han, S. S., Bae, T. H., Jang, G. G., & Tak, T. M. Influence of sludge retention time on membrane fouling and bioactivities in membrane bioreactor system. *Process Biochem.,* **2005**, *40*(7), 2393–2400.

Haq, I., Kumar, S., Kumari, V., Singh, S. K., & Raj, A. Evaluation of bioremediation potentiality of ligninolytic *Serratia liquefaciens* for detoxification of pulp and paper mill effluent. *Journal of Hazardous Materials*, **2016**, *305*, 190–199.

Harms, G., Layton, A. C., Dionisi, H. M., Gregory, I. R., Garrett, V. M., Hawkins, S. A., Robinson, K. G., & Sayler, G. S. Real-time PCR quantification of nitrifying bacteria in a municipal wastewater treatment plant. *Environ. Sci. Technol.*, **2003**, *37*(2), 343–351.

Hassard, F., Biddle, J., Cartmell, E., Jefferson, B., Tyrrel, S., & Stephenson, T. Rotating biological contactors for wastewater treatment–a review. *Process Saf. Environ. Prot.*, **2014**, *94*, 285–306.

He, S., Gu, A. Z., & McMahon, K. D. Progress toward understanding the distribution of *Accumuli bacter* among full-scale enhanced biological phosphorus removal systems. *Microb. Ecol.*, **2008**, *55*(2), 229–236.

Higa T. *US Patent No, 4839051*. Washington, DC, USA, **1989**.

Higa, T. Effective microorganisms: A biotechnology for mankind. In: *Proceedings of the First International Conference on Kyusei Nature Farming* (pp. 8–14). US Department of Agriculture, Washington, DC, USA, **1991**.

Higa, T. Effective microorganisms: A new dimension for nature farming. In: *Proceedings of the Second International Conference on Kyusei Nature Farming* (pp. 20–22). US Department of Agriculture, Washington, DC, USA, **1994**.

Higa, T., Chinen N. *EM Treatments of Odor, Waste Water, and Environment Problems.* University of Ryukyus, Okinawa, Japan, **1998**.

Huijun, J., & Qiuyan, Y. Removal of nitrogen from wastewater using microalgae and microalgae–bacteria consortia. *Cogent Environmental Science*, **2016**, *2*, 1275089, http://dx.doi.org/10.1080/23311843.2016.1275089 (Accessed on 7 October 2019).

Ike, M., Jin, C. S., & Fujit, M. Isolation and characterization of a novel bisphenol A-degrading bacterium *Pseudomonas paucimobilis* strain FJ-4. Japanese. *J. Water Treat. Biol.*, **1995**, *31*, 203–212.

Jarboui, R., Baati, H., Fetoui, F., Gargouri, A., Gharsallah, N., & Ammar, E. Yeast performance in wastewater treatment: Case study of *Rhodotorula mucilaginosa*. *Environ. Technol.*, **2012**, *33*(7–9), 951–960.

Jiang, X., Ma, M., Li, J., Lu, A., & Zhong, Z. Bacterial diversity of active sludge in wastewater treatment plant. *Earth Sci. Front*, **2008**, *15*, 163–168.

Jianlong, W., Xiangchun, Q., Libo, W., Yi, Q., & Hegemann, Q. Bioaugmentation as a tool to enhance the removal of refractory compound in coke plant wastewater. *Process Biochemistry*, **2002**, *38*(50), 777–781.

Jin, M., Wang, X. W., Gong, T. S., Gu, C. Q., Zhang, B., Shen, Z. Q., & Li, J. W. A novel membrane bioreactor enhanced by effective microorganisms for the treatment of domestic wastewater. *Applied Microbiology and Biotechnology*, **2005**, *69*(2), 229–235.

Jogdand, V. G., Chavan, P. A., Ghogare, P. D., & Jadhav, A. G. Remediation of textile industry waste water using immobilized *Aspergillus terreus*. *European Journal of Experimental Biology*, **2012**, *2*(5), 1550–1555.

Kantachote, D., Torpee, S., & Umsakul, K. The potential use of anoxygenic phototrophic bacteria for treating latex rubber sheet wastewater. *Electronic Journal of Biotechnology*, **2005**, *8*(3), http://www.ejbiotechnology.info/index.php/ejbiotechnology/article/view/244 (Accessed on 7 October 2019).

Kapagiannidis, A. G., Zafiriadis, I., & Aivasidis, A. Effect of basic operating parameters on biological phosphorus removal in a continuous-flow anaerobic–anoxic activated sludge system. *Bioprocess Biosyst. Eng.*, **2012**, *35*(3), 371–382.

Karamalidis, A. K., Evangelou, A. C., Karabika, E., Koukkou, A. I., Drainas, C., & Voudrias, E. A. Laboratory scale bioremediation of petroleum-contaminated soil by indigenous microorganisms and added *Pseudomonas aeruginosa* strain Spet. *Bioresource Technology*, **2010**, *101*(16), 6545–6552.

Kaur, R., Dhir, G., Kumar, P., Laishram, G., Ningthoujam, D., & Sachdeva, P. Constructed wetland technology for treating municipal wastewaters. *ICAR News*, **2012**, *18*(1), 7–8.

Ke, B., Xu, Z., Ling, Y., Qiu, W., Xu, Y., Higa, T., & Aruoma, O. I. Modulation of experimental osteoporosis in rats by the antioxidant beverage effective microorganism-X (EM-X). *Biomed. Pharmacother*, **2009**, *63*(2), 114–119.

Kim, S. U., Cheong, Y. H., Seo, D. C., Hur, J. S., Heo, J. S., & Cho, J. S. Characterization of heavy metal tolerance and biosorption capacity of bacterium strain CPB4 (*Bacillus* spp.). *Water Science and Technology*, **2007**, *55*(1/2), 105–111.

Kowalska, E., Paturej, E., & Zielińska, M. Use of *Lecane inermis* for control of sludge bulking caused by the *Haliscomenobacter* genus. *Desalination and Water Treatment*, **2015**, *57*(23), 10916–10923.

Kumar, B. L., & Sai, G. D. V. R. Effective role of indigenous microorganisms for sustainable environment. *3 Biotech.*, **2015**, *5*, 867–876.

Lau, P. S., Tam, N. F. Y., & Wonga, Y. S. Effect of algal density on nutrient removal from primary settled wastewater. *Environmental Pollution*, **1995**, *89*(1), 59–66.

Lettinga, G., Rebac, S., & Zeeman, G. Challenge of psychrophilic anaerobic wastewater treatment. *Trends in Biotechnology*, **2001**, *19*(9), 363–370.

Linich, M. *Microbial Processes and Practical Guidance for On-Site Assessment* (pp. 253–260). University of New England, Armidale, **2001**.

Lu, Y., Wu, X., & Guo, J. Characteristics of municipal solid waste and sewage sludge co-composting. *Waste Manage*, **2009**, 29, 1152–1157.

Lücker, S., Schwarz, J., Gruber-Dorninger, C., Spieck, E., Wagner, M., & Daims, H. *Nitrotoga*-like bacteria are previously unrecognized key nitrite oxidizers in full-scale wastewater treatment plants. *ISME J.*, **2015**, *9*(3), 708–720.

Ma, Q., Qu, Y., Shen, W., Zhang, Z., Wang, J., Liu, Z., Li, D., Li, H., & Zhou, J. Bacterial community compositions of coking wastewater treatment plants in steel industry revealed by Illumina high-throughput sequencing. *Bioresour. Technol.*, **2015**, *179*, 436–443.

Manyuchi, M. M., & Ketiwa, E. Distillery effluent treatment using membrane bioreactor technology utilising *Pseudomonas fluorescens*. *Inter. J. Sci. Eng. Tech.*, **2013**, *2*, 1252–1254.

Mara, D., & Horan, N. J. *The Handbook of Water and Wastewater Microbiology* (p. 4). Academic Press, London, UK, **2003**.

Marshall, F. M., Holden, J., Ghose, C., Chisala, B., Kapungwe, E., Volk, J., Agarwal, M., Agarwal, R., Sharma, R. K., & Singh, R. P. *Contaminated Irrigation Water and Food*

Safety for the Urban and Peri-Urban Poor: Appropriate Measures for Monitoring and Control from Field Research in India and Zambia. Inception Report, DFID Enkar R**8160**, SPRU, University of Sussex, **2007**.

Marti, E., Monclús, H., Jofre, J., Rodriguez-Roda, I., Comas, J., & Balcázar, J. L. Removal of microbial indicators from municipal wastewater by a membrane bioreactor (MBR). *Bioresour. Technol.*, **2011**, *102*, 5004–5009.

Mbouobda, H. D., Djeuani, C. A., Fai, K., & Omokolo, N. D. Impact of effective and indigenous microorganisms manures on *Colocassia esculenta* and enzymes activities. *African Journal of Agricultural Research*, **2013**, *8*(12), 1086–1092.

Mbouobda, H. D., Fotso, Djeuani, C. A., Baliga, M. O., & Omokolo, D. N. Comparative evaluation of enzyme activities and phenol content of Irish potato (*Solanum tuberosum*) grown under EM and IMO manures Bokashi. *Int. J. Biol. Chem. Sci.*, **2014**, *8*(1), 157–166.

Melamane, X. L., Strong, P. J., & Burgess, J. E. Treatment of wine distillery wastewater: A review with emphasis on anaerobic membrane reactors. *S Afr. J. Enol. Vitic.*, **2007**, *28*, 25–36.

Mielczarek, A. T., Kragelund, C., Eriksen, P. S., & Nielsen, P. H. Population dynamics of filamentous bacteria in Danish wastewater treatment plants with nutrient removal. *Water Res.*, **2012**, *46*(12), 3781–3795.

Mohapatra, D. P., Brar, S. K., & Tyagi, R. D. Degradation of endocrine disrupting bisphenol A during pre-treatment and biotransformation of wastewater sludge. *Chem. Eng. J.*, **2010**, *163*, 273–283.

Monica, S., Karthik, L., Mythili, S., & Sathiavelu, A. Formulation of effective microbial consortia and its application for sewage treatment. *Journal of Microbial and Biochemical Technology*, **2011**, *3*, 51–55.

Mueller, J. G., Middaugh, D. P., Lantz, S. E., & Chapman, P. J. Biodegradation of creosote and pentachlorophenol in contaminated groundwater: Chemical and biological assessment. *Applied and Environmental Microbiology*, **1991**, *57*(5), 1277–1285.

Narmadha, D., & Mary, S. K. V. J. Treatment of domestic waste water using natural flocculants. *Environmental Science: An Indian Journal*, **2012**, *7*(5), 173–178.

Nasirpour, N., Mousavi, S., & Shojaosadati, S. Biodegradation potential of hydrocarbons in petroleum refinery effluents using a continuous anaerobic-aerobic hybrid system. *Korean J. Chem. Eng.*, **2015**, *32*, 874–881.

Nguyen, H. T., Le, V. Q., Hansen, A. A., Nielsen, J. L., & Nielsen, P. H. High diversity and abundance of putative polyphosphate-accumulating *Tetrasphaera*-related bacteria in activated sludge systems. *FEMS Microbiol. Ecol.*, **2011**, *76*(2), 256–267.

Nielsen, P. H., Kragelund, C., Seviour, R. J., & Nielsen, J. L. Identity and ecophysiology of filamentous bacteria in activated sludge. *FEMS Microbiol. Rev.*, **2009**, *33*, 969–998.

Nielsen, P. H., Mielczarek, A. T., Kragelund, C., Nielsen, J. L., Saunders, A. M., Kong, Y., Hansen, A. A., & Vollertsen, J. A conceptual ecosystem model of microbial communities in enhanced biological phosphorus removal plants. *Water Res.*, **2010**, *44*, 5070–5088.

Oehmen, A., Lemos, P. C., Carvalho, G., Yuan, Z., Keller, J., Blackall, L. L., & Reis, M. A. Advances in enhanced biological phosphorus removal: From micro to macro scale. *Water Res.*, **2007**, *41*(11), 2271–2300.

Okuda, A., & Higa, T. *Purification of Waste Water with Effective Microorganisms and its Utilization in Agriculture.* University of the Ryukyus, Okinawa, Japan, **1995**.

Olawale, A. M. Bioremediation of waste water from an industrial effluent system in Nigeria using *Pseudomonas aeruginosa*: Effectiveness tested on albino rats. *J. Pet. Environ. Biotechnol.*, **2014**, *5*, 167.

Oshiman, K., Tsutsumi, Y., Nishida, T., & Matsumura, Y. Isolation and characterization of a novel bacterium, *Sphingomonas bisphenolicum* strain AO1, that degrades bisphenol A. *Biodegradation*, **2007**, *18*(2), 247–255.

Paillard, D., Dubois, V., Thiebaut, R., Nathier, F., Hogland, E., & Caumette, P. Q. C. Occurrence of *Listeria* spp. in effluents of French urban wastewater treatment plants. *Appl. Environ. Microbiol.*, **2005**, *71*(11), 7562–7566.

Pathak, A., Dastidar, M. G., & Sreekrishnan, T. R. Bioleaching of heavy metals from sewage sludge by indigenous iron-oxidizing microorganisms using ammonium ferrous sulfate and ferrous sulfate as energy sources: A comparative study. *Journal of Hazardous Materials*, **2009**, *171*(1–3), 273–278.

Peng, X., Guo, F., Ju, F., & Zhang, T. Shifts in the microbial community, nitrifiers and denitrifiers in the biofilm in a full-scale rotating biological contactor. *Environ. Sci. Technol.*, **2014**, *48*(14), 8044–8052.

Philip, L., & Venkobachr, C. An insight into mechanism of biosorption of Cu by *B. polymyxa*. *Indian J. Environ. Pollut.*, **2001**, *15*, 448–460.

PSDC. *Penang Skills Development Centre*, **2009**, http://www.psdc.com.my/ (Accessed on 7 October 2019).

Rajasulochana, P., & Preethy, V. Comparison on efficiency of various techniques in treatment of waste and sewage water-a comprehensive review. *Resource-Efficient Technologies*, **2016**, *2*, 175–184.

Rastegar, S. O., Mousavi, S. M., Shojaosadati, S. A., & Sheibani, S. Optimization of petroleum refinery effluent treatment in a UASB reactor using response surface methodology. *J. Hazard Mater.*, **2011**, *197*, 26–32.

Rossello-Mora, R. A., Wagner, M., Amann, R., & Schleifer, K. H. The abundance of *Zoogloea ramigera* in sewage treatment plants. *Applied and Environmental Microbiology*, **1995**, *61*(2), 702–705.

Russell, N. J., & Fukunaga, N. A comparison of thermal adaptation of membrane lipids in psychrophilic and thermophilic bacteria. *FEMS Microbiology Reviews*, **1990**, *75*(2), 1–2.

Sahoo, N. K., Ghosh, P. K., & Pakshirajan, K. Treatment of refinery wastewater using *Arthrobacter chlorophenolicus* A6 in an upflow packed bed reactor. *Desalination and Water Treatment*, **2014**, *55*(7), 1762–1770.

Salla, A. K., Abu-Alteen, K. H., & Jafri, A. M. Enumeration of *Pseudomonas* species and *Pseudomonas aeruginosa* bacteriophages in domestic sewage. *Micriobios*, **1989**, *60*(242), 35–43.

Saravanan, P., Pakshirajan, K., & Saha, P. Growth kinetics of an indigenous mixed microbial consortium during phenol degradation in a batch reactor. *Bioresource Technology*, **2008**, *99*(1), 205–209.

Satyawali, Y., & Balakrishnan, M. Wastewater treatment in molasses based alcohol distilleries for COD and color removal a review. *Journal of Environmental Management*, **2008**, *86*(3), 481–497.

Saunders, A. M., Albertsen, M., Vollertsen, J., & Nielsen, P. H. The activated sludge ecosystem contains a core community of abundant organisms. *The ISME Journal*, **2016**, *10*, 11–20.

Schuler, A. J., Jenkins, D., & Ronen, P. Microbial storage products, biomass density, and settling properties of enhanced biological phosphorus removal activated sludge. *Water Sci. Technol.*, **2001**, *43*, 173–180.

Schultz, T. E. Biological wastewater treatment. *Chemical Engineering Magazine*, **2005**, *112*, 44–51.

Shalaby, E. A. Prospects of effective microorganisms technology in wastes treatment in Egypt. *Asian Pac. J. Trop. Biomed.*, **2011**, *1*(3), 243–248.

Sheng, Z., Chaohai, W., Chaodeng, L., & Haizhen, W. Damage to DNA of effective microorganisms by heavy metals: Impact on wastewater treatment. *Journal of Environmental Sciences*, **2008**, *20*(12), 1514–1518.

Shivajirao, P. A. Treatment of distillery wastewater using membrane technologies. *Int. J. Adv. Eng. Res. Stud.*, **2012**, *1*(3), 275–283.

Siezen, R. J., & Galardini, M. Genomics of biological wastewater treatment. *Microb. Biotechnol.*, **2008**, *1*(5), 333–340.

Siripong, S., & Rittmann, B. E. Diversity study of nitrifying bacteria in full-scale municipal wastewater treatment plants. *Water Res.*, **2007**, *41*(5), 1110–1120.

Spina, F., Anastasi, A., Prigione, V., Tigini, V., & Varese, G. C. Biological treatment of industrial wastewaters: A fungal approach. *Chemical Engineering Transactions*, **2012**, *27*, 175–180.

Srinath, T., Garg, S. K., & Ramteke, P. W. Biosorption and elution of Cr from immobilized *Bacillus coagulens* biomass. *Indian Journal of Experimental Biology*, **2003**, *41*, 986–990.

Szymanski, N., & Patterson, R. A. Effective microorganisms (EM) and wastewater systems-future directions for on-site systems: Best management practice. In: Patterson, R. A., & Jones, M. J., (eds.), *Proceedings of On-Site '03 Conference* (pp. 347–354). Lanfax Laboratories, Armidale, **2003**.

Tadkaew, N., Sivakumar, M., Khan, S. J., McDonald, J. A., & Nghiem, L. D. Effect of mixed liquor pH on the removal of trace organic contaminants in a membrane bioreactor. *Biores. Technol.*, **2010**, *101*, 1494–1500.

Tawfik, A., Temmink, H., Zeeman, G., & Klapwijk, B. Sewage treatment in a rotating biological contactor (rbc) system. *Water Air Soil Pollut.*, **2006**, *175*, 275–289.

Tchobanoglous, G., Burton, F. L., & Stensel, H. D. *Wastewater Engineering: Treatment and Resource Recovery* (5th edn.). McGraw-Hill Education, New York, **2014**.

Templeton, M. R., & Butler, D. *Introduction to Wastewater Treatment* (pp. 66–67). Ventus Publishing, UK, **2011**.

Teruo, H., & James, F. P. *Beneficial and Effective Microorganisms for a Sustainable Agriculture and Environment* (pp. 1–16). International Nature Farming Research Center Press, Atami, Japan, **1994**.

Thakur, I. S. Screening and identification of microbial strains for removal of color and adsorbable organic halogens in pulp and paper mill effluent. *Process Biochemistry*, **2004**, *39*(11), 1693–1699.

Thawale, P. R., Juwarkar, A. A., & Singh, S. K. Resource conservation through land treatment of municipal wastewater. *Current Science*, **2006**, *90*, 704–711.

Tripathi, M., & Garg, S. K. Co-remediation of pentachlorophenol and Cr^{6+} by free and immobilized cells of native *Bacillus cereus* isolate: spectrometric characterization of PCP dechlorination products, bioreactor trial and chromate reductase activity. *Process Biochemistry*, **2013**, *48*(3), 496–509.

Tyagi, R. D., Blais, J. F., Auclair, J. C., & Meunier, N. Bacterial leaching of toxic metals from municipal sludge: Influence of sludge characteristics. *Water Environ Res.*, **1993**, *65*, 196–204.

Van Der Star, W. R., Abma, W. R., Blommers, D., Mulder, J. W., Tokutomi, T., Strous, M., Picioreanu, C., & Van Loosdrecht, M. C. Startup of reactors for anoxic ammonium oxidation: Experiences from the first full-scale anammox reactor in Rotterdam. *Water Res.*, **2007**, *41*(18), 4149–4163.

Verma, A. K., Dash, R. R., & Bhuniam P. A review on chemical coagulation/flocculation technologies for removal of color from textile wastewaters. *Journal of Environmental Management*, **2012**, *93*(1), 154–168.

Wagner, M. *In situ* analysis of nitrifying bacteria in sewage treatment plants. *Water Science and Technology*, **1996**, *1*, 237–244.

Waites, M. J., Morgan, N. L., Rockey, J. S., & Highton, G. *Industrial Microbiology* (pp. 24–25). Blackwell Science Limited, Oxford, UK, **2001**.

Wang, B., Yi, W., Yingxin, G., Guomao, Z., Min, Y., Song, W., & Jianying, H. Occurrences and behaviors of naphthenic acids in a petroleum refinery wastewater treatment plant. *Environ. Sci. Technol.*, **2015**, *49*, 5796–5804.

Wang, J., Li, Q., Qi, R., Tandoi, V., & Yang, M. Sludge bulking impact on relevant bacterial populations in a full-scale municipal wastewater treatment plant. *Process Biochem.*, **2014**, *49*, 2258–2265.

Wood, M. T., Tabora, P., Gabert, L., Hernandez, C., & Miles, R. Sustainable treatment of banana industry and crop residue wastes for crop production using effective microorganisms. *Proceedings of the 12th International IFOAM Scientific Conference* (pp. 212–218). Mar del Plata, Argentina, **1999**.

Xiang, L., Chan, L. C., & Wong, J. W. C. Removal of heavy metals from anaerobically digested sewage sludge by isolated indigenous iron-oxidizing bacteria. *Chemosphere*, **2000**, *41*(1/2), 283–287.

Zakaria, Z., Gairola, S., & Shariff, N. M. Effective microorganisms (EM) technology for water quality restoration and potential for sustainable water resources and management. In: Swayne, D. A., Yang, W., Voinov, A. A., Rizzoli, A., & Filatova, T., (eds.), *Proceeding of International Congress on Environmental Modeling and Software Modeling for Environment's Sake*, http://scholarsarchieve.byu.edu/iemssconference/2010/all/142 (Accessed on 7 October 2019).

Zhang, T., Shao, M. F., & Ye, L. 454 pyrosequencing reveals bacterial diversity of activated sludge from 14 sewage treatment plants. *ISME J.*, **2012**, *6*(6), 1137–1147.

Zhang, X., Li, A., Szewzyk, U., & Ma, F. Improvement of biological nitrogen removal with nitrate-dependent Fe(II) oxidation bacterium *Aquabacterium parvum* B6 in an up-flow bioreactor for wastewater treatment. *Bioresource Technology*, **2016**, *219*, 624–631.

Zhao, X., Wang, Y., Ye, Z. F., & Ni, J. R. Kinetics in the process of oil field wastewater treatment by effective microbe B350. *China Water Wastewater*, **2006**, *11*, 350–357.

Zou, X. L. Treatment of heavy oil wastewater by UASB–BAFs using the combination of yeast and bacteria. *Environ. Tech.*, **2015**, *36*(18), 2381–2389.

CHAPTER 12

EARTHWORMS AND MICROBES IN ENVIRONMENTAL MANAGEMENT THROUGH VERMITECHNOLOGY-MEDIATED ORGANIC FARMING

ABDULLAH ADIL ANSARI

Department of Biology, University of Guyana, Georgetown, Guyana

12.1 INTRODUCTION

Earth has a number of plants that have been providing food, promoting health, and some form of shelter towards civilization throughout the years. The soil is considered to be a major component of plant growth that helps to provide homes for many organisms (Ismail, 2005). Soil microbiology influences the above-ground ecosystem by contributing to plant nutrition, plant health, and soil structure, and soil fertility. However, they also play a pivotal role in various biogeochemical cycles and cycling organic compounds (Kirk et al., 2004). Plant growth is improved when beneficial microbes that increase nutrient availability, also stimulates plant growth without actually increasing nutrient availability to plants (Ismail, 2005). Microorganisms are tiny one-celled organisms found in the micro-biotic layer of the soil, responsible for building fertile soil for plant growth.

Bio-fertilizers, referred to as the use of soil microorganisms to increase the availability and uptake of mineral nutrients for the plant (Ismail, 2005), they are a substance that is added to the soil to enhance the microorganisms, in order to increase the nutrient status. Various Composting techniques are used for the production of biofertilizers vermicompost, biodung, tank compost, and microbe enhancers-Biodynamic preparation 500.

Vermicompost helps to promote humidification, increases microbial activity, and enhances enzyme production. It also facilitates the soil aggregation resulting in better aeration when applied to the soil. It is prepared through the application of earthworms. It has excellent structure, porosity, aeration drainage, and moisture-holding capacity, and helps to improve the physical, chemical, and biological properties of the soil (Ansari, 2008). Biodung compost and tank compost is similar to that of Vermicompost, except that earthworms, is used in the degradation of organic materials, and cow dung slur is used to aid in the degradation of organic waste in the tank and biodung compost. Biodynamic preparation 500, also known as cow horn manure, is fermented cow dung, which upon application aid in the fertility of soil. This technique helps to enhance and improve the structure of the soil quickly and effectively. Cow horns with the cow dung buried in the soil produce humus, which is to be beneficial to soil (Proctor, 2006).

Excessive use of chemical fertilizers and pesticides in agricultural lands over a long period of time has resulted in poor soil health with a combined effect on crop production and increase incidences of pests and diseases. These concerns have led to a greater economic impact on farmers. Over the last few years, the problems associated with food security has led to thinking in terms of organic agriculture by soil management techniques and microbial innovations. Soil microbiology influences the above-ground ecosystem by contributing to plant nutrition, health, soil structure, and fertility. They also play a pivotal role in various biogeochemical cycles and cycling of organic compounds (Kirk et al., 2004). Plant growth is improved when beneficial microbes increase nutrient availability and stimulate plant growth (Ismail, 2005). Biofertilizers referred to the use of soil microorganisms to increase the availability and uptake of mineral nutrients for the plant (Ansari, 2008); they are substances added to the soil to enhance the microorganisms, in order to increase the nutrient status. Vermicompost is one of the biofertilizers that helps to promote humification, increased microbial activity, and enzyme production, which subsequently helps to increase the aggregate stability of soil particles resulting in better aeration when applied to the soil. The material has excellent structure, porosity, aeration drainage, and moisture-holding capacity, and helps to improve the physical, chemical, and biological properties of the soil (Ansari, 2008).

The biocomposting method is made up of two phases (breakdown and buildup phase). In the breakdown phase, biodegradable wastes are decomposed into smaller particles. Proteins are broken down into amino acids and finally to ammonia, nitrates, and free nitrogen. Similarly, urea, uric acids, and other non-protein nitrogen-containing compounds are reduced to form different plant nutrients. In the build-up phase, there is the re-synthesis of simple compounds into complex humic substances. The organisms responsible for transformation to humus are aerobic and facultative aerobic, sporing, and non-sporing and nitrogen-fixing bacteria of the *Azotobacter* and *Nitrosomonas* group. *Actinomycetes* also play an important role. There are two major reasons why vermicomposting is better. Waste is converted faster, and conventional composting takes weeks to months to convert organic matter to compost, which are very labor-intensive. By using earthworms, waste is rapidly turned into vermicompost. The vermicompost is far superior to conventional compost. The worm castings in the vermicompost have nutrients that are highly utilizable by plants, and the castings have a mucous coating that allows the nutrients to "time-release." Vermicompost forms fine stable granular organic matter that assists in the aeration, released mucus that are hygroscopic absorbs water and prevents waterlogging, and improves water-holding capacity. Vermicompost added to the soil releases nutrients slowly and consistently and enables the plant to absorb these nutrients more readily. Soils enriched with vermicompost provide additional substances that are not found in the chemicals (Kale, 1998; Ansari and Ismail, 2001). Biofertilizers contribute both macro and micronutrients in amounts that are required by the plant and upon application have an emphatic effect on plant growth parameters and production.

Vermicomposting is the biological degradation and stabilization of organic waste by earthworms and microorganisms to form vermicompost. This is an essential part in organic farming today. It can be easily prepared, has excellent properties, and is harmless to plants. The earthworms fragment the organic waste substrates, stimulate microbial activity greatly, and increase rates of mineralization, they release a coelomic fluid that has antibacterial properties and destroys all pathogens in the media in which it inhibits. These rapidly convert the waste into humus-like substances with finer structure than thermophilic composts but possessing a greater and more diverse microbial activity (Ansari, 2008). The material has excellent structure, porosity, aeration, drainage, and moisture-holding capacity, and helps to improve the physical, chemical, and biological properties of soil (Ansari, 2008).

Biodynamic agriculture is a system of organic agriculture that has proved to be very effective throughout the world. The results are better soils, better quality food, healthy plants, healthy, and contented animals, and enthusiastic farmers and consumers. Cow horns and cow dung is used to produce this fertilizer, and after being buried together in the earth, make the most wonderful hummus to spread on the land. This help to improve the structure of the soil dramatically and quickly. Good soil structure means better water-holding capacity, which means better control, with consequently less loss of topsoil, which means better and deeper roots and less need for irrigation, which means deeper soil and more natural fertility (Proctor, 2006).

Biodynamic agriculture falls into the category of general organic agriculture, the main differences being in the use of biodynamic preparations for soil (preparation 500), plants (preparation 501), and compost (preparations 502–507). The agronomic system is based on a universal approach of the whole farm, and under this aspect, the simple use of preparations could not completely be defined as a biodynamic approach. However, from a scientific point of view, the only way to compare results from different farming systems is to study the effect of preparations. Even if some efforts have been made to explain the mechanism by which these preparations act, their effect on plant physiology, soil microbiology, and compost characteristics is still not explained. Only a few papers compare the results obtained with biodynamic agriculture with those from conventional farming (Heimler et al., 2012). Many organic practices are scientifically testable and can result in improved soil and plant health parameters (Scott, 2005).

Since many experiments were done with regards to biodynamic agriculture, the researcher found it quite interesting in trying this technique, along with the others, to investigate their effect on plant parameters. Organic waste possesses a serious environmental problem globally. This can be solved by vermitechnology, including vermiwash and vermicompost, and also biodynamic preparation (500), which is an essential component of biodynamic farming. Many researches over the years have been conducted, whereby solid waste were used and recycled to produce organic fertilizers using different technologies. In many developing countries, there is a serious organic solid waste problem; preparing these organic fertilizers will be cost-effective, and beneficial for farming (Ansari, 2008). The use of organic processes and materials in agriculture also helps to prevent environmental hazards, soil damage, and nutrients loss due to the excess use of toxic chemical fertilizers and pesticides (Nath et al., 2009).

12.2   EARTHWORMS IN SOIL FERTILITY AND MICROBIAL MANAGEMENT

Earthworms are key to maintaining soil fertility and nutrient cycling. Earthworms process organic nutrients for the efficient growth of plants.

Earthworms also contribute to the physical and chemical changes in the soil, transforming in terms of soil fertility and affect plant growth. Earthworms release casts into the soil, which is enriched with beneficial microorganisms. Earthworms are classified into three ecological types. Epigeics (*Eisenia fetida, Eudrilus eugeniae*) are surface dwellers serving as efficient agents of comminuting and fragmentation of leaf litter. They are phytophagous and generally have no effect on the soil structure as they cannot dig into the soil. Anecics (*Lampito mauritii*) feed on the leaf litter mixed with the soil of the upper layers and are said to be geophytophagous. They may also produce surface casts generally depending on the bulk density of the soil. Endogeic earthworms (*Octochaetona thurstoni*) are geophagous and live within the soil deriving nutrition from the organically rich soil they ingest (Ismail, 2005).

12.3   VERMITECHNOLOGY AND ORGANIC FARMING

Vermitechnology is a method of converting all the biodegradable wastes into useful product, i.e., vermicompost, through the action of earthworms. Vermicompost is a sustainable bio-fertilizer regenerated from organic wastes using earthworm, which contains 1.2 to 6.1% more nitrogen, 1.8 to 2.0% more phosphate, and 0.5 to 0.75% more potassium compared to farmyard manure. It also contains hormones like auxins and cytokinins, enzymes, vitamins, and useful microorganisms like bacteria, actinomycetes, protozoans, and fungi (Ansari and Ismail, 2001). This process of decomposition results in the production of vermicompost. Vermicompost or castings is worm manure. It is considered by many in the farming arena to be the very good soil improver. The nutrient content of castings is dependent on the material fed to the worms, and worms are commonly fed materials with high nutrient content (Ismail, 1997). It is the worm castings that provide these nutrients in a form that is readily available to plants. The biology of the worm's gut facilitates the growth of fungus and bacteria that are beneficial to plant growth.

12.4   VERMICOMPOSTING

Vermicomposting is a simple biotechnological process of composting, in which epigeic species of earthworms are used to enhance the process of waste conversion and produce a better end product. Vermicompost is a nutrient-rich organic soil conditioner that can be applied to improve soil conditions for a wide range of soil types. The use of earthworms is very essential in this process, as the worms act for the composing of organic matter into a stable nontoxic material with good structure, which has a potentially high economic value and also acts as a soil conditioner for plant growth. Vermicomposting has many environmental benefits is proven to be an easy way of getting rid of garbage waste. This technique is also beneficial to the soil and results in a lower use of synthetic fertilizers.

Vermicomposting units can be set up in many ways. This system can be set up in a large box, a bucket, a bin, a basket, and even in a pit in the soil. It is very important to keep in mind that a vermicomposting unit should be more than 1 meters in depth, but maybe as long as preferred in width. It is also very important to note that such a unit is set up in the shade. Organic matter that is added to the unit should be dry to prevent an increase of temperature in the unit. The unit should be kept moist; therefore, watering is very essential. The amount of materials which are layered during the building of the unit depends on the size of the unit, which is set up. The basic layering in a vermicomposting bin is as given in Figure 12.1.

FIGURE 12.1   Layering in the vermicomposting unit.

The basal layer of the vermi-bed comprises of broken bricks followed by a layer of coarse sand (10 cm thick) in-order to ensure proper drainage. A layer (10 cm) of loamy soil should be placed at the top. 100 locally collected earthworms were introduced into the soil. Fresh cattle dung is scattered over the soil, and then it was covered with a 10 cm layer of dried grasses. Water is sprinkled on the unit in order to keep it moist. The dried grasses, along with cattle dung, is turned once a week. After 60 days, vermicompost units are regularized for the harvesting of vermicompost every 45 days. When the layering is completed, the unit should be covered with dried leaves and left for 60 days. During the period of these 60 days, organic material and cow dung should be added on a weekly basis, while watering every other day, depending on the moisture content of the material in the bin.

Vermicompost should be ready for harvesting in a maximum of 40–45 days. When the organic material in the unit is changed completely in structure and smells soil-like, it is ready for harvest (Figure 12.2). The compost should be pressed in hand to check on moisture content. Before harvesting, no water should be added to the unit for 3–4 days, and a heap of the compost should be formed after harvesting. These actions will derive the earthworms in the deeper layers of the unit, where the moisture content is slightly higher. On the fourth day, the compost can be harvested and is ready to be used for agricultural purposes. This compost can be used directly in the soil and can be stored for 3 months if disposed of well in a plastic bag.

FIGURE 12.2   Vermicompost at harvest.

12.5   BENEFITS OF VERMICOMPOST

Vermicompost not only benefits soil, but also impact soil economics. It improves the physical structure, enriches the soil with microorganisms. Microbial activity in worm castings is 10 to 20 times higher than in the soil and organic matter that the worm ingests thereby attract deep-burrowing earthworms already present in the soil to enhance burrowing activity. It also improves water holding capacity. Vermicompost plays a major role in improving the growth and yield of different field crops, vegetables, flowers, and fruit crops and enhances germination, plant growth, and crop yield. It Improves root growth and structure (rhizosphere) and enriches the soil with microorganisms (adding plant hormones such as auxins and gibberellic acid). It is a good quality organic soil additive that enhances the water holding capacity and nutrient supplying capacity of the soil and also brings about the development of resistance in plants to pests and diseases, thereby providing a sustainable environment in the soil. A waste recycling through vermicomposting causes no pollution, as it becomes valuable raw materials for enhancing the soil health. It helps to close the "metabolic gap" through recycling waste onsite and thereby cause a reduction in greenhouse gas emissions such as methane and nitric oxide (NO) (produced in landfills or incinerators when not composted or through methane harvest).

12.6   VERMIWASH

Vermiwash is one of the materials produced by vermicomposting, which is an "ecobiotechnological process that transforms energy-rich and complex organic substances into a stabilized vermicomposts" primarily through the action of earthworms but with support of other microorganisms. Vemiwash contains the soluble nutrients that were released in the vermicomposting process (Nath et al., 2009). Organic fertilizers such as vermiwash provide a relatively cost-effective and safe alternative to chemical fertilizers. According to Ansari and Sukhraj (2010), the use of chemical fertilizers, which is widespread in many developing countries, can lead to soil damage and reduced soil health and production levels while increasing the incidence of pests and disease and environmental pollution. Vermiwash is a liquid that is obtained when water is left to flow slowly through a vermicomposting-like unit. Vermiwash has fertilizing abilities and has also proven to have a pesticidal action when applied as a foliar spray. The layering of a vermiwash bin is the same as a vermicomposting unit, with the exception that this unit consists of a bucket to which a tap is attached at the lowest point to collect the vermiwash when ready. The organic matter that is added to this unit varies from ordinary grass clippings to plant material with pesticidal properties. The organic matter should be dried for 3 to 4 days to accelerate the composting action and regulate the temperature in the bin.

The vermiwash unit is set up using buckets (Figure 12.3). A tap is fixed on the lower side of each bucket. The bucket is placed on a stand to facilitate the collection of vermiwash. About 5 cm of broken pebbles are placed at the bottom of the buckets followed by 5 cm layer of coarse sand. Water is then allowed to flow through these layers to enable the settling of the basic filter unit. A 15 cm layer of loamy soil is placed on top of the filter bed. Approximately 300 earthworms are introduced into the soil. Dried grass and cattle dung are placed on top of the soil. The vermiwash unit is left to regularize after 60 days for the collection of vermiwash every day. Approximately 0.5 liters can be collected on a daily basis. After layering the different material to the bin, the unit is left for 60 days to regulate with the tap open. Organic matter and cattle dung should be added on a weekly basis as needed. The unit should be watered every other day, depending on the moisture content in the bin. Access water should be left to flow through the open tap. Vermiwash will be ready to collect when the liquid that is flowing through the tap gets pale yellow in color. When the color change is seen, the tap should be closed, and water should be allowed to drip through the unit overnight. The following day the tap should be opened, and the vermiwash should be collected in a plastic container. The color intensity of the vermiwash will differ according to the organic material that is added to the bin. After the first collection, vermiwash can be collected on a daily basis by repeating the same process of adding water to the unit. The vermiwash that is collected can be kept stored for 3 months in plastic containers. Vermiwash can be used by a dilution of 10% of the vermiwash with water and spray to the desired plant/crop.

12.7   BIODYNAMIC PREPARATION 500

Cow horns (about 20–40) are collected, cleaned by removal of residues from within the horns. Fresh cattle dung is collected from a healthy female lactating cow. A pit of dimension 12 × 18 × 12 inches is dug in fertile land (shady area will be preferred). Soil dug out of the pit is mixed with dried cattle dung with little added moisture to make the soil enriched (Figure 12.4). Cow horns are then filled with fresh cow dung and place in upright position tip of horn pointing upwards, in the pit. The soil mixed with dried cattle dung and moisture is placed into the pit and covered with natural materials to prevent the soil from getting dry. The moisture in the pit is maintained at an appropriate level by sprinkling water on and around the pit once a week except during the rainy periods, till the time of harvest (November-December). This is continued until the six month period is up. After six months (March-April), the horns are retrieved from the pit, and biofertilizer is harvested (Proctor, 2006). BD 500 acts inoculum of soil microbes and can be added to the soil (at the rate of 1 gram per hectare) by diluting with chemical-free water along with the use of organic compost for the cultivation of crops.

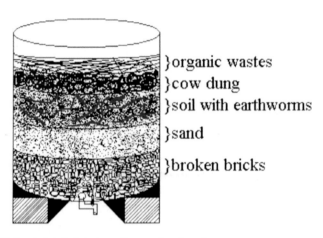

FIGURE 12.3   A detailed design of a vermiwash unit.

12.8   BIODUNG COMPOSTING

Organic waste such as Water hyacinth and Bermuda grass (used as waste); along with dried cattle dung, are collected. Cattle dung is mixed with water to make a liquid solution. Water hyacinth is placed on a flat surface longitudinal about 48 inches in length and 5 inches high, then another layer of Bermuda grass with the same measurement as the water hyacinth. Cow dung solution is then added uniformly after each layer. Step 3 is repeated two more times to make a total of three layers, and then the compost material is covered with a plastic sheet, ensuring all edges are kept down. Temperature is monitored every week for a period of three months. After every 15 days, the compost heap is turned so that the top layer will be at the bottom, and the bottom layer on top, also the heaps are sprayed with cattle dung solution after turning of each layer. Biodung compost is harvested after three months.

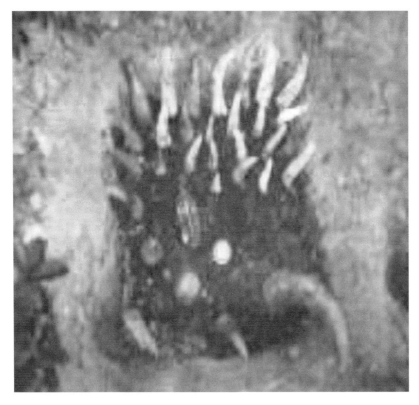

FIGURE 12.4   Biodynamic preparation 500 set-ups.

12.9   ORGANIC INPUTS IN SOIL AND IMPACT

Organic amendments like vermicompost, BD 500 and biodung compost promote humification, increased microbial activity and enzyme production, which, in turn, increase the aggregate stability of soil particles, resulting in better aeration (Tisdale and Oades, 1982; Dong et al., 1983; Haynes and Swift, 1990; Perucci, 1990). Organic matter has a property of binding mineral particles like calcium, magnesium, and potassium in the form of colloids of humus and clay, facilitating stable aggregates of soil particles for desired porosity to sustain plant growth (Haynes, 1986). Soil microbial biomass and enzyme activity are important indicators of soil improvement as a result of the addition of organic matter (Perucci, 1990). Apart from these, earthworm castings are reported to contain plant growth promoters, such as auxins and cytokinins (Krishnamoorthy and Vajranabhaiah, 1986). Vermiwash, a liquid fertilizer produced by the action of earthworms, contains soluble plant nutrients, some organic acids, mucus, and microbes, that have proved to be effective, both as a biological fertilizer (as a foliar spray) as well as a pesticide (Pramoth, 1995; Ismail, 1997; Kale, 1998).

The high content of organic matter in compost and the resultant effects of the organic matter on the humic fractions and nutrients in soil effectively increase the microbial population, activity, and enzyme production, which in turn increases aggregate stability (Tisdale and Oades, 1982; Dong et al., 1983; Haynes and Swift, 1990; Perucci, 1990). Humic acid and fulvic acid are important as persistent binding agents in mineral organic complexes, and 52 to 92% of soil organic matter may be involved in these complexes (Edwards and Bremner, 1967; Hamblin, 1977). Increased plant litter incorporation, improved aggregation, better aeration, and water relationships and the development of mull characteristics can be observed soils amended with organic inputs. These improvements in soil structure were confirmed by soil morphological studies, as illustrated by Rogaar and Boswinkel (1978). On the contrary there was reduction in organic carbon in plots treated with chemical fertilizers which may be due to negligible organic matter as input, moreover chemical inputs cause degradation of the soil structure resulting in unfavorable conditions for crop growth in an already difficult soil (Pagliai et al., 1983a, b; Shipitalo and Protz, 1988).

Vermicompost, one of the important types of compost, contains earthworm casts that are reported to be higher in available nitrogen (De Vleeschauwer and Lal, 1981) which enhance the activity and number of microorganisms (Stewart and Chaney, 1975; Satchell and Martin, 1984; Satchell et al., 1984). An increase in soil nitrogen through the application of vermicompost is likely to be due to the stimulation of microbial activity specifically through increase in the colonization of nitrogen fixers and actinomycetes (Kale 1998; Borken et al., 2002). Much of the effect of application of compost on crop yield and productivity is derived from the plant nutrients, particularly nitrogen in composts (Woodbury, 1992; Maynard, 1993; Ozores-Hampton et al., 1994). Reports indicate that adequate quantities of phosphorus and potassium were supplied by compost application to the soil (Smith, 1992; Maynard, 1993; Ozores-Hampton et al., 1994). Vermicompost is reported to contain desired quantity of phosphorus (De Vleeschauwer and Lal, 1981) which enhances the activity and number of microorganisms producing acid-phosphatases in the soil (Satchell and Martin, 1984; Satchell et al., 1984). Synergistically, these specific effects appear to raise phosphorus availability in soils amended with vermicompost (Buchanan and Gliessman, 1990).

Vermicompost application in the wheat-paddy cropping system has been reported to increase crop yield (Sharma and Mittra, 1991; Ismail, 1997). This is because nutrients present in vermicompost are readily available to the plants (Ismail, 1997; Rajkhowa et al., 2000). The effect of application of organic amendments like vermicompost on crop yield and production is derived from the plant nutrients, particularly nitrogen (Woodbury, 1992; Maynard, 1993; Ozores-Hampton et al., 1994). Organic phosphorus solubilized by microbial activity in composts like the vermicompost is more effective for plant absorption (Mishra and Banger, 1986; Singh et al., 1987). The reduced cost of cultivation, less cost-benefit ratio and higher net income has been recorded in wheat and paddy cultivation through vermitech compared with the use of chemical fertilizers along with the other economically important crops like peanut (*Arachis hypogaea*) and brinjal (*Solanum melongena*) by organic methods (Ismail, 1997). Organic farming has proved to be environment-friendly, sustainable, and cost-effective (Reganold et al., 2001).

Experiments on the effect of earthworms and vermicompost on the cultivation of vegetables like tomato (*Lycopersicum esculentum*), brinjal (*Solanum melongena*), and okra (*Abelmoschus esculentus*) have yielded significant results (Ismail, 1997). Vermicompost, as an organic input, has been applied to grow vegetables and other crops successfully (Ismail, 1997). The application of composts like vermicompost could contribute to the increased availability of food (Ouédraogo et al., 2001). This is attributed to better growth of plants and higher yield by slow release of nutrients for absorption with additional nutrients like gibberellin, cytokinin, and auxins, by the application of organic inputs like vermicompost in combination with vermiwash (Raviv et al., 1998; Subler et al., 1998; Lalitha et al., 2000). The yield of potato and the average weight of potato tubers were significantly higher in plots treated with vermicompost. This may be attributed to the increased bioavailability of phosphorus by the application of the organic amendment in the form of vermicompost (Erich et al., 2002).

Organic manure like vermicompost and vermiwash, when added to soil, augment crop growth and yield (Lalitha et al., 2000). The yields of spinach and onion in response to diluted vermiwash along with vermicompost was highly significant which may be due to increased availability of more exchangeable nutrients in the soil by the application of vermiwash along with vermicompost (Ponomareva, 1950; Finck, 1952; Nijhawan and Kanwar, 1952; Nye, 1955; Atlavinyte and Vanagas, 1973, 1982; Czerwinski et al., 1974; Watanabe, 1975; Cook et al., 1980; Tiwari et al., 1989). Concern about the environment and the economic and social impacts of chemical or conventional agriculture has led to many thinking groups seeking alternative practices that will make agriculture more suitable. Biodynamic farming practices and systems have shown promise in mitigating some of the detrimental effects of chemical-dependent, conventional agriculture on the environment (Reganold et al., 1993).

12.10 CONCLUSION

Soils are critical to productivity of both agriculture and natural ecosystems. Soil is an integral system, which is to be maintained through the sustainability of nutrient resources. The continuous worldwide soil degradation by erosion, chemicals, acidification, and physical abuse requires management in terms of soil quality. The use of organic amendments augmented with vermitechnology could be adopted as a means for crop production and soil stability. The use of combinations of organic amendments such as vermiwash, and vermicompost can effectively bring about an improvement in soil quality; enhance microbial population and impact crop productivity, thereby bringing about long term sustainability. Considering all aspects, such as studies on soil health, the yield of crops, and the cost-effectiveness of vermitechnology as a means of microbial innovation, it is concluded such technology could be applied for sustainable soil enrichment and crop productivity.

KEYWORDS

- **biodung composting**
- **biodynamic farming**
- **biofertilizer**
- **chemical fertilizer**
- **microbial activity**
- **microorganism**

- **organic input**
- **organic matter**
- **soil fertility**
- **soil health**
- **soil quality**
- **vermicompost**
- **vermitechnology**
- **vermiwash**

REFERENCES

Ansari, A. A. Effect of vermicompost on the productivity of Potato (*Solanum tuberosum*), Spinach (*Spinach oleracea*) and Turnip (*Brassica campestris*). *World Journal of Agricultural Sciences*, **2008**, *4*(3), 333–336.

Ansari, A. A., & Sukhraj, K. Effect of vermiwash and vermicompost on soil parameters and productivity of okra (*Abelmoschus esculentus*) in Guyana. *Pakistan Journal of Agricultural Research*, **2010**, *23*, 3–4.

Ansari, A. A., Ismail, S. A. A case study on organic farming in Uttar Pradesh. *Journal of Soil Biology*, **2001**, *27*, 25–27.

Atlavinyte, O., & Vanagas, J. Mobility of nutritive substances in relation to earthworm numbers in the soil. *Pedobiologia*, **1973**, *13*, 344–352.

Atlavinyte, O., & Vanagas, J. The effect of earthworms on the quality of barley and rye and grain. *Pedobiologia*, **1982**, *23*, 256–262.

Borken, W., Muhs, A., & Beese, F. Changes in microbial and soil properties following compost treatment of degraded temperate forest soils. *Soil Biol. Biochem.*, **2002**, *34*, 403–412.

Buchanan, R. A., & Gliessman, S. R. The influence of conventional and compost fertilization on phosphorus use efficiency by broccoli in a phosphorus deficient soil. *Am. J. Alt. Agric.*, **1990**, *5*(1), 38–46.

Cook, A. G., Critchley, B. R., & Critchley, U. Effects of cultivation and DDT on earthworm activity in a forest soil in the sub-humid tropics. *J. Appl. Ecol.*, **1980**, *17*(1), 21–29.

Czerwinski, Z., Jakubczyk, H., & Nowak, E. Analysis of sheep pasture ecosystem in the Pieniny Mountains (The Carpathians). XII. The effect of earthworms on pasture soil. *Ekol. Pol.*, **1974**, *22*, 635–650.

De Vleeschauwer, D. D., & Lal, R. Properties of worm casts under secondary tropical forest regrowth. *Soil Sci.*, **1981**, *132*, 175–181.

Dong, A., Chester, G., & Simsiman, G. V. Soil dispersibility. *J. Soil Sci.*, **1983**, *136*(4), 208–212.

Edwards, A. P., & Bremmer, J. M. Microaggregates in soils. *J. Soil Sci.*, **1967**, *18*(1), 64–73.

Erich, M. S., Fitzgerald, C. B., & Porter, G. A. The effect of organic amendments on phosphorus chemistry in a potato cropping system. *Agric. Ecosys. Environ.*, **2002**, *88*(1), 79–88.

Finck, A., Ökologische und bodenkundliche studien über die leistungen der regenwürmer für die bodenfruchtbarkeit. *Z. PflErnähr Düng*, **1952**, *58*, 120–145.

Hamblin, A. P. Structural features of aggregates in some East Anglian silt soils. *J. Soil Sci.*, **1977**, *28*(1), 23–28.

Haynes, R. J. The decomposition process mineralization, immobilization, humus formation and degradation. In: Haynes, R. J., (eds.), *Mineral Nitrogen in the Plant-Soil System* (pp. 52–186). Academic Press, New York, **1986**.

Haynes, R. J., & Swift, R. S. Stability of soil aggregates in relation to organic constituents and soil water content. *J. Soil Sci.*, **1990**, *41*(1), 73–83.

Heimler, D., Vignolini, P., Arfaioli, P., Isolani, L., & Romani, A. Conventional organic and biodynamic farming: differences in polyphenol content and antioxidant activity of Batavia lettuce. *J. Sci. Food Agric.*, **2012**, *92*(3), 551–556.

Ismail, S. A. *The Earthworm Book* (pp. 1–101). Other India Press, Mapusa, Goa, **2005**.

Ismail, S. A. *Vermicology: The Biology of Earthworms* (pp. 1–92). Orient Longman Press, Hyderabad, **1997**.

Kale, R. D. *Earthworm Cinderella of Organic Farming* (pp. 1–88). Prism Book Pvt. Ltd., Bangalore, India, **1998**.

Kirk, J. L., Beandette, L. A., Hart, M., Moutoglis, P., Klironomos, J. N., Lee, H., & Trevors, J. T. Methods of studying soil microbial diversity. *Journal of Microbiological Methods*, **2004**, *58*, 169–188.

Krishnamoorthy, R. V., & Vajranabhaiah, S. N. Biological activity of earthworm casts: An assessment of plant growth promoter levels in the casts. *Proc. Indian Acad. Sci. (Anim. Sci.)*, **1986**, *95*(3), 341–351.

Lalitha, R., Fathima, K., & Ismail, S. A. Impact of biopesticides and microbial fertilizers on productivity and growth of *Abelmoschus esculentus*. *Vasundhara – The Earth*, **2000**, *1 & 2*, 4–9.

Maynard, A. Evaluating the suitability of MSW compost as a soil amendment in field growth tomatoes. Part A: Yield of tomatoes. *Compost Sci. Util.*, **1993**, *1*, 34–36.

Mishra, M. M., & Banger, K. C. Rock phosphate comprising: transformation of phosphorus forms and mechanisms of solubilization. *Biol. Agric. Hort.*, **1986**, *3*, 331.

Nath, G., Singh, K., & Singh, D. Chemical analysis of vermicomposts/vermiwash of different combinations of animal, agro and kitchen wastes. *Australian Journal of Basic and Applied Sciences*, **2009**, *3*(4), 3672–3676.

Nijhawan, S. D., & Kanwar, J. S. Physicochemical properties of earthworm castings and their effect on the productivity of soil. *Ind. J. Agric. Sci.*, **1952**, *22*, 357–373.

Nye, P. H. Some soil-forming processes in the humid tropics. IV. The action of soil fauna. *J. Soil Sci.*, **1955**, *6*, 78.

Ouédraogo, E., Mando, A., & Zombré, N. P. Use of compost to improve soil properties and crop productivity under low input agricultural system in West Africa. *Agric. Ecosys. Environ.*, **2001**, *84*, 259–266.

Ozores-Hampton, M., Schaffer, B., Bryan, H. H., & Hanlon, E. A. Nutrient concentrations, growth and yield of tomato and squash in municipal solid-waste-amended soil. *Hort. Sci.*, **1994**, *29*(7), 785–788.

Pagliai, M., Bisdom, E. B. A., & Ledin, S. Changes in surface structure (crusting) after application of sewage sludges and pig slurry to cultivated agricultural soils in northern Italy. *Geoderma*, **1983a**, *30*, 35–53.

Pagliai, M., La Marca, M., & Lucamante, G. Micromorphometric and micromorphological investigations of a clay loam soil in viticulture under zero and conventional tillage. *J. Soil Sci.*, **1983b**, *34*, 391–403.

Perucci, P. Effect of the addition of municipal solid-waste compost on microbial biomass and enzyme activities in soil. *Biol. Fertil. Soils*, **1990**, *10*(3), 221–226.

Ponomareva, S. I. The role of earthworms in the creation of a stable structure in ley rotations. *Pochvovedenie*, **1950**, 476–486.

Pramoth, A. *Vermiwash-a Potent Bio-Organic Liquid "Ferticide."* (pp. 1–29). MSc Dissertation, University of Madras, Chennai, India, **1995**.

Proctor, P. *Why Biodynamic Agriculture – from Grasp the Nettle*. Bio-Dynamic Association of India, Bengaluru, **2006**.

Rajkhowa, D. J., Gogoi, A. K., Kandal, R., & Rajkhowa, K. M. Effect of vermicompost on Greengram nutrition. *J. Ind. Soc. Soil Sci.*, **2000**, 48(1), 207–208.

Raviv, M., Zaidman, B. Z., & Kapulnik, Y. The use of compost as a peat substitute for organic vegetable transplants production. *Compost Science and Utilization*, **1998**, *6*(1), 46–52.

Reganold, J. P., Glover, J. D., Andrews, P. K., & Hinman, H. R. Sustainability of three apple production systems. *Nature*, **2001**, *410*, 926–925.

Reganold, J. P., Palmer, A. S., Lockhart, J. C., & Macgrogor, A. N. Soil quality and financial performance of biodynamic and conventional farms in New Zealand. *Science*, **1993**, *260*, 344–349.

Rogaar, H., & Boswinkel, J. A. Some soil morphological effects of earthworm activity, field data and X-ray radiography. *Neth. J. Agric. Sci.*, **1978**, *26*, 145–160.

Satchell, J. E., & Martin, K. Phosphatase activity in earthworm species. *Soil Biol. Biochem.*, **1984**, *16*(2), 191–194.

Satchell, J. E., Martin, K., & Krishnamoorthy, R. V. Stimulation of microbial phosphatase production by earthworm activity. *Soil Biol. Biochem.*, **1984**, *16*(2), 195.

Scott, L. C. *The Myth of Biodynamic Agriculture*. Puyallup Research and Extension Centre, Washington State University, USA, **2005**.

Sharma, A. R., & Mittra, B. N. Effect of different rates of application of organic and nitrogen fertilizers in a rice-based cropping system. *J. Agric. Sci.*, **1991**, *117*(3), 313–318.

Shipitalo, M. J., & Protz, R. Factors influencing the dispersibility of clay in worm casts. *Soil Sci. Soc. Am. J.*, **1988**, *52*(3), 764–769.

Singh, C. P., Singh, Y. P., & Singh, M. Effect of different carbonaceous compounds on the transformation of soil nutrients. II. Immobilization and mineralization of phosphorus. *Biol. Agric. Hort.*, **1987**, *4*(4), 301–307.

Smith, S. R., **1992**. Sewage sludge and refuse composts as peat alternatives for conditioning impoverished soils: Effects on the growth response and mineral status of *Petunia grandiflora*. *J. Hort. Sci.*, **1992**, *67*(5), 703–716.

Subler, S., Edwards, C. A., & Metzer, J. Comparing vermicomposts and composts. *Biocycle*, **1998**, *39*, 63–66.

Tisdale, J. L., & Oades, J. M. Organic matter and water-stable aggregates in soil. *J. Soil Sci.*, **1982**, *33*(2), 141–163.

Tiwari, S. C., Tiwari, B. K., & Mishra, R. R. Microbial populations, enzyme activities and nitrogen-phosphorus-potassium enrichments in earthworm casts and in the surrounding soil of a pineapple plantation. *Biol. Fertil. Soils*, **1989**, *8*, 178–182.

Watanabe, H. On the amount of cast production by the megascolecid earthworm *Pheretima hupeiensis*. *Pedobioligia*, **1975**, *15*, 20–28.

Woodbury, P. B. Trace elements in municipal solid waste composts: A review of potential detrimental effects on plants, soil biota, and water quality. *Biomass and Bioenergy*, **1992**, *3*(3/4), 239–259.

INDEX

Non-trophic interface mechanisms, 131
Non-water stressed (NWS), 6
Novosphingobium sediminicola, 84, 90, 99
Nuclear weapons, 304
Nucleoid
  condensation, 291
  like structure, 306
Nucleotide
  excision repair, 296
  sequence, 125
Nutrient
  availability, 15, 99, 340, 341, 353, 354
  content, 6, 9, 357
  cycling, 201, 270, 356
  deficient conditions, 276
  diffusion, 341
  source, 266, 339
  uptake, 6, 8, 14, 19, 96, 324
    transport by mycorrhizae, 14
Nutritional
  analysis, 180
  elements, 5
  status, 8, 9
  value, 179
Nylon membranes, 122
Nypa fruticans, 87, 95, 99

O

Ochrobactrum
  anthropi, 96, 99
  intermedium, 99
Octochaetona thurstoni, 357
Oil refining effluent, 172
Oil-palm pericarp, 181, 182
Oligonucleotide, 32, 123
  probes, 32, 123
Onygena sp, 168
Open-air burning, 143
Operational taxonomic units (OTUs), 88
Operon profiles, 29
Orange peel waste, 164
Organic
  acids, 42, 131, 160, 162, 169, 364
  agriculture, 354–356
  amendments, 363, 365, 366
  compost, 362

compounds, 38, 47, 54, 159, 324, 338, 353, 354
farming, 355
fertilizers, 356
inorganic
  compounds, 241
  substances, 289
input, 363, 365, 367
material, 115, 116, 120, 158, 359, 361
matter, 19, 115, 116, 118, 162, 165, 171–173, 179, 201, 212, 223, 266, 272, 329–332, 334, 337, 338, 340, 355, 358–361, 363, 364, 367
metabolites, 2
nitrogen, 14, 272
nutrients, 130, 356
peroxides, 241
pollutant, 39, 57, 145, 174, 227, 228, 324, 338
polymer, 160
solid waste problem, 356
waste, 18, 160–161, 163, 164, 170, 177, 180, 182, 228, 324, 340, 354–357, 362
  possesses, 356
  substrates, 355
Organogenesis, 34
Ortho-dihydroxy phenols, 17
Orthologous genes, 296
Oryza sativa, 41, 83, 84, 90, 94
Oscillateria tenuis, 230
Osmosis, 225
Osmotolerant, 120
Osteomalacia, 268
Overcoming *penicillium* infections in plants, 129
Oxidation
  ditches, 260
  molecule, 227
  processes, 151, 227
Oxidative
  deamination of 3-chloroaniline, 212
  stress, 16, 266, 267, 269, 288, 300–302, 306
Oxidoreductases, 238, 241, 248
Oxygen
  demand, 219, 245, 338, 340, 342, 343